AF691077

stochastic processes and functional analysis

PURE AND APPLIED MATHEMATICS

A Program of Monographs, Textbooks, and Lecture Notes

LECTURE NOTES IN PURE AND APPLIED MATHEMATICS

1. *N. Jacobson,* Exceptional Lie Algebras
2. *L.-Å. Lindahl and F. Poulsen,* Thin Sets in Harmonic Analysis
3. *I. Satake,* Classification Theory of Semi-Simple Algebraic Groups
4. *F. Hirzebruch, W. D. Newmann, and S. S. Koh, Differentiable Manifolds and Quadratic Forms*
5. *I. Chavel,* Riemannian Symmetric Spaces of Rank One
6. *R. B. Burckel,* Characterization of C(X) Among Its Subalgebras
7. *B. R. McDonald, A. R. Magid, and K. C. Smith,* Ring Theory: Proceedings of the Oklahoma Conference
8. *Y.-T. Siu,* Techniques of Extension on Analytic Objects
9. *S. R. Caradus, W. E. Pfaffenberger, and B. Yood,* Calkin Algebras and Algebras of Operators on Banach Spaces
10. *E. O. Roxin, P.-T. Liu, and R. L. Sternberg,* Differential Games and Control Theory
11. *M. Orzech and C. Small,* The Brauer Group of Commutative Rings
12. *S. Thomier,* Topology and Its Applications
13. *J. M. Lopez and K. A. Ross,* Sidon Sets
14. *W. W. Comfort and S. Negrepontis,* Continuous Pseudometrics
15. *K. McKennon and J. M. Robertson,* Locally Convex Spaces
16. *M. Carmeli and S. Malin,* Representations of the Rotation and Lorentz Groups: An Introduction
17. *G. B. Seligman,* Rational Methods in Lie Algebras
18. *D. G. de Figueiredo,* Functional Analysis: Proceedings of the Brazilian Mathematical Society Symposium
19. *L. Cesari, R. Kannan, and J. D. Schuur,* Nonlinear Functional Analysis and Differential Equations: Proceedings of the Michigan State University Conference
20. *J. J. Schäffer,* Geometry of Spheres in Normed Spaces
21. *K. Yano and M. Kon,* Anti-Invariant Submanifolds
22. *W. V. Vasconcelos,* The Rings of Dimension Two
23. *R. E. Chandler,* Hausdorff Compactifications
24. *S. P. Franklin and B. V. S. Thomas,* Topology: Proceedings of the Memphis State University Conference
25. *S. K. Jain,* Ring Theory: Proceedings of the Ohio University Conference
26. *B. R. McDonald and R. A. Morris,* Ring Theory II: Proceedings of the Second Oklahoma Conference
27. *R. B. Mura and A. Rhemtulla,* Orderable Groups
28. *J. R. Graef,* Stability of Dynamical Systems: Theory and Applications
29. *H.-C. Wang,* Homogeneous Branch Algebras
30. *E. O. Roxin, P.-T. Liu, and R. L. Sternberg,* Differential Games and Control Theory II
31. *R. D. Porter,* Introduction to Fibre Bundles
32. *M. Altman,* Contractors and Contractor Directions Theory and Applications
33. *J. S. Golan,* Decomposition and Dimension in Module Categories
34. *G. Fairweather,* Finite Element Galerkin Methods for Differential Equations
35. *J. D. Sally,* Numbers of Generators of Ideals in Local Rings
36. *S. S. Miller,* Complex Analysis: Proceedings of the S.U.N.Y. Brockport Conference
37. *R. Gordon,* Representation Theory of Algebras: Proceedings of the Philadelphia Conference
38. *M. Goto and F. D. Grosshans,* Semisimple Lie Algebras
39. *A. I. Arruda, N. C. A. da Costa, and R. Chuaqui,* Mathematical Logic: Proceedings of the First Brazilian Conference
40. *F. Van Oystaeyen,* Ring Theory: Proceedings of the 1977 Antwerp Conference
41. *F. Van Oystaeyen and A. Verschoren,* Reflectors and Localization: Application to Sheaf Theory
42. *M. Satyanarayana,* Positively Ordered Semigroups
43. *D. L Russell,* Mathematics of Finite-Dimensional Control Systems
44. *P.-T. Liu and E. Roxin,* Differential Games and Control Theory III: Proceedings of the Third Kingston Conference, Part A
45. *A. Geramita and J. Seberry,* Orthogonal Designs: Quadratic Forms and Hadamard Matrices
46. *J. Cigler, V. Losert, and P. Michor,* Banach Modules and Functors on Categories of Banach Spaces
47. *P.-T. Liu and J. G. Sutinen,* Control Theory in Mathematical Economics: Proceedings of the Third Kingston Conference, Part B

48. *C. Byrnes,* Partial Differential Equations and Geometry
49. *G. Klambauer,* Problems and Propositions in Analysis
50. *J. Knopfmacher,* Analytic Arithmetic of Algebraic Function Fields
51. *F. Van Oystaeyen,* Ring Theory: Proceedings of the 1978 Antwerp Conference
52. *B. Kadem,* Binary Time Series
53. *J. Barros-Neto and R. A. Artino,* Hypoelliptic Boundary-Value Problems
54. *R. L. Sternberg, A. J. Kalinowski, and J. S. Papadakis,* Nonlinear Partial Differential Equations in Engineering and Applied Science
55. *B. R. McDonald,* Ring Theory and Algebra III: Proceedings of the Third Oklahoma Conference
56. *J. S. Golan,* Structure Sheaves Over a Noncommutative Ring
57. *T. V. Narayana, J. G. Williams, and R. M. Mathsen,* Combinatorics, Representation Theory and Statistical Methods in Groups: YOUNG DAY Proceedings
58. *T. A. Burton,* Modeling and Differential Equations in Biology
59. *K. H. Kim and F. W. Roush,* Introduction to Mathematical Consensus Theory
60. *J. Banas and K. Goebel,* Measures of Noncompactness in Banach Spaces
61. *O. A. Nielson,* Direct Integral Theory
62. *J. E. Smith, G. O. Kenny, and R. N. Ball,* Ordered Groups: Proceedings of the Boise State Conference
63. *J. Cronin,* Mathematics of Cell Electrophysiology
64. *J. W. Brewer,* Power Series Over Commutative Rings
65. *P. K. Kamthan and M. Gupta,* Sequence Spaces and Series
66. *T. G. McLaughlin,* Regressive Sets and the Theory of Isols
67. *T. L. Herdman, S. M. Rankin III, and H. W. Stech,* Integral and Functional Differential Equations
68. *R. Draper,* Commutative Algebra: Analytic Methods
69. *W. G. McKay and J. Patera,* Tables of Dimensions, Indices, and Branching Rules for Representations of Simple Lie Algebras
70. *R. L. Devaney and Z. H. Nitecki,* Classical Mechanics and Dynamical Systems
71. *J. Van Geel,* Places and Valuations in Noncommutative Ring Theory
72. *C. Faith,* Injective Modules and Injective Quotient Rings
73. *A. Fiacco,* Mathematical Programming with Data Perturbations I
74. *P. Schultz, C. Praeger, and R. Sullivan,* Algebraic Structures and Applications: Proceedings of the First Western Australian Conference on Algebra
75. *L Bican, T. Kepka, and P. Nemec,* Rings, Modules, and Preradicals
76. *D. C. Kay and M. Breen,* Convexity and Related Combinatorial Geometry: Proceedings of the Second University of Oklahoma Conference
77. *P. Fletcher and W. F. Lindgren,* Quasi-Uniform Spaces
78. *C.-C. Yang,* Factorization Theory of Meromorphic Functions
79. *O. Taussky,* Ternary Quadratic Forms and Norms
80. *S. P. Singh and J. H. Burry,* Nonlinear Analysis and Applications
81. *K. B. Hannsgen, T. L. Herdman, H. W. Stech, and R. L. Wheeler,* Volterra and Functional Differential Equations
82. *N. L. Johnson, M. J. Kallaher, and C. T. Long,* Finite Geometries: Proceedings of a Conference in Honor of T. G. Ostrom
83. *G. I. Zapata,* Functional Analysis, Holomorphy, and Approximation Theory
84. *S. Greco and G. Valla,* Commutative Algebra: Proceedings of the Trento Conference
85. *A. V. Fiacco,* Mathematical Programming with Data Perturbations II
86. *J.-B. Hiriart-Urruty, W. Oettli, and J. Stoer,* Optimization: Theory and Algorithms
87. *A. Figa Talamanca and M. A. Picardello,* Harmonic Analysis on Free Groups
88. *M. Harada,* Factor Categories with Applications to Direct Decomposition of Modules
89. *V. I. Istrăţescu,* Strict Convexity and Complex Strict Convexity
90. *V. Lakshmikantham,* Trends in Theory and Practice of Nonlinear Differential Equations
91. *H. L. Manocha and J. B. Srivastava,* Algebra and Its Applications
92. *D. V. Chudnovsky and G. V. Chudnovsky,* Classical and Quantum Models and Arithmetic Problems
93. *J. W. Longley,* Least Squares Computations Using Orthogonalization Methods
94. *L. P. de Alcantara,* Mathematical Logic and Formal Systems
95. *C. E. Aull,* Rings of Continuous Functions
96. *R. Chuaqui,* Analysis, Geometry, and Probability
97. *L. Fuchs and L. Salce,* Modules Over Valuation Domains
98. *P. Fischer and W. R. Smith,* Chaos, Fractals, and Dynamics
99. *W. B. Powell and C. Tsinakis,* Ordered Algebraic Structures

100. *G. M. Rassias and T. M. Rassias,* Differential Geometry, Calculus of Variations, and Their Applications
101. *R.-E. Hoffmann and K. H. Hofmann,* Continuous Lattices and Their Applications
102. *J. H. Lightbourne III and S. M. Rankin III,* Physical Mathematics and Nonlinear Partial Differential Equations
103. *C. A. Baker and L. M. Batten,* Finite Geometrics
104. *J. W. Brewer, J. W. Bunce, and F. S. Van Vleck,* Linear Systems Over Commutative Rings
105. *C. McCrory and T. Shifrin,* Geometry and Topology: Manifolds, Varieties, and Knots
106. *D. W. Kueker, E. G. K. Lopez-Escobar, and C. H. Smith,* Mathematical Logic and Theoretical Computer Science
107. *B.-L. Lin and S. Simons,* Nonlinear and Convex Analysis: Proceedings in Honor of Ky Fan
108. *S. J. Lee,* Operator Methods for Optimal Control Problems
109. *V. Lakshmikantham,* Nonlinear Analysis and Applications
110. *S. F. McCormick,* Multigrid Methods: Theory, Applications, and Supercomputing
111. *M. C. Tangora,* Computers in Algebra
112. *D. V. Chudnovsky and G. V. Chudnovsky,* Search Theory: Some Recent Developments
113. *D. V. Chudnovsky and R. D. Jenks,* Computer Algebra
114. *M. C. Tangora,* Computers in Geometry and Topology
115. *P. Nelson, V. Faber, T. A. Manteuffel, D. L. Seth, and A. B. White, Jr.,* Transport Theory, Invariant Imbedding, and Integral Equations: Proceedings in Honor of G. M. Wing's 65th Birthday
116. *P. Clément, S. Invernizzi, E. Mitidieri, and I. I. Vrabie,* Semigroup Theory and Applications
117. *J. Vinuesa,* Orthogonal Polynomials and Their Applications: Proceedings of the International Congress
118. *C. M. Dafermos, G. Ladas, and G. Papanicolaou,* Differential Equations: Proceedings of the EQUADIFF Conference
119. *E. O. Roxin,* Modern Optimal Control: A Conference in Honor of Solomon Lefschetz and Joseph P. Lasalle
120. *J. C. Díaz,* Mathematics for Large Scale Computing
121. *P. S. Milojevič,* Nonlinear Functional Analysis
122. *C. Sadosky,* Analysis and Partial Differential Equations: A Collection of Papers Dedicated to Mischa Cotlar
123. *R. M. Shortt,* General Topology and Applications: Proceedings of the 1988 Northeast Conference
124. *R. Wong,* Asymptotic and Computational Analysis: Conference in Honor of Frank W. J. Olver's 65th Birthday
125. *D. V. Chudnovsky and R. D. Jenks,* Computers in Mathematics
126. *W. D. Wallis, H. Shen, W. Wei, and L. Zhu,* Combinatorial Designs and Applications
127. *S. Elaydi,* Differential Equations: Stability and Control
128. *G. Chen, E. B. Lee, W. Littman, and L. Markus,* Distributed Parameter Control Systems: New Trends and Applications
129. *W. N. Everitt,* Inequalities: Fifty Years On from Hardy, Littlewood and Pólya
130. *H. G. Kaper and M. Garbey,* Asymptotic Analysis and the Numerical Solution of Partial Differential Equations
131. *O. Arino, D. E. Axelrod, and M. Kimmel,* Mathematical Population Dynamics: Proceedings of the Second International Conference
132. *S. Coen,* Geometry and Complex Variables
133. *J. A. Goldstein, F. Kappel, and W. Schappacher,* Differential Equations with Applications in Biology, Physics, and Engineering
134. *S. J. Andima, R. Kopperman, P. R. Misra, J. Z. Reichman, and A. R. Todd,* General Topology and Applications
135. *P Clément, E. Mitidieri, B. de Pagter,* Semigroup Theory and Evolution Equations: The Second International Conference
136. *K. Jarosz*, Function Spaces
137. *J. M. Bayod, N. De Grande-De Kimpe, and J. Martínez-Maurica*, *p*-adic Functional Analysis
138. *G. A. Anastassiou*, Approximation Theory: Proceedings of the Sixth Southeastern Approximation Theorists Annual Conference
139. *R. S. Rees*, Graphs, Matrices, and Designs: Festschrift in Honor of Norman J. Pullman
140. *G. Abrams, J. Haefner, and K. M. Rangaswamy*, Methods in Module Theory
141. *G. L. Mullen and P. J.-S. Shiue*, Finite Fields, Coding Theory, and Advances in Communications and Computing
142. *M. C. Joshi and A. V. Balakrishnan*, Mathematical Theory of Control: Proceedings of the International Conference

143. *G. Komatsu and Y. Sakane*, Complex Geometry: Proceedings of the Osaka International Conference
144. *I. J. Bakelman*, Geometric Analysis and Nonlinear Partial Differential Equations
145. *T. Mabuchi and S. Mukai*, Einstein Metrics and Yang–Mills Connections: Proceedings of the 27th Taniguchi International Symposium
146. *L. Fuchs and R. Göbel*, Abelian Groups: Proceedings of the 1991 Curaçao Conference
147. *A. D. Pollington and W. Moran*, Number Theory with an Emphasis on the Markoff Spectrum
148. *G. Dore, A. Favini, E. Obrecht, and A. Venni*, Differential Equations in Banach Spaces
149. *T. West*, Continuum Theory and Dynamical Systems
150. *K. D. Bierstedt, A. Pietsch, W. Ruess, and D. Vogt*, Functional Analysis
151. *K. G. Fischer, P. Loustaunau, J. Shapiro, E. L. Green, and D. Farkas*, Computational Algebra
152. *K. D. Elworthy, W. N. Everitt, and E. B. Lee*, Differential Equations, Dynamical Systems, and Control Science
153. *P.-J. Cahen, D. L. Costa, M. Fontana, and S.-E. Kabbaj*, Commutative Ring Theory
154. *S. C. Cooper and W. J. Thron*, Continued Fractions and Orthogonal Functions: Theory and Applications
155. *P. Clément and G. Lumer*, Evolution Equations, Control Theory, and Biomathematics
156. *M. Gyllenberg and L. Persson*, Analysis, Algebra, and Computers in Mathematical Research: Proceedings of the Twenty-First Nordic Congress of Mathematicians
157. *W. O. Bray, P. S. Milojevič and Č. V. Stanojevič*, Fourier Analysis: Analytic and Geometric Aspects
158. *J. Bergen and S. Montgomery*, Advances in Hopf Algebras
159. *A. R. Magid*, Rings, Extensions, and Cohomology
160. *N. H. Pavel*, Optimal Control of Differential Equations
161. *M. Ikawa*, Spectral and Scattering Theory: Proceedings of the Taniguchi International Workshop
162. *X. Liu and D. Siegel*, Comparison Methods and Stability Theory
163. *J.-P. Zolésio*, Boundary Control and Variation
164. *M. Křížek, P. Neittaanmäki, and R. Stenberg*, Finite Element Methods: Fifty Years of the Courant Element
165. *G. Da Prato and L. Tubaro*, Control of Partial Differential Equations
166. *E. Ballico*, Projective Geometry with Applications
167. *M. Costabel, M. Dauge, and S. Nicaise*, Boundary Value Problems and Integral Equations in Nonsmooth Domains
168. *G. Ferreyra, G. R. Goldstein, and F. Neubrander*, Evolution Equations
169. *S. Huggett*, Twistor Theory
170. *H. Cook, W. T. Ingram, K. T. Kuperberg, A. Lelek, and P. Minc*, Continua: With the Houston Problem Book
171. *D. F. Anderson and D. E. Dobbs*, Zero-Dimensional Commutative Rings
172. *K. Jarosz*, Function Spaces: The Second Conference
173. *V. Ancona, E. Ballico, and A. Silva*, Complex Analysis and Geometry
174. *E. Casas*, Control of Partial Differential Equations and Applications
175. *N. Kalton, E. Saab, and S. Montgomery-Smith*, Interaction Between Functional Analysis, Harmonic Analysis, and Probability
176. *Z. Deng, Z. Liang, G. Lu, and S. Ruan*, Differential Equations and Control Theory
177. *P. Marcellini, G. Talenti, and E. Vesentini*, Partial Differential Equations and Applications: Collected Papers in Honor of Carlo Pucci
178. *A. Kartsatos*, Theory and Applications of Nonlinear Operators of Accretive and Monotone Type
179. *M. Maruyama*, Moduli of Vector Bundles
180. *A. Ursini and P. Aglianò*, Logic and Algebra
181. *X. H. Cao, S. X. Liu, K. P. Shum, and C. C. Yang*, Rings, Groups, and Algebras
182. *D. Arnold and R. M. Rangaswamy*, Abelian Groups and Modules
183. *S. R. Chakravarthy and A. S. Alfa*, Matrix-Analytic Methods in Stochastic Models
184. *J. E. Andersen, J. Dupont, H. Pedersen, and A. Swann*, Geometry and Physics
185. *P.-J. Cahen, M. Fontana, E. Houston, and S.-E. Kabbaj*, Commutative Ring Theory: Proceedings of the II International Conference
186. *J. A. Goldstein, N. E. Gretsky, and J. J. Uhl, Jr.*, Stochastic Processes and Functional Analysis

Additional Volumes in Preparation

stochastic processes and functional analysis

in celebration of M.M. Rao's 65th birthday

edited by

Jerome A. Goldstein
University of Memphis
Memphis, Tennessee

Neil E. Gretsky
University of California-Riverside
Riverside, California

J. J. Uhl, Jr.
University of Illinois

Marcel Dekker, Inc. **New York•Basel•Hong Kong**

Library of Congress Cataloging-in-Publication Data

Stochastic processes and functional analysis : in celebration of M. M. Rao's 65th birthday / edited by Jerome A. Goldstein, Neil E. Gretsky, John Jerry Uhl, Jr.
p. cm. — (Lecture notes in pure and applied mathematics : v. 186)
Held at the Univ. of Calif. —Riverside, Nov. 18-20, 1994.
"Published writings of M. M. Rao": p.
ISBN 0-8247-9801-5 (pbk. : alk. paper)
1. Stochastic processes—Congresses. 2. Functional analysis--Congresses. I. Rao, M. M. (Malempati Madhusudana). II. Goldstein, Jerome A. III. Gretsky, Neil E. IV. Uhl, J. J. (J. Jerry) V. Series.
QA274.A1S7665 1997
515'.7—dc21

96-48137
CIP

The publisher offers discounts on this book when ordered in bulk quantities. For more information, write to Special Sales/Professional Marketing at the address below.

This book is printed on acid-free paper.

MARCEL DEKKER, INC.
270 Madison Avenue, New York, New York 10016

Current printing (last digit):
10 9 8 7 6 5 4 3 2 1

PRINTED IN THE UNITED STATES OF AMERICA

Preface

A conference in Modern Analysis and Probability on the occasion of his 65th birthday was held in honor of M. M. Rao at the University of California, Riverside. Over a hundred friends, colleagues, students, and other mathematicians attended during the three-day meeting. The Mathematics Department provided copious amounts of coffee and doughnuts. An on-campus dinner was held on a Friday night and an off-campus dinner was held Saturday night following a gala reception at the home of M. M. and his wife, Durgamba. Support for the conference was supplied by the College of Natural and Agricultural Sciences at the University of California–Riverside, the Mathematics Department at the University of California–Riverside, and the National Science Foundation.

This festschrift volume contains most of the talks given at the conference as well as several that were contributed later. The beginning portions of the book include a biography of M. M. Rao, a bibliography of his published writings, an ancestral mathematical family tree, and a list of Ph.D. theses written under Rao and his students.

The talks at the conference included four keynote addresses by Rao, Jean Bourgain, S. R. S. Varadhan, and Michael Crandall. All but Crandall's talk are contained here; Crandall's talk will appear in a paper that will be published elsewhere. Rao's paper is an account of that portion of his work which originated in problems arising in applications. It is organized by area and features the work of his students as much as his own. The breadth and depth of Rao's mathematical work and its impact on analysis, probability, and stochastic processes can be seen not only by what is included in this paper but also by the portion of his bibliography which is *not* in this paper. The editors enjoyed this paper immensely. Even as good as the paper is, it does not capture the charm and the emotion with which the talk was given.

Jean Bourgain's paper is a long, densely written survey (an "exposé" in his terminology) of persistency of quasi-periodic solutions of linear or integrable partial differential equations after Hamiltonian perturbation. Much of the original work is due to Bourgain and is not in print elsewhere. The talk given by Varadhan reported on joint work with H. T. Yau concerning scaling limits for lattice gas models. This provides a way to give a simplified description of the state of a large system of interacting particles which is evolving in time. The results typify recent deep research involving hydrodynamic limits, which establish that nonlinear partial differential equations govern many large particle systems in the limit.

The remaining eighteen papers are original contributions in probability and statistics, stochastic processes, Banach space theory, measure theory, and differential equations—both deterministic and stochastic.

Many other people attended the conference who did not give talks for one reason or another. Although we cannot list all of them (our sincere apologies) we would like to mention two esteemed intellectual colleagues of M. M. Rao, Mannie Parzen and Howard Tucker, as well as two former students, William Kraynek and Marc Mehlman.

Jerome A. Goldstein
Neil E. Gretsky
J. J. Uhl, Jr.

Contents

Contributors

Barry C. Arnold Department of Statistics, University of California, Riverside, California

Jean Bourgain School of Mathematics, Institute for Advanced Study, Princeton, New Jersey

Bruce L. Chalmers Department of Mathematics, University of California, Riverside, California

Terry Cheng Department of Mathematics, Irvine Valley Community College, Irvine, California

John S. Chipman Department of Economics, University of Minnesota, Minneapolis, Minnesota

Angelo Favini Dipartimento di Matematica, Università di Bologna, Bologna, Italy

Gisèle Ruiz Goldstein Department of Mathematics, Louisiana State University, Baton Rouge, Louisiana, and CERI and Department of Mathematical Sciences, University of Memphis, Memphis, Tennessee

Jerome A. Goldstein Department of Mathematical Sciences, University of Memphis, Memphis, Tennessee

Neil E. Gretsky Department of Mathematics, University of California, Riverside, California

S. L. Hobbs Naval Command Control and Ocean Surveillance Center, San Diego, California

Yûichirô Kakihara Department of Mathematics, University of California, Riverside, California

Dan Kalman Department of Mathematics and Statistics, American University, Washington, DC

Alan Krinik Department of Mathematics, California State Polytechnic University, Pomona, California

Hui-Hsiung Kuo Department of Mathematics, Louisiana State University, Baton Rouge, Louisiana

Chin-Yuan Lin Department of Mathematics, University of South Carolina, Columbia, South Carolina, and Department of Mathematics, National Central University, Chang-Li, Republic of China

Sergey Lototsky Center for Applied Mathematical Sciences, University of Southern California, Los Angeles, California

Daniel Marcus Department of Mathematics, California State Polytechnic University, Pomona, California

Frederic T. Metcalf Department of Mathematics, University of California, Riverside, California

Joseph M. Ostroy Department of Economics, University of California, Los Angeles, California

Roger N. Pedersen Department of Mathematics, Carnegie Mellon University, Pittsburgh, Pennsylvania

Balram S. Rajput Department of Mathematics, The University of Tennessee, Knoxville, Tennessee

Malempati M. Rao Department of Mathematics, University of California, Riverside, California

Zhongdau Ren Department of Mathematics, University of California, Riverside, California

Silvia Romanelli Dipartimento di Matematica, Università di Bari, Bari, Italy

Boris L. Rozovskii Center for Applied Mathematical Sciences, University of Southern California, Los Angeles, California

V. V. Sazonov Steklov Mathematical Institute, Moscow, Russia, and Department of Mathematics, Hong Kong University of Science and Technology, Hong Kong

Boris Shektman Department of Mathematics, University of South Florida, Tampa, Florida

S. S. Sritharan Naval Command Control and Ocean Surveillance Center, San Diego, California

Randall J. Swift Department of Mathematics, Western Kentucky University, Bowling Green, Kentucky

Makato Tsukada Department of Information Sciences, Toho University, Funabashi City, Chiba, Japan

S. R. S. Varadhan Courant Institute of Mathematical Sciences, New York University, New York, New York

Kunyang Wang Department of Mathematics, Louisiana State University, Baton Rouge, Louisiana

Jacek Wesolowski Mathematical Institute, Warsaw University of Technology, Warsaw, Poland

H. T. Yau Courant Institute of Mathematical Sciences, New York University, New York, New York

William R. Zame Department of Mathematics, University of California, Los Angeles, California

Biography of M. M. Rao

M. M. Rao was born Malempati Madhusudana Rao in the village of Nimmagadda in the state of Andhra Pradesh in India on June 6, 1929. After studying at the College of Andhra University and the Presidency College of Madras University, he came to the United States and obtained his Ph.D in 1959 at the University of Minnesota under the supervision of Monroe Donsker (as well as Bernard R. Gelbaum, Leonid Hurwicz, and I. Richard Savage). His first academic appointment was at Carnegie Institute of Technology (now called Carnegie Mellon University) in 1959. In 1972, he accepted an offer from University of California, Riverside where he remains even today. Along the way, he has held visiting positions at the Institute for Advanced Study (Princeton), the Indian Statistical Institute, University of Vienna, University of Strassbourg, and the Mathematical Sciences Research Institute (Berkeley). In 1966 he married Durgamba Kolluru in India. They have twin daughters Leela and Uma.

M.M. started out in probability and mathematical statistics but his voracious mathematical appetite found him branching out into a wide range of mathematical analysis including stochastic processes, functional analysis, ergodic theory and related asymptotics, differential equations and difference equations. His breadth of interest is mirrored by his students, many of whom are recognized as experts in diverse fields such as measure theory, operator theory, partial differential equations, etc.

M.M. has always strived for complete understanding and full generality and has seldom been willing to accept less from others. This attitude has always affected his teaching as well. Many generations of students have found him to be a demanding but truly inspiring teacher and mentor. He has always championed the open door policy for his Ph. D. students and has always put their welfare above his own. It's no accident that, throughout his career, he has had his share of the best available graduate students lined up to work with him.

His article in this volume is a remarkable summary of his broad research program and how he is able to view many diverse topics in a naturally unified way. Mathematicians rarely get to see in print such an intimate view of the thought processes of a leading mathematician and how he involved his students in his work. We are delighted that M.M. wrote this article for all of us to share.

M.M. continues to work on and we hope that he will be around twenty years from now to write another such article, to enjoy his beloved mathematics, his beloved students, and his beloved family.

M. M. Rao

Published Writings of M. M. Rao

[1] Note on a remark of Wald, Amer. Math. Monthly 65 (1958), 277-278.

[2] Lower bounds for risk functions in estimation, Proc. Nat'l Acad. of Sciences 45 (1959), 1168-1171.

[3] Estimation by periodogram, Trabajos de Estadistica 11 (1960), 123-137.

[4] Two probability limit theorems and an application, Indagationes Mathematicae 23 (1961), 551-559.

[5] Theory of lower bounds for risk functions in estimation, Mathematische Annalen 143 (1961), 379-398.

[6] Consistency and limit distributions of estimators of parameters in explosive stochastic difference equations, Annals of Math. Stat. 32 (1961), 195-218.

[7] Some remarks on independence of statistics, Trabajos de Estadistica 12 (1961), 19-26.

[8] Remarks on a multivariate gamma distribution, Amer. Math. Monthly 68 (1961), (with P. R. Krishnaiah, 342-346).

[9] Theory of order statistics, Mathematische Annalen 147 (1962), 298-312.

[10] Nonsymmetric projections in Hilbert Space, Pacific J. Math. 12 (1962), 343-357, (with V. J. Mizel).

[11] Characterizing normal law and a nonlinear integral equation, J. Math. & Mech. 12 (1963), 869-880.

[12] Inference in stochastic processes I, Teoria Veroyatnastei i ee Primcneniya 8 (1963), 282-298.

[13] Some inference theorems in stochastic processes, Bull. Amer. Math. Soc. 68 (1963), 72-77.

[14] Discriminant analysis, Annals of Inst. of Stat. Math. 15 (1963), 11-24.

[15] Bayes estimation with convex loss, Annals of Math. Stat. 34 (1963), 839-846, (with M. H. DeGroot).

[16] Stochastic give-and-take, J. Math. Anal. & Appl. 7 (1963), 489-498, (with M.H. DeGroot).

[17] Projections, generalized inverses, and quadratic forms, J. Math. Anal. & Appl. 9 (1964), 1-11, (with J. S. Chipman).

[18] Decomposition of vector measures, Proceedings of Nat'l. Acad. of Sciences 51 (1964), 771-774.

[19] Linear functionals on Orlicz spaces, Nieuw Archief voor Wiskunde 312 (1964), 77-98.

[20] The treatment of linear restrictions in regression analysis, Econometrica 32 (1964), 198-209, (with J.S. Chipman).

[21] Conditional expectations and closed projections, Indagationes Mathematicae 27 (1965), 100-112.

[22] Smoothness of Orlicz spaces-I and II, Indagationes Mathematicae 27 (1965), 671-680, 681-690.

[23] Existence and determination of optimal estimators relative to convex loss, Annals of Inst. of Stat. Math 17 (1965), 113-147.

[24] Interpolation, ergodicity, and martingales, J. of Math. & Mech. 16 (1965), 543-567.

[25] Inference in stochastic processes-II, Zeitschrift fur Wahrscheinlichkeitstheorie 5 (1966), 317-335.

[26] Approximations to some statistical tests, Trabajos de Estadistica 17 (1966), 85-100.

[27] Multidimensional information inequalities and prediction, Proceedings of Int'l. Symposium on Multivariate Anal., Academic Press, (1966) 287-313, (with M.H. DeGroot).

[28] Convolutions of vector fields and interpolation, Proceedings of Nat'l. Acad. Sciences 57 (1967), 222-226.

[29] Abstract Lebesgue-Radon-Nikodym theorems, Annali di Matematica Pura ed Applicata (4) 76 (1967), 107-132.

[30] Characterizing Hilbert space by smoothness, Indagationes Mathematicae 29 (1967), 132-135.

[31] Notes on pointwise convergence of closed martingales, Indagationes Mathematicae 29 (1967), 170-176.

[32] Inference in stochastic processes-III, Zeitschrift fur Wahrscheinlichkeitstheorie 8 (1967), 49-72.

[33] Characterization and extension of generalized harmonizable random fields, Proceedings Nat'l. Acad. Sciences 58 (1967), 1213-1219.

[34] Local functionals and generalized random fields, Bull. Amer. Math. Soc. 74 (1968), 288-293.

[35] Extensions of the Hausdorff-Young theorem, Israel J. of Math. 6 (1968), 133-149.

[36] Linear functionals on Orlicz spaces: General theory, Pacific J. Math. 25 (1968), 553-585.

[37] Almost every Orlicz space is isomorphic to a strictly convex Orlicz space, Proceedings Amer. Math. Soc. 19 (1968), 377-379.

[38] Predictions nonlineares et martingales d'operateurs, Comptes rendus (Academie des Sciences, Paris), Ser. A, 267 (1968) 122-124.

[39] Representation theory of multidimensional generalized random fields, Proceedings 2d Int'l. Sympt. Multivariate Anal., Academic Press (1969), 411-436.

[40] Operateurs de moyennes et moyennes conditionnelles, C.R. Acad. Sciences, Paris, Ser. A, 268 (1969), 795-797.

[41] Produits tensoriels et espaces de fontions, C.R. Acad. Sci., Paris 268 (1969), 1599-1601.

[42] Stone-Weierstrass theorems for function spaces, J. Math. Anal. 25 (1969), 362-371.

[43] Contractive projections and prediction operators, Bull. Amer. Math. Soc. 75 (1969), 1369-1373.

[44] Generalized martingales, Proceedings 1st Midwestern Symp. on Ergodic Theory & Prob., Lecture Notes in Math., Springer-Verlag, 160 (1970), 241-261.

[45] Linear operations, tensor products and contractive projections in function spaces, Studia Math. 38, 131-186, Addendum 48 (1970), 307-308.

[46] Approximately tame algebras of operators, Bull. Acad. Pol. Sci., Ser. Math. 19 (1971), 43-47.

[47] Abstract nonlinear prediction and operator martingales, J. Multivariate Anal. 1 (1971), 129-157, Erratum, 9. p. 646.

[48] Local functionals and generalized random fields with independent values, Teor. Verojatnost., Prem. 16 (1971), 466-483.

[49] Projective limits of probability spaces, J. Multivariate Anal. 1 (1971), 28-57.

[50] Contractive projections and conditional expectations, J. Multivariate Anal. 2 (1972), 262-381, (with N. Dinculeanu).

[51] Prediction sequences in smooth Banach spaces, Ann. Inst. Henri Poincare, Ser. B, 8 (1972), 319-332.

[52] Notes on characterizing Hilbert space by smoothness and smooth Orlicz spaces, J. Math. Anal. & Appl. 37 (1972), 228-234.

[53] Abstract martingales and ergodic theory, Proc. 3rd Symp. on Multivariate Anal., Academic Press (1973), 100-116.

[54] Remarks on a Radon-Nikodym theorem for vector measures, Proc. Symp. on Vector & Operator Valued Measures and Appl., Academic Press (1973), 303-317.

[55] Inference in stochastic processes-IV: Predictors and projections, Sankhya, Ser. A 36 (1974), 63-120.

[56] Inference in stochastic processes-V: Admissible means, Sankhya, Ser. A. 37 (1974), 538-549.

[57] Extensions of stochastic transformations, Trab. Estadistica 26 (1975), 473-485.

[58] Conditional measures and operators, J. Multivariate Anal. 5 (1975), 330-413.

[59] Compact operators and tensor products, Bull. Acad. Pol. Sci. Ser. Math. 23 (1975), 1175-1179.

[60] Two characterizations of conditional probability, Proc. Amer. Math. Soc. 59 (1976), 75-80.

[61] Conjugate series, convergence and martingales, Rev. Roum. Math. Pures et Appl. 22 (1977), 219-254.

[62] Inference in stochastic processes-VI: Translates and densities, Proc. 4th Symp. Multivariate Anal., North Holland, (1977), 311-324.

[63] Bistochastic operators, Commentationes Mathematicae, Vol. 21 March, (1978), 301-313.

[64] Asymptotic distribution of an estimator of the boundary parameter of an unstable process, Ann. Statistics 6 (1978), 185-190.

[65] Convariance analysis of nonstationary time series, Developments in Statistics 1 (1978), 171-225.

[66] Non L1-bounded martingales, Stochastic Control Theory and Stochastic Differential Systems, Lecture Notes in Control and Information Sciences, 16 (1979), 527-538, Springer Verlag.

[67] Processus lineaires sur Coo(G), C. R. Acad. Sci., Paris, 289 (1979), 139-141.

[68] Convolutions of vector fields-I, Math. Zeitschrift, 174 (1980), 63-79.

[69] Local Functionals on Coo(G) and probability, J. Functional Analysis 39 (1980), 23-41.

[70] Local functionals, Proceedings of Oberwolfach Conference on Measure Theory, Lecture Notes in Math. 794, Springer-Verlag (1980), 484-496.

[71] Structure and convexity of Orlicz spaces of vector fields, Proceedings of the F.B. Jones Conference on General Topology and Modern Analysis, University of California, Riverside (1981), 457-473.

[72] Representation of weakly harmonizable processes, Proc. Nat. Acad. Sci., 79, No. 9 (1981), 5288-5289.

[73] Stochastic processes and cylindrical probabilities, Sankhya, Ser. A (1981), 149-169.

[74] Application and extension of Cramer's Theorem on distributions of ratios, In Contributions to Statistics and Probability, North Holland (1981), 617-633.

[75] Harmonizable processes: structure theory, L'Enseignement Mathematique, 28 (1982), 295-351.

[76] Domination problem for vector measures and applications to non-stationary processes, Oberwolfach Measure Theory Proceedings, Springer Lecture Notes in Math. 945 (1982), 296-313.

[77] Bimeasures and sampling theorems for weakly harmonizable processes, Stochastic Anal. Appl. 1 (1983), 21-55, (with D.K. Chang).

[78] Filtering and smoothing of nonstationary processes, Proceedings of the ONR workshop on "Signal Processing", Marcel-Dekker Publishing (1984), 59-65.

[79] The spectral domain of multivariate harmonizable processes, Proc. Nat. Acad. Sci., 81 (1984), 4611-4612.

[80] Harmonizable, Cramer, and Karhunen classes of processes, Handbook in Statistics, Vol. 5 (1985), 279-310.

[81] Bimeasures and nonstationary processes, Real and Stochastic Analysis, Wiley & Sons (1986), 7-118, (with D.K. Chang).

[82] A commentary on "On equivalence of infinite product measures", in S. Kakutani's selected works, Birkhauser Boston Series (1986), 377-379.

[83] Probability, Academic Press, Inc., New York, Encyclopedia of Physical Science and Technology, Vol. 11 (1987), pp. 290-310.

[84] Special representations of weakly harmonizable processes, Stochastic Anal. Appl., 6 (1988), 169-189, (with D.K. Chang).

[85] Paradoxes in conditional probability, J. Multivariate Anal., 27, (1988), pp. 434-446.

[86] Harmonizable signal extraction, filtering and sampling Springer-Verlag, New York, Topics in Non-Guassian Signal Processing, Vol. II (1989), pp. 98-117.

[87] A view of harmonizable processes, North-Holland, New York, in Statistical Data Analysis and Inference (1989), pp. 597-615.

[88] Bimeasures and harmonizable processes; (analysis, classification, and representation), Springer-Verlag Lecture Notes in Math., 1379, (1989), pp. 254-298.

[89] Sampling and prediction for harmonizable isotropic random fields, J. Combinatorial Analysis, Information & System Sciences, Vol 16 (1991), pp. 207-220.

[90] L2,2 - boundedness, harmonizability and filtering, Stochastic Anal. Appl., Vol 10 (1992), pp. 323-342.

[91] Probability (expanded for 2nd ed.), Encyclopedia of Physical Science and Technology, Vol 13 (1992), pp. 491-512.

[92] Stochastic integration: a unified approach, C. R. Acad. Sci., Paris, Vol 314 (Series 1), (1992), pp. 629-633.

[93] A projective limit theorem for probability spaces and applications, Theor. Prob. and Appl., Vol 38 (1993), (with V. V. Sazonov, in Russian), pp. 345-355.

[94] Exact evaluation of conditional expectations in the Kolmogonov model, Indian J. Math., Vol 35 (1993) pp 57-70.

[95] An approach to stochastic integration (a generalized and unified treatment), in Multivariate Analysis: Future Directions, Elsivier Science Publishers, The Netherlands (1993), pp. 347-374.

[96] Harmonizable processes and inference: unbiased prediction for stochastic flows, J. Statistic. Planning and Inf., Vol 39 (1994), pp. 187-209.

[97] Packing in Orlicz sequence spaces, (1995), (with Z. D. Ren), 18 pages (to appear).

[98] Characterization of isotropic harmonizable covariances and representations, (1995), 14 pages, (to appear).

Books Edited

[1] General Topology and Modern Analysis. Proceedings of the F.B. Jones Conference, Academic Press, Inc., New York (1981), 514 pages, (Edited jointly with L.F. McCauley).

[2] Handbook in Statistics, Volume 5, Time Series in the Time Domain, (Edited jointly with E.J. Hannan, P.R. Krishnaiah), North-Holland Publishing Co., Amsterdam (1985), 484

pages.

[3] Real and Stochastic Analysis, (Editor), Wiley & Sons, New York (1986), 347 pages.

[4] Multivariate Statistics and Probability, (Edited jointly with C.R. Rao), Academic Press Inc., Boston (1989), 565 pgs.

Books Written

[5] Stochastic Processes and Integration. Sijthoff & Noordhoff International Publishers, Alpehn aan den Rijn, The Netherlands, (1979), 460 pages.

[6] Foundations of Stochastic Analysis, Academic Press, Inc., New York, (1981), 295 pages.

[7] Probability Theory with Applications, Academic Press, Inc. New York, (1984), 495 pages.

[8] Measure Theory and Integration, Wiley-Interscience, New York (1987), 540 pages.

[9] Theory of Orlicz Spaces (jointly with Z. D. Ren), Marcel Dekker Inc., New York (1991), 449 pages.

[10] Conditional Measures and Applications, Marcel Dekker Inc., New York (1993), 417 pages.

[11] Stochastic Processes: General Theory, Kluwer Academic Publishers, The Netherlands (1995), 620 pages.

A Family Tree of Mathematical Ancestors

Godfrey H. Hardy
⇓
Norbert Wiener
⇓
Robert H. Cameron
⇓
Munroe D. Donsker
⇓
Malempati M. Rao

Ph.D. Theses Completed Under the Direction of M. M. Rao and His Students

M. M. Rao at Carnegie-Mellon University

Dietmar R. Borchers (1964), "Second order stochastic differential equations and related Ito processes."

J. Jerry Uhl. Jr (1966), "Orlicz spaces of additive set functions and set martingales."

Jerome A. Goldstein (1967), "Stochastic differential equations and nonlinear semi-groups."

Neil E. Gretsky (1967), "Representation theorems on Banach function spaces."

William T. Kraynek (1968), "Interpolation of sub-linear operators on generalized Orlicz and Hardy spaces."

Robert L. Rosenberg (1968), "Compactness in Orlicz spaces based on sets of probability measures."

George Y. H. Chi (1969), "Nonlinear prediction and multiplicity of generalized random processes."

M. M. Rao at University of California, Riverside

Vera Darlean Briggs (1973), "Densities for infinitely divisible processes."

Stephen V. Noltie (1975), "Integral representations of chains and vector measures."

Theodore R. Hillmann (1977), "Besicovitch - Orlicz spaces of almost periodic functions."

Michael D. Brennan (1978), "Planar semi-martingales and stochastic integrals."

James P. Kelsh (1978), "Linear analysis of harmonizable time series."

Alan C. Krinik (1978), "Stroock - Varadhan theory of diffusion in a Hilbert space and likelihood ratios."

Derek K. Chang (1983), "Bimeasures, harmonizable process and filtering."

Marc H. Mehlman (1990), "Moving average representation and prediction for multidimensional strongly harmonizable process."

Randall J. Swift (1992), "Structural and sample path analysis of harmonizable random fields."

Michael L. Green (1995), "Multi-parameter semi-martingale integrals and boundedness principles."

Heroe Soedjak (1996), "Estimation problems for harmonizable random processes and fields."

Jerome A. Goldstein at Tulane University

Charles J. Monlezun (1972), "Temporally inhomogeneous scattering theory."

James T. Sandefur, Jr. (1974), "Higher order abstract Cauchy problems."

Joseph H. Hendrickson (1974), "Temporally inhomogeneous scattering with modified wave operators."

Bruno J. Wichnoski (1974), "Temporally inhomogeneous scattering theory."

B. Clay Burch (1975), "A semigroup treatment of the Hamilton-Jacobi equation in several space variables." (co-advised with E. D. Conway)

Janet M. Hughes (Diem) (1976), "Estimation in a non-stationary Markov chain." (co-advised with C.B. Bell and A. Levine)

Kuong Lin OuYoung (1978), "Periodic solutions of conservation laws."

Michael E. Ballotti (1983), "Modern versions of the theorems of Kneser and Wiener."

Gisèle Ruiz Rieder (1986), "Mathematical contributions to Thomas-Fermi Theory."

Lige Li (1986), "Positive solutions of some predator-prey interacting systems." (co-advised with E.D. Conway)

Chin-Yuan Lin (1987), "Degenerate non-linear parabolic boundary value problems."

Douglas D. Pickett (1989), "Scattering theory for higher order equations."

Mi Ai Park (1990), "Model equations in fluid dynamics."

Gabriella Segall (1991), "Linear integro-differential equations in Banach spaces."

Denise Kirschner (1991), "Epidemiology and immunology in AIDS."

Chien-an Lung (1992), "The nuclear cusp condition in spin polarized Thomas-Fermi theory."

Jerome A. Goldstein at Louisiana State University

Genbao Shi (1995), "Mathematical contributions to elasticity and quantum theory."

Kunyang Wang (1995), "The generalized Kompaneets equation."

Andrei Breazna, in progress.

Aurora Breazna, in progress.

Radu Cascaval, in progress.

Matthew Cliff, in progress.

Neil Gretsky at University of California, Riverside

Joseph J. Sroka (1987), "Factorization of pNA games through Banach spaces."

Bahman Soltani (1989), "A game-theoretic analysis of hostile takeovers."

Zhongdau Ren (1994), "Packing in Orlicz function spaces."

Timothy Watson (1994), "Potential in infinite games."

Jaime Londono, in progress.

J.J. Uhl, Jr. at University of Illinois, Champaign-Urbana

Barry Turett (1976)

Elias Saab (1979)

Paulette Saab (1979)

Robert Geitz (1980)

Kevin Andrews (1980)

Frank Page (1981)

Larry Riddle (1982)

Minos Petrakis (1985)

Russ Gordon (1986)

Maria Girardi (1989)

Lige Li at Kansas State University

Afshin Ghoreishi (1990), "Positive solutions of interacting models in a heterogeneous enviroment under mixed boundary conditions."

Roger Logan (1990), "A study of a two-species competing interaction model in mathematical biology."

Inkying Ahn (1992), "Elliptic interacting systems with nonlinear diffusions."

Mufid Abudiab (1993), "Dynamic analysis of AIDS infection: Reaction-diffusion modeling."

Yaping Liu (1993), "Positive solutions to general ellipic interacting systems."

Elias Saab at University of Missouri, Columbia
Musbah Abdul-Al Salam
Narcisse Randianantoanina

Paulette Saab at University of Missouri, Columbia
Brenda Smith

Some Problems of Real and Stochastic Analysis Arising from Applications

MALEMPATI M. RAO Department of Mathematics, University of California, Riverside, CA 92521-0135

Introduction The following is a brief account of some problems coming from applications, chiefly economics, meteorology, physics and statistical practice. They have led to substantial studies in real and stochastic analysis with a nontrivial interaction between these areas leading to further understanding of the subjects. The ensuing questions have helped increase the interests of several of my colleagues and me in those areas of mathematics. More often the problems that came to our attention are those for which one has to go beyond the usual limits of the classical areas that were considered by many people in the past, and this appears often enough to keep the curiosity, as seen from the accounts described below.

I. Idempotent matrices and non-symmetric projections In some studies of economic analysis, for instance in the work with Cobb-Douglas production functions, linear restrictions play an important role after a suitable conversion of the data. The problem here, after a logarithmic transformation, leads to linear regression with restrictions which involved a use of non-symmetric idempotent matrices. This was treated in detail by John Chipman and me in the latter part of 1950's and the results were first issued as a report and then written up in two papers separating the applications and structural analysis in 1964. A typical result is the following:

Theorem 1. *Let X be an n-vector which is normally distributed with mean μ and covariance Σ. Let A be an n-by-n matrix. Then the quadratic form $Q = X^tAX$ is distributed as a non-central chi-squared variable with r degrees of freedom (=rank of A) with the non-centrality parameter $\lambda = \frac{1}{2}\mu^t A\mu$, if and only if the (not necessarily symmetric) matrix $A\Sigma$ is idempotent of rank r. [Here X^t, μ^t are transposes of the vectors X, μ.]*

The non-symmetric idempotent (=projection) operators were not treated thoroughly in the literature until then. The problem was analyzed by Victor Mizel and myself (1962, 1963) in the Hilbert space context, first for (non-symmetric) projections which have several curious properties (regarding their spectral and convergence aspects), and then for operators satisfying a quadratic equation $ax^2 + bx + c = 0$. A basic assertion on the spectral analysis is given by the following:

Theorem 2. *A continuous (non-symmetric) linear operator $T : \mathcal{H} \to \mathcal{H}$, a Hilbert space, satisfies the equation $T^2 - bT + cI = 0$ if and only if there is a selfadjoint positive definite operator $S : \mathcal{H} \to \mathcal{H}$ such that (i) $S^2 - bS + cI = 0$, (ii) a unitary operator U which is invariant on the closure of the range $\bar{R}_{S^2-bS+cI}$, and (iii) its restriction to the latter space satisfies: (a) $U^2 = -I$, (b) $SU = U(bI - S)$, (c) $T = S + U(S^2 - bS + cI)^{\frac{1}{2}}$.*

When $b = 1$ and $c = 0$, this was first studied in (1962) and then the work was extended to the general case stated in the above theorem. Later these results led to a structural analysis of Reynolds operators and conditional expectations. A detailed account of some of this analysis is given in my paper (1975).

1. J. S. Chipman and M. M. Rao (1964), "The treatment of linear restrictions in regression analysis," *Econometrica,* **32**, 198–209.
2. ______ ______ (1964),"Projections, generalized inverses and quadratic forms," *J. Math. Anal. Appl.,* **9**, 1–11.
3. V. J. Mizel and M. M. Rao (1962), "Non-symmetric projections in Hilbert space," *Pacific J. Math.,* **12**, 343–357.
4. ______ ______ (1963), "Averaging and quadratic equations in operators," *Technical Report No.***9**, Carnegie-Mellon Univ., Pittsburgh, PA.
5. M. M. Rao (1975), "Conditional measures and operators," *J. Multivar. Anal.,* **5**, 330–413.

II. Generalized Cramér-Rao inequalities and function spaces Suppose X is a random variable (or vector) having an absolutely continuous distribution F with density $f(\cdot, \theta)$ depending on a parameter $\theta \in A$, an open (nonempty) interval. If $\frac{\partial f}{\partial \theta}$ exists and is dominated by an F-integrable function, and if $\hat{\theta}(= \hat{\theta}(X))$ is a Borel function of X, called an estimator of θ, let $E_\theta(\hat{\theta}) = \int_\Omega \hat{\theta}(X)dP = \theta + b(\theta)$. Then

$$(*) \qquad E_\theta[(\hat{\theta} - \theta)^2] \geq (1 + \frac{db}{d\theta})^2/[E_\theta(\frac{\partial \log f(X, \theta)}{\partial \theta})^2]$$

holds, and is called the *Cramér-Rao inequality.* This depends on a clever use of the CBS (=Cauchy-Buniyakowski-Schwarz) inequality presented in Cramér's book (cf. H. Cramér (1946), p.479), and was also derived independently at about the same time by C. R. Rao (hence the name C-R inequality) who published it in a paper in Calcutta. The left side of (*) can be replaced by $E_\theta(|\hat{\theta}-\theta|^p), 1 < p < \infty$, and the corresponding result based on Hölder's inequality is valid, and it was given by E. W. Barankin (1949). Unaware of Barankin's work, I obtained the same result ten years later. (I. R. Savage informed me of Barankin's results on seeing my manuscript.) After analyzing the computations and applications, I realized that the convexity of a function φ is actually at work, and hence sought a lower bound for $E_\theta(\varphi(\hat{\theta}-\theta))$ where $\varphi : \mathbb{R} \to \mathbb{R}^+$ is a symmetric Young function, with $\varphi(x) = 0$ if and only if $x = 0$, so that $\varphi(x) = |x|^p, p \geq 1$ includes the above results. This led to a set of inequalities, a typical one of which is as follows. Let $\varphi(x) = \varphi_1(x)^k$ for some $k \geq 1$ where φ_1 satisfies the same conditions as φ. Let $D_1 = \frac{\partial f}{\partial \theta}(X)$ as above. Then the inequality below obtained in (1959) reduces to Barankin's when $\varphi_1(x) = |x|$, and to (*) if $k = 2$ in addition. A multidimensional extension was given in (1963), after a series type bound (extension of Bhattacharyya bounds) in (1961), and an application to Bayes estimation was then obtained by Morris DeGroot and me (1963).

Theorem 1. *Under the regularity conditions and the integrability of $\varphi(D_1)$, as in (*), one has with $b' = \frac{db}{d\theta}$,*

$$E_\theta(\varphi(\hat{\theta}-\theta)) \geq \varphi_1\Big(\frac{1+b'(\theta)}{E_\theta(|D_1|)}\Big)^k \sup_{k\geq 1}\Big[\frac{E_\theta(|D_1|)}{\|D_1\|_{k'}}\Big]^{k'},$$

where $k' = k/(k-1)$ and if $k' = \infty$, then $\|D_1\|_{k'}$ is the essential supremum norm.

1. E. W. Barankin (1949), "Locally best unbiased estimates," *Ann. Math. Statist.*, **20**, 477–501.
2. H. Cramér (1946), *Mathematical Methods of Statistics*, Princeton Univ. Press, Princeton, NJ.
3. M. H. DeGroot and M. M. Rao (1963), "Bayes estimation with convex loss," *Ann. Math. Statist.*, **34**, 839–846.
4. M. M. Rao (1959), "Lower bounds for risk functions in estimation," *Proc. Nat. Acad. Sci.*, **45**, 1168–1171.
5. ________ (1961), "Theory of lower bounds for risk functions in estimation," *Math. Annalen*, **143**, 379–398.
6. ________ (1963), "Some inference theorems in stochastic processes," *Bulletin Am. Math. Soc.*, **69**, 72–77.

(i) *Dual spaces $(L^\varphi)^*$ on the scene.* The preceding work on convex (loss) functions was not sufficient for constructing the best $\hat{\theta}$ which gives equality in the above lower bounds, as was possible in Barankin's case. The corresponding study demands an analysis of L^φ-spaces (called Orlicz spaces), and their adjoint spaces $(L^\varphi)^*$. The structure of the latter space was an open problem at the time, as indicated in the book by Krasnosel'skii and Rutickii

(1960). This problem was solved after much work (1964, 1968). It can be stated as follows. Let $\psi : x \mapsto \psi(x) = \sup\{|x|y - \varphi(y) : y \geq 0\}$, the complementary Young function to φ. An additive set function $\nu : \mathcal{A} \to \mathbb{R}$, where $\mathcal{A}$ is an algebra of subsets of a set Ω, has finite ψ-bounded variation relative to μ if

$$\|\nu\|_\psi = \sup\{\sum_{i=1}^{n} \psi(\frac{\nu(A_i)}{\mu(A_i)})\mu(A_i) : A_i \quad disjoint, \quad A_i \in \mathcal{A}\} < \infty.$$

Then one has on writing $\Sigma = \sigma(\mathcal{A})$:

Theorem 2. $\left(L^\varphi(\mu)\right)^* = \mathfrak{A}_\psi(\mu) = \{\nu : \Sigma \to \mathbb{R} | \quad \|\nu\|_\psi < \infty\}$.

For the construction problem of optimal estimators, it is also necessary to analyze $\mathfrak{A}_\psi(\mu)$ for smooth φ. This was done in (1965a) and the desired construction of the estimator $\hat{\theta}$ was accomplished in (1965b). However, the structure of the space $(\mathfrak{A}_\psi(\mu))^* = (L^\varphi(\mu))^{**}$ was also needed in its generality for some related applications. If $\varphi(x) = |x|^p, p \geq 1$, the corresponding space was analyzed by S. Leader (1953), but the more general $(L^\varphi(\mu))^{**}$ has not been studied before. The problem was solved by J. Jerry Uhl (1967) who also considered the vector valued function space $(L^\varphi_{\mathcal{X}}(\mu))^{**}$ and analyzed some of its subspaces when $\mathcal{X}$ is a Banach space. But here a new Radon-Nikodým theorem for $\mathcal{X}$-valued additive set functions was needed. He settled these questions and went on to apply the results for (additive) vector set martingales (1969) and then chose to classify Banach spaces $\mathcal{X}$ with the Radon-Nikodým property as the next project.

1. M. A. Krasnosel'skii and Ya. B. Rutickii (1960), *Convex Functions and Orlicz Spaces*, P. Noordhoff Ltd., Groningen, The Netherlands.
2. S. Leader (1953), "The theory of L^p spaces for finitely additive set functions," *Ann. Math.*, **58**, 528–543.
3. M. M. Rao (1964) "Linear functionals on Orlicz spaces," *Nieuw Arch. Wisk.*, **12**, 77–98.
4. ________ (1965a), "Smoothness of Orlicz spaces," *Indag. Math.*, **27**, 671–690.
5. ________ (1965b), "Existence and determination of optimal estimators relative to convex loss," *Ann. Inst. Statist. Math.*, **17**, 133–147.
6. ________ (1968), "Linear functionals on Orlicz spaces: general theory," *Pacific J. Math.*, **25**, 553–585.
7. J. J. Uhl, Jr (1967), "Orlicz spaces of additive set functions," *Studia Math.*, **29**, 19–58.
8. ________ (1969), "Martingales of vector valued set functions," *Pacific J. Math.*, **30**, 533–548.

(ii) *Some extensions and applications.* In this study a great many arguments appear to generalize for Banach function spaces $L^\rho(\mu)$, studied by several people at the time, especially by A. C. Zaanen and his associates in the Netherlands. Here $\rho(\cdot)$ is a function norm, and $L^\rho(\mu)$ is also called a Riesz space. The corresponding problem of a characterization of

$(L^{\rho}_{\mathcal{X}}(\mu))^*$ has been analyzed in detail by Neil Gretsky and his results appear in a memoir (1968). Later he and Uhl (1972, 1981) used these ideas in obtaining various other results.

During this time, certain applications to ergodic theory based on the Lebesgue spaces $L^p, 1 < p < \infty$, were being studied by R. V. Chacón, and a corresponding analysis in L^{φ}-spaces was natural but was dependent on an interpolation theorem. The desired extension (even with changes of measures) and applications to ergodic theory and martingales, especially for Reynolds operators considered by Rota (1964) in the context of L^2-spaces, were obtained in (1966). A further generalization of some of the latter results to sublinear operators and a detailed analysis of it in Hardy-Orlicz spaces (being initiated by a student of W. Orlicz at that time) were undertaken in his thesis by William Kraynek (1970, 1972).

1. N. E. Gretsky (1968), "Representation theorems for Banach function spaces," *Memoirs Am. Math. Soc.*, **84**, 1–56.

2. ________ and J. J. Uhl, Jr (1972) "Bounded linear operators on Banach function spaces of vector valued functions," *Trans. Am. Math. Soc.*, **167**, 263–277.

3. ______ ______ (1981), "Carleman and Korotkov operators on Banach spaces," *Acta Sci. Math.*, **43**, 207–218.

4. W. T. Kraynek (1970), "Interpolation of multilinear functionals on $L^{\varphi\psi}$-spaces," *J. Math. Anal. Appl.*, **31**, 414–430.

5. ________ (1972), "Interpolation of sublinear operators on generalized Orlicz and Hardy-Orlicz spaces," *Studia Math.*, **43**, 93–123.

6. M. M. Rao (1966), "Interpolation, ergodicity and martingales," *J. Math. Mach.*, **16**, 543–567..

7. G.-C. Rota (1964), "Reynolds operators," *Proc. Am. Math. Soc. Symp. Appl. Math.*, **16**, 70–83.

(iii) *Nonlinear prediction and related problems.* During these investigations some applications to probability theory have been made. The smoothness of $L^{\varphi}(\mu)$-space analysis has led to a study of prediction problems on these spaces. I considered the problem on real L^{φ}-spaces (1965) and also the $L^p_{\mathcal{X}}$-case for $1 < p < \infty$ with $\mathcal{X}$ as a smooth Banach space (1967). A further analysis has been undertaken to determine the structure of projection operators, and their relation to the conditional expectation operators, by Nicolae Dinculeanu and me (1974). The work led to an abstract formulation of prediction problems in general smooth Banach spaces (1972). Thus a familiarity with Orlicz spaces provided a strong motivation to obtain several results that can be studied on the L^p-spaces to the more inclusive L^{φ}-versions. Such an extension in most cases is not straightforward since unlike the former case with $1 < p < \infty$, the reflexivity, uniform convexity, smoothness and other properties are all distinguished and separated in Orlicz spaces. For this reason, these are useful "test spaces" to try new ideas and methods. In this way, Theodore Hillmann (1977) has studied the Besicovitch-Orlicz spaces of almost periodic functions. In a similar manner, there is an analysis of Fenchel-Orlicz spaces of vector functions by J. Barry Turett (1980), a thesis written under the supervision of Uhl.

Thus most of the advances since the publication of Krasnosel'skii-Rutickii volume noted earlier, obtained by my associates, me, and several other researchers abroad, have recently been compiled in a monograph in collaboration with Z. D. Ren who came from P. R. China to spend time here at Riverside with us. He brought the work on Orlicz spaces from his side, since much of it was published in the Chinese language, and fortunately he can read Russian to verify the corresponding contributions from that side. Also further results on sphere packing in Orlicz spaces were recently completed by him (1994), in a thesis under Gretsky's guidance. Thus at Riverside an important phase of this work has been accomplished.

1. N. Dinculeanu and M. M. Rao (1974) "Contractive projections and conditional expectations," *J. Multivar. Anal.*, **2**, 362–381.
2. T. R. Hillmann (1977), "Besicovitvh-Orlicz spaces of almost periodic functions," Ph.D. thesis at UCR, (see also Chapter 2 in *Real and Stochastic Analysis*, Wiley, New York, (1986), 119–167).
3. M. M. Rao (1965) "Conditional expectations and closed projections," *Indag. Math.*, **27**, 100–112.
4. ________ (1967), "Inference in stochastic processes-III," *Z. Wahrs.*, **8**, 49–72.
5. ________ (1972), "Prediction sequences in smooth Banach spaces," *Ann. Inst. H. Poincaré, Ser. B*, **8**, 319–332.
6. M. M. Rao and Z. D. Ren (1991), *Theory of Orlicz Spaces*, Marcel Dekker, Inc., New York.
7. Z. D. Ren (1994), "Packing in Orlicz function spaces," Ph.D. thesis at UCR, Riverside, CA.
8. J. B. Turett (1980), "Fenchel-Orlicz spaces," *Dissertationes Math.*, **181**, 1–55.

III. Higher order stochastic differential equations In an early 1960 visit to Carnegie Tech, Prof. S. Kakutani remarked to me that there are still some fresh ideas in the classical paper on applications of probability to physics, by S. Chandrasekhar (1943), and I should read it. There is a treatment of Langevin's equation which is of first order and then there is also a description of the motion of a simple harmonic oscillator driven by a random disturbance. The latter is a second order linear equation. Several works exist in the literature extending the Langevin's linear equation to nonlinear problems using Itô's theory of stochastic integration. For a detailed modern treatment of the subject, see, for instance, Stroock and Varadhan (1979), and for infinite dimensional versions Yor (1974) and Krinik (1986). It seemed to me that the corresponding equation of a simple harmonic oscillator should be similarly generalized and studied, using a suitable definition of (higher order) stochastic derivative. When the work on this problem was started, Dietmar Borchers, a graduate student at Carnegie, was looking for a dissertation topic and I gave this to him when he wanted to work on his thesis with me. The equation is:

$$(+) \qquad d\dot{X}_t + q(t, X_t, \dot{X}_t)dt = \sigma(t, X_t, \dot{X}_t)dZ_t,$$

with $X_0 = A, \dot{X}_0 = B$ as initial conditions when in fact all these are finite vector (or suitable matrix) functions, and Z_t is a martingale process with independent increments. Under some suitable Lipschitz conditions and using an appropriate mean derivative concept, Borchers (1964) established the existence and uniqueness of the solutions of (+)

and showed that the vector process $V_t = (X_t, \dot{X}_t)$ is Markovian. He also calculated the generator of the associated semigroup of linear operators when the Z-process is Brownian motion and q, σ of (+) are independent of the time variable t. The solution X was termed the *second order Itô process*. After completing his work Borchers returned to his native land (Germany) and did not pursue or publish his work. At about this time, Jerome Goldstein started as a graduate student moving into this area, considered first the existence problem for (higher order) linear vector equations in a general setting (1967), and turned to continue his studies concentrating on equation (+).

Thus, Goldstein worked on the operator theoretical treatment related to the general second order Itô process as well as their path properties. The infinitesimal operator of the above V_t-process, with Z as Brownian motion, turns out to be a degenerate elliptic operator, on a certain continuous function space on $\mathbb{R}^2$ made into a Banach space relative to a weighted uniform norm, and it belongs to a class of partial differential operators not yet fully explored at that time as described by O. A. Oleinik at the ICM congress at Nice in 1970. Several interesting path properties of these Itô processes were obtained by Goldstein (1969) under some restrictions. The associated operator family of the V_t-processes, in general, does not form a semigroup (unless q, σ of (+) are independent of t), but it satisfies an abstract evolution equation. This is closely related to the (abstract) Cauchy problem and after presenting an analysis of this family (1969), he has been spending most of his time in this area of PDE and nonlinear analysis ever since. There is still a great need for further study of the path properties, extension of the Yosida-Kakutani theory of the operator treatment of Markov processes with stationary transitions, and a generalization of the work if Z is, for instance, a martingale.

For the linear and nonlinear equations noted above one has two types of stochastic integrals to work with. The first one is the Cramér-Kolmogorov-Wiener type and the second one (nonlinear case) is a martingale or a generalized Itô type integral. Both of these can be unified by utilizing and extending a boundedness principle formulated by Bochner (1956). A simplified version of it can be stated as: $X = \{X_t, t \in [a, b]\}$ on (Ω, Σ, P) is $L^{p,q}$*-bounded* if there is a σ-finite measure μ on $\Sigma \otimes \mathcal{B}([a, b])$, such that for simple functions $f \mapsto f(t, \omega) = \sum_{i=1}^{n} a_i(\omega)\chi_{[t_i, t_{i+1})}(t)$, a_i is $\sigma(X_s, s \leq t)$-adapted, one has

$$E\Big(\int_a^b f(t, \cdot)dX_t\Big)^p \leq K \int_{\Omega\times[a,b]} |f(t, \omega)|^q d\mu(t, \omega),$$

where $K > 0$ is an absolute constant depending only on $p > 0, q > 0$. The first integral in the above for simple functions is the standard sum. Now the $L^{p,q}$-bounded X can be used to integrate more general functions, and the thus obtained stochastic integrals include semimartingale, Stratonovich as well as Skorokhod integrals. Interestingly, Orlicz spaces play a role in this extended principle for showing that such a result is essentially optimal since then one has to replace p, q above with Young functions $\varphi_i, i = 1, 2$, and I indicated these ideas in (1992) and will include a detailed account with an extension of Borcher's work from this point of view in the second edition of my book (1979), under preparation. Much remains and much can be done in this area as well.

1. S. Bochner (1956), "Stationarity, boundedness, almost periodicity of random valued functions," *Proc. Third Berkeley Symp. Math. Statist. Prob.*, **2**, 7–27.

2. D. R. Borchers (1964), "Second order stochastic differential equations and related Itô processes," Ph.D. thesis at Carnegie-Mellon Univ., Pittsburgh, PA.
3. S. Chandrasekhar (1943), "Stochastic problems in physics and astronomy," *Rev. Modern Phy.*, **15**, 1–89. (Reprinted in *Selected Papers on Noise and Stochastic Processes,* (Ed. N. Wax), Dover Publications, New York.
4. J. A. Goldstein (1967), "An existence theorem for linear stochastic differential equations," *J. Diff. Eqns.*, **3**, 78–87.
5. ________ (1969), "Abstract evolution equations," *Trans. Am. Math. Soc.*, **141**, 159–185.
6. ________ (1969), "Second order Itô processes," *Nagoya Math. J.*, **36**, 27–63.
7. A. Krinik (1986), "Diffusion processes in Hilbert Space and likelihood ratios," in *Real and Stochastic Analysis,* Wiley, New York, 168–220. [Condensed from a Ph.D. thesis at UCR, Riverside, CA, 1978.]
8. M. M. Rao (1979) *Stochastic Processes and Integration,* Sijthoff and Noordhoff, Alphen aan den Rijn, The Netherlands, (2nd revised and enlarged edition, published by Kluewer in 1995).
9. ________ (1992), "Stochastic Integration: a unified approach," *C. R. Acad. Sci., Ser. A,* **314**, 629–633.
10. D. W. Stroock and S. R. S. Varadhan (1979), *Multidimensional Diffusion Processes,* Springer-Verlag, Berlin.
11. M. Yor (1974), "Existence et unicité de diffusions à valeurs dans un espace de Hilbert," *Ann. Inst. H. Poincaré, Ser. B,* **10**, 55–88.

IV. Unstable stochastic difference equations Stochastic fluctuations pertaining to stock market and other economic behavior can be modelled with difference equations. A first order equation of the type $X_t = X_{t-1} + \varepsilon_t$ is sometimes termed a random walk model. Here X_t is the value of a stock at time t, and it is the same as the previous observation with an additive random disturbance ε_t. Generalizing this, one can consider the kth order equation:

$$X_t + \alpha_1 X_{t-1} + \cdots + \alpha_k X_{t-k} = \varepsilon_t, \quad t = 0, 1, 2, \ldots \tag{++}$$

where ε_t's are independent identically distributed random variables with mean zero and variance one. The α_k's are constants. If the characteristic equation of (++), namely, $\lambda^k + \alpha_1 \lambda^{k-1} + \cdots + \alpha_k = 0$, has all its roots inside the unit circle, then the X_t-process is called *stable.* If at least one root is on the circumference of the circle (and the rest inside), then X_t is *unstable* and if at least one root is outside (and at least one inside) the circle, it is called an *explosive* process. These terms describe, among other things, the behavior of the variance of X_t as $t \to \infty$. Since usually the α_i are unknown, one estimates them by the least squares method, based on a set of N observations; and the limit distributions of the estimators, as $N \to \infty$, are of interest for economic and other studies.

The first rigorous treatment of such problems from a general point of view was given by H. B. Mann and A. Wald (1943), if all the roots of the characteristic equation are distinct, inside the unit circle, and all moments of the ε_t exist. If $k = 1$ and the ε_t are standard normal random variables, then John White (1958) studied the asymptotic behavior of the estimator when $|\alpha| \geq 1$, and T. W. Anderson (1959) for general k if all the roots are either outside or all are inside the unit circle, and that the ε_t have two moments finite. I have considered the case that one root is outside and the rest are inside the unit circle (1961). It turns out that the Donsker invariance principle is not generally valid for the unstable cases, and in fact the limit distribution depends on that of the ε_t's in the explosive case. The problem was later studied in some generality by T. L. Lai and C. Z. Wei (1983) for the strong consistency when the ε_t's are martingale differences but having $2 + \delta(\delta > 0)$ moments. The case with $\delta = 0$, as well as the limit distributions of the estimators in that context are still unresolved. If $k = 1$ and $|\alpha| = 1$, then the limit distribution is already complicated, as shown by me (1978), and the case that $k > 1$ with some roots on the unit circle is yet to be solved.

1. T. W. Anderson (1959), "On asymptotic distribution of estimates of parameters of stochastic difference equations," *Ann. Math. Statist.*, **30**, 676–687.
2. T. L. Lai and C. Z. Wei (1983), "Asymptotic properties of general autoregressive models and strong consistency of least squares estimators of the parameters," *J. Multivar. Anal.*, **13**, 1–23.
3. H. B. Mann and A. Wald (1943), "On the statistical treatment of linear stochastic difference equations," *Econometrica*, **11**, 173–220.
4. M. M. Rao (1961), "Consistency and limit distributions of estimators of parameters in explosive stochastic difference equations," *Ann. Math. Statist.*, **32**, 195–218.
5. ________ (1978), "Asymptotic distribution of an estimator of the boundary parameter of an unstable process," *Ann. Statist.*, **6**, 185–190.
6. J. S. White (1958), "The limiting distribution of the serial correlation coefficient in the explosive case," *Ann. Math. Statist.*, **29**, 1188–1197.

V. Harmonizable processes In the 1940's Loève defined a class of second order processes, called harmonizable, whose covariance is the Fourier transform of a continuous function (necessarily positive definite) of finite Vitali variation in the complex plane. This generalization of (Khintchine's) stationarity prompted (and he asked for) a characterization of the corresponding harmonizable covariance. This is messy and in 1967, I gave a solution which reduces to the classical Bochner-Khintchine case for stationary processes. In 1956 in the third Berkeley symposium paper, quoted in Section III, Bochner introduced a concept called "V-boundedness" and noted that it includes the Loève harmonizability, and then presented some of its properties. Later Rozanov (1959) defined a related notion, also termed "harmonizable", and made some applications. It is found that Rozanov's definition is more general than Loève's and in fact it coincides with Bochner's concept of V-boundedness. In his thesis Niemi (1975) studied the Fourier transforms of certain Hilbert space valued measures directly. To understand and classify these classes, I have called the Loève concept *strong* harmonizability, and the Bochner-Rozanov concept *weak* harmonizability, and presented a structure theory of these processes (1982). To clarify the distinctions noted above, it was necessary to use the Morse-Transue theory of integration of scalar functions relative to a bimeasure of finite Fréchet variation. This integral is weaker than the Lebesgue-Stieltjes concept and coincides with the latter iff the Fréchet variation is replaced by the Vitali variation.

A slightly restricted form of the MT-integral was useful for this work. This with applications to linear filtering were then developed, jointly with D. K. Chang (1986), and further extensions to random fields in (1989). The area of harmonizable processes is now suitable for analysis by extending (nontrivially) many results from the stationary case. Some applications of these processes were considered in studies begining with J. Kelsh (1978), and continuing the work by Marc Mehlman (1991). It was noted in the book by Yadrenko (1983) that only the trivial (i.e., constant) stationary isotropic random fields satisfy the Laplacian. This motivated an introduction of isotropy for harmonizable classes for which nontrivial solutions of the above equation exist, and the desired concept is presented and studied in (1991). Further structural analysis and applications of this class are given by R. J. Swift (1994). Much can be (and remains to be) done here, and in fact homogeneous (= stationary) isotropic random currents introduced by Itô (1956) and recently analyzed in more detail by Wong and Zakai (1992) can be extended to the harmonizable class and this is actively pursued in the present research. The area is an interesting place in which a (not necessarily abelian) harmonic analysis of vector (or Hilbert space valued) measures and probability theory meet, and a fruitful interaction results (1989). It is also possible to consider some of the ideas by generalizing strict stationarity (i.e., invariance of finite dimensional distributions under transtations) with vector harmonic analysis. This is somewhat more restricted than the preceding class, as seen from Hosoya (1982), but can be studied for its own interest.

1. D. K. Chang and M. M. Rao (1986), "Bimeasures and nonstationary processes," in *Real and Stochastic Analysis*, Wiley, New York, 7–118.
2. Y. Hosoya (1982), "Harmonizable stable processes," *Z. Wahrs.*, **60**, 517–533.
3. K. Itô (1956), "Isotropic random current," *Proc. Third Berkeley Symp. Math. Statist. Prob.*, **2**, 125–132.
4. J. P. Kelsh (1978), "Linear analysis of harmonizable time series," Ph.D. thesis at UCR, Riverside, CA.
5. M. H. Mehlman (1991), "Structure and moving average representation for multidimensional strongly harmonizable processes," *Stochastic Anal. Appl.*, **9**, 323–361.
6. H. Niemi (1975), "Stochastic processes as Fourier transforms of stochastic measures," *Ann. Acad. Sci. Fenn. AI. Math.*, **591**, 1–47.
7. M. M. Rao (1967), "Characterization and extension of generalized harmonizable random fields," *Proc. Nat. Acad. Sci.*, **58**, 1213–1219.
8. ________ (1982), "Harmonizable processes: structure theory," *L'Ensign. Math.*, **28**, 295–351.
9. ________ (1989), "Bimeasures and nonstationary process," in *Probability on Groups-IX*, Lect. Notes Math., **1379**, 254–298.
10. ________ (1991), "Sampling and prediction for harmonizable isotropic random fields," *J. Comb. Info. Syst. Sci.*, **16**, 207–220.
11. Yu. A. Rozanov (1959), "Spectral analysis of abstract functions," *Theor. Prob. Appl.*, **4**, 271–287.
12. R. J. Swift (1994), "The structure of harmonizable isotropic random fields," *Stochastic Anal. Appl.*, **12**, 583–616.

13. M. I. Yadrenko (1983), *Spectral Theory of Random Fields*, Optimization Software Inc., New York.

VI. Precise measurements and generalized random fields Stochastic models describing a random phenomenon can be interpreted and verified when observations on random variables are measured at specified instants. If the process is continuous, the value of $X_t(\omega)$ at the instant t cannot exactly be observed without affecting a nearby instance. Consequently one can only have at best an averaged version, i.e., $\int_T X_t\varphi(t)dt$ where φ is a smooth function of compact support describing the measuring instrument. Thus one is led to consider a functional $\Phi(\varphi) = \int_T X_t\varphi(t)dt$, on the space of such (real) functions $\mathcal{D}$. This motivates a study of such $\Phi : \mathcal{D} \to L^2(P)$, called *generalized random functions or functionals.* K. Itô and I. M. Gel'fand independently introduced these objects in the middle 1950's. In their seminal work, Gel'fand and Vilenkin (1964) gave various properties of these functionals and presented representation theorems for them. They also considered Φ to have independent values, i.e., if $\varphi_j \in \mathcal{D}, j = 1, 2$, with $\varphi_1 \cdot \varphi_2 = 0$, then $\Phi(\varphi_1), \Phi(\varphi_2)$ are mutually independent random variables. If $L(\varphi) = E(\exp i\Phi(\varphi))$, then $L : \mathcal{D} \to \mathbb{C}$ is called a *local functional* whenever $L(\varphi_1+\varphi_2) = L(\varphi_1)L(\varphi_2)$ for $\varphi_1 \cdot \varphi_2 = 0$. These authors raised the question of characterizing such functionals when $\mathcal{D}$ is the Schwartz space of infinitely differentiable (real) functions with compact supports and $T = \mathbb{R}$.

While teaching a course based on the Generalized Functions noted above, I became interested in this question and first considered the representation of generalized harmonizable random fields (1969). Then the Gel'fand-Vilenkin problem was solved for $\mathcal{D}$ if it is taken as $C_{00}(G)$, the space of real continuous compactly based functions on a locally compact group G, or if it is $C_{00}^m(\mathbb{R}^n)$, the space of m-times continuously differentiable real functions with compact supports on $\mathbb{R}^n$, in (1981) and (1971) respectively. But the problem in the latter case with $m = \infty$ is still not satisfactorily settled, and it seems to present some interesting new challenges. During these studies, a prediction question for generalized random functions when $\mathcal{D} = C_{00}^\infty(\mathbb{R})$, has been considered by George Chi (1969). Also these functionals and their representations are of interest in the study of second order processes whose n^{th} order increments form a stationary or harmonizable process, as noted by Yaglom (1987) and used in the recent work of Swift (1992).

1. G. Y. H. Chi (1969), "Nonlinear prediction and multiplicity of generalized random processes," Ph.D. thesis at Carnegie-Mellon Univ., Pittsburgh, PA.

2. I. M. Gel'fand and N. Ya. Vilenkin (1964), *Generalized Functions,* Vol.**4**, Academic Press, Inc., New York.

3. M. M. Rao (1969), "Representation theory of multidimensional generalized random fields," *Proc. Second Internat. Symp. Multivariate Analysis'*, Academic Press, Inc., New York, 411–436.

4. ________ (1971), "Local functionals and generalized random fields with independent values," *Theor. Prob. Appl.,* **16**,457–473.

5. ________ (1981), "Local functionals on $C_{00}(G)$ and probability," *J. Functional Anal.,* **39**, 23–41.

6. R. J. Swift (1992), "Structure and sample path analyses of harmonizable random

fields," Ph.D. thesis at UCR, Riverside, CA.
7. A. M. Yaglom (1987), *Correlation Theory of Stationary and Related Random Functions*, Vols.**1-2**, Springer-Verlag, Berlin.

VII. Sufficiency studies If $f(\cdot, \theta) : \mathbb{R}^n \to \mathbb{R}^+$ is a probability density function of a random vector $X = (X_1, \ldots, X_n)$ where $\theta \in I \subset \mathbb{R}$, one estimates θ by a Borel function of X, denoted $\hat{\theta}(X)$, and if $\tilde{\theta}(X)$ is another such estimator, then R. A. Fisher (1922) suggested that $\hat{\theta}$ be called *sufficient* for θ whenever the conditional distribution of $\tilde{\theta}$ given $\hat{\theta}$ does not involve the parameter θ, so that all the "information" about θ is contained in $\hat{\theta}$ which is thus preferable to $\tilde{\theta}$ in any statistical inference problem about θ; and he presented a substantial theory of it. This useful concept was generalized, and an analytical expression obtained, by Halmos and Savage (1949) as follows. If $\{P_\theta, \theta \in A\}$ is a family of probability measures on (Ω, Σ), then a σ-algebra $\mathcal{B} \subset \Sigma$ is termed *sufficient* for that class iff for each bounded measurable function $f : \Omega \to \mathbb{R}$, there is a bounded $\mathcal{B}$-measurable F_f such that

$$E_\theta^{\mathcal{B}}(f) = F_f, \qquad a.e.[P_\theta],\ \theta \in A.$$

Then, among others, the following result was established by these authors:

Theorem 1. *If $\{P_\theta, \theta \in A\}$ is dominated by some measure $\lambda : \Sigma \to \bar{\mathbb{R}}^+$, such that $\lambda|\mathcal{B}$ is σ-finite, then $\mathcal{B}$ is sufficient for the P_θ-family iff $\frac{dP_\theta}{d\lambda}$ is $\mathcal{B}$-measurable for all $\theta \in A$.*

If the (σ-finite) domination condition is simply omitted, then Burkholder (1961) showed that this result does not hold and, moreover, discussed the many serious difficulties that can result in the ensuing analysis. Then a more general condition than domination was introduced by T. S. Pitcher (1965) and this was generalized by Robert Rosenberg (1970) using the theory of Orlicz spaces. His work leads to a study of martingales relative to families of measures. It was also found later that Pitcher's condition is essentially equivalent to the existence of a "localizable" (which is more general than σ-finiteness) dominating measure λ for the given family. There are still many interesting problems in this area for investigation. The undominated case, for instance, is related to the Radon-Nikodým theory of vector valued measures which do not have finite variation. Using the order topology on real functions $\mathcal{M}$ on (Ω, Σ), integrable relative to the family $\{P_\theta, \theta \in A\}$, one can prove some vector R-N theorems which are useful in this context. On the other hand, many basic unresolved questions also exist on proper constructions and evaluations of conditional expectations or probabilities. This is of central importance for a satisfactory analysis of such areas as Bayesian statistics and calculation of transition probabilities in general Markov Processes. A detailed account of these questions and available rigorous results have been given in my recent work (1993a,b).

1. D. L. Burkholder (1961), "Sufficiency in the undominated case," *Ann. Math. Statist.*, **32**, 1191–1200.
2. R. A. Fisher (1922), "On the mathematical foundations of theoretical statistics," *Roy. Soc. Phil. Trans., Ser. A*, **222**, 309–368.

3. P. R. Halmos and L. J. Savage (1949), "Application of Radon-Nikodým theorem to the theory of sufficient statistics," *Ann. Math. Statist.*, **20**, 225–241.

4. T. S. Pitcher (1965), "A more general property than domination for sets of probability measures," *Pacific. J. Math.*, **15**, 597–611.

5. M. M. Rao (1993a), "Exact evaluation of conditional expectations in the Kolmogorov model," *Indian J. Math.*, **35**, 57–70.

6. ________ (1993b), *Conditional Measures and Applications*, Marcel Dekker, Inc., New York.

7. R. L. Rosenberg (1970), "Orlicz spaces based on families of measures," *Studia Math.*, **35**, 15–49.

VIII. Stochastic processes and inference The classical statistical inference questions for sequences of independent random variables with common distributions is first extended to stochastic processes by Grenander (1950), opening up a vastly new and important area for research. He also solved many problems there and introduced new methods which employ deep results from functional analysis. For instance, the nonlinear prediction based on the past and present observations leads to a structural analysis of prediction operators.

If the index set of the process is one dimensional, usually identified as time, and the error criterion is a nonnegative convex function which need not be a square, then one has to consider the L^p or even the Orlicz spaces L^φ for this work. In the latter case, the structure of prediction and projection operators have been studied in detail (1974). On the other hand if the process satisfies an nth order linear stochastic differential equation with the driving force as a harmonizable process, an unbiased prediction problem of the following type is important in some applications (1994). The work leads to systems of Fredholm integral equations. Solutions of these equations have not yet been adequately analyzed. A somewhat simplified version of a typical statement will be given as an illustration.

Proposition 1. *Let $\{X_t = Y_t + Z_t, t \in [a,b]\}$ be a process with Y_t as a random signal and Z_t as a noise process. Suppose both are of second order, orthogonal to each other and $E(Y_t) = \sum_{j=1}^{m} a_j g_j(t), a_j \in \mathbb{R}$. If r_y is the covariance function of the Y-process, Z_t has orthogonal increments, $E(Z_t) = 0$, with the variance function $H(\cdot)$, let X_t be observed on $[a,b]$ and g_j be smooth. Then an unbiased weighted linear least squares predictor $\hat{X}_{t_0}$ of Y_{t_0} for $t_0 > b$, of the form $\hat{X}_{t_0} = \int_a^b X_t dp(t, t_o)$, where $p(\cdot, t_0)$ is a (weight) function of bounded variation on $[a,b]$, exists whenever $p(,\cdot, t_0)$ is a solution of the Fredholm integral equation:*

$$r_y(t, t_0) + \sum_{j=1}^{m} \lambda_j(t_0) g_j(t) = \int_a^b k(s,t) dp(s, t_0).$$

Here $k(s,t) = r_y(s,t) + H(min(s,t))$, and the $\lambda_j(t_0)$ are the Lagrange multipliers for the system's constraints,

$$\int_a^b g_j(t) dp(t, t_0) = g_j(t_0), 1 \le j \le m.$$

If the index set is multidimensional, i.e., X_t, Y_t, Z_t are random fields, the corresponding study involves multiple stochastic integrals. These problems lead to a general analysis of multi-parameter processes or fields. Here the basic results of Cairoli and Walsh (1975) have to be employed and extended. Some of it has been done by Michael Brennan (1979) and much further research is needed. The desired stochastic calculus involving many new questions are subjects of current research interests.

1. M. D. Brennan (1979), "Planar semi-martingales," *J. Multivar. Anal.*, **9**, 465–486.
2. R. Cairoli and J. B. Walsh (1975), "Stochastic integrals in the plane," *Acta Math.*, **134**, 111–183.
3. U. Grenander (1950), "Stochastic processes and statistical inference," *Ark. Mat.*, **1**, 195–277.
4. M. M. Rao (1974), "Inference in stochastic processes-IV," *Sankhyā, Ser. A*, **36**, 63–120.
5. ________ (1994), "Harmonizable processes and inference: unbiased prediction for stochastic flows," *J. Statist. Plan. Inf.*, **39**, 187–209.

IX. The existence problem for processes In all the above work the existence of various types of processes, to be used in modeling, is a nontrivial question to settle. The basic result that one invokes in such problems is due to Kolmogorov (1933). It says that if a family $\{F_{t_1,\dots,t_n}, t_i \in T, n \geq 1\}$ of compatible finite dimensional distributions is given, then there exists a probability space (Ω, Σ, P) and a stochastic process $\{X_t, t \in T\}$ on it having the given family as its finite dimensional distributions. The compatibility condition which is both necessary and sufficient can be stated as follows: (i) $\lim_{x_n \to \infty} F_{t_1,\dots,t_n}(x_1, \cdots, x_n) = F_{t_1,\dots,t_{n-1}}(x_1, \cdots, x_{n-1})$, and (ii)$F_{t_{i_1},\dots,t_{i_n}}(x_{i_1}, \cdots, x_{i_n}) = F_{t_1,\dots,t_n}(x_1, \cdots, x_n)$, for all permutations $(i_1, \dots, i_n)$ of $(1, \dots, n)$, and for all $n \geq 1$. The mysterious condition (ii) above is better understood when it is recognized as part of a projective limit problem that is involved here. The latter identification and its structure of the basic theorem is due to Bochner (1955) who then extended it. It was further generalized by Prokhorov (1956) and several others (e.g., Choksi, Métivier, Sion and his students). There is also a dual problem, called the direct limits of measure spaces. I have discussed various forms of both these problems in my book (1981). An application of the direct limit result is moreover found useful in solving an open question of Whitney's in geometric integration theory, generalizing it to infinite dimensions, by Stephen Noltie (1975) in his thesis. A comprehensive result for the projective limits, which includes and refines several previous cases, has recently been given in collaboration with Vjačeslav Sazonov (1993). This has close relations with set martingales and other parts of analysis. The duality of direct and projective limit theory has interesting connections with topology. Further work is possible in this class of ideas.

1. S. Bochner (1955), *Harmonic Analysis and the Theory of Probability*, Univ. Calif. Press, Berkeley, CA.
2. A. N. Kolmogorov (1933), *Grundbegriffe der Wahrschienlichkeitsrechnung*, Springer-Verlag, Berlin. [English translation (1956), Chelsea Publishing Co., New York.]

3. S. V. Noltie (1975), "Integral representations of chains and vector measures," Ph.D. thesis at UCR, Riverside, CA. (See also, Chapter 4 of *Real and Stochastic Analysis*, Wiley, New York, (1986), 211–248.)
4. Yu. V. Prokhorov (1956), "Convergence of random processes and limit theorems in probability," *Theor. Prob. Appl.*, **1**, 157–214.
5. M. M. Rao (1981), *Foundations of Stochastic Analysis*, Academic Press, Inc., New York.
6. ________ and V. V. Sazonov (1993), "A projective limit theorem for probability spaces and applications," *Theor. Prob. Appl.* **38**, 307–315.

Quasi-periodic Solutions of Hamiltonian Evolution Equations

JEAN BOURGAIN School of Mathematics, Institute for Advanced Study, Princeton, NJ 08540

The problem discussed here is the persistency of quasi-periodic solutions of linear or integrable equations after Hamiltonian perturbation. This subject is closely related to the well known "KAM-theory" on invariant tori in smooth dynamical systems. Most of the research has been achieved within the last ten years by various authors (including J. Fröhlich, T. Spencer, E. Wayne, S. Kuksin, and more recently W. Craig, E. Wayne and myself). The main difference with KAM is the fact that one considers (finite dimensional) tori in an infinite dimensional phase space. Considering PDE's in space dimension > 1, the most significant new difficulty arising is the presence of large clusters of normal frequencies. This feature which seems incompatible with the use of a "standard" KAM technology as applied by E. Wayne [W] and S. Kuksin [K1] has lead to the development of a new method to approach the persistency problems, also in finite dimensional phase space. The first PDE-results obtained along these lines appear in the work of W. Craig and E. Wayne [C-W] on time-periodic solutions of 1D-nonlinear wave equations. Later on the author extended their method to cover the full setting of quasi-periodic solutions, leading in particular to a new proof of the KAM theorem and Melnikov theorem in finite dimensional phase space under weaker non-resonance hypothesis (not excluding multiplicities in the normal frequencies). This understanding lead to significant progress on the PDE problems. Presently the persistency problem is in 1D (1 space dimension) in satisfactory way understood (under the Dirichlet and periodic boundary conditions). In higher space dimension several cases can be treated (especially the nonlinear Schrödinger equation (NSE) and the case of time periodic solutions). Many problems remain and are today an active research topic.

The purpose of this exposé is to give a brief account of the model problems, the techniques and some of the results.

Consider an equation of the form

$$iu_t + Au + \varepsilon \frac{\partial H}{\partial \bar{u}} = 0 \tag{1}$$

where A is a selfadjoint operator and $H = H(u, \bar{u})$ is polynomial (or real analytic). We assume that A is diagonal in an eigenfunction basis which is "well localized" *wrt* the exponentials. The following are the main examples

(i)

$$A = -\Delta + M_\sigma \qquad \text{where} \qquad M_\sigma\, e^{i\langle n,x\rangle} = \sigma_n\, e^{i\langle n,x\rangle},\ \sigma_n \in \mathbb{R}$$

obtained by adding to the Laplacian a Fourier multiplier M_σ. In this case, the eigenfunction basis are the exponentials and the eigenvalues

$$\lambda_n = |n|^2 + \sigma_n. \tag{2}$$

We assume

$$\lim_{|n|\to\infty} \sigma_n = 0 \tag{3}$$

or

$$\lim_{|n|\to\infty} |\sigma_{n_1} - \sigma_{n_2}| = 0. \tag{4}$$

(ii) Consider the 1D Sturm Liouville operator

$$A = -\frac{d^2}{dx^2} + V(x) \tag{5}$$

where V is a real analytic 1-periodic potential.

The periodic spectrum of A is a sequence

$$\lambda_0 < \lambda_1 \le \lambda_2 < \lambda_3 \le \lambda_4 < \ldots < \lambda_{2n-1} \le \lambda_{2n} < \ldots \tag{6}$$

which satisfies

$$\lambda_{2n-1}, \lambda_{2n} = \pi^2 n^2 + \int V dx + 0(n^{-2}) \tag{7}$$

$$\lambda_{2n} - \lambda_{2n-1} \to 0 \qquad \text{rapidly}. \tag{8}$$

The corresponding eigenfunction s $\varphi_{2n-1}, \varphi_{2n}$ are periodic or anti-periodic according to the parity of n. Expanding φ_j in a Fourier series

$$\varphi_j(x) = \sum_m \widehat{\varphi}_j(m)\ e^{i\pi mx}. \tag{9}$$

There is the localization property *wrt* exponentials

$$|\widehat{\varphi}_j(m)| < C\ e^{-c|\ |m|-j/2\ |} \tag{10}$$

for some constants $c > 0,\ C < \infty$ depending on V.

Conversely, one has an expansion

$$e^{i\pi mx} = \sum\nolimits_j\ \hat{e}_m(j)\ \varphi_j(x) \tag{11}$$

with again

$$|\hat{e}_m(j)| < C\ e^{-c|\ |m|-j/2\ |}. \tag{12}$$

Assume V even i.e. $V(x) = V(-x)$. Consider the spectrum of $A = -\frac{d^2}{dx^2} + V$ subject to Dirichlet boundary conditions on [0,1]. This gives a sequence $\{\mu_n\}$ interlacing the periodic spectrum

$$\lambda_0 < \lambda_1 \leq \mu_1 \leq \lambda_2 < \ldots < \lambda_{2n-1} \leq \mu_n < \lambda_{2n} < \ldots \tag{13}$$

and the corresponding eigenfunction s $\{\psi_n\}$ form a basis for the 2-periodic odd functions. Thus

$$\psi_n(x) = \sum_{m=1}^{\infty} \widehat{\psi}_n(m)\ \sin\ \pi mx \tag{14}$$

$$\sin\ \pi nx = \sum_{m=1}^{\infty} \widehat{s}_n(m)\ \psi_m(x) \tag{15}$$

where again

$$|\widehat{\psi}_n(m)|,\ |\widehat{s}_n(m)| < C\ e^{-c|n-m|}. \tag{16}$$

(iii) The analogue of the theory in 1D described above fails in higher dimension. One may however, consider the special case of a potential of the form

$$V(x) = V_1(x_1) + \cdots + V_d(x_d). \tag{17}$$

For periodic boundary conditions, one gets for eigenvalues and eigenfunction s

$$\{\lambda^1_{j_1} + \cdots + \lambda^d_{j_d},\ \varphi^1_{j_1}(x_1) \ldots \varphi^d_{j_d}(x_d)\}. \tag{18}$$

In the Dirichlet case, one gets

$$\{\mu^1_{n_1} + \cdots + \mu^d_{n_d},\ \psi^1_{n_1}(x_1)\ldots\psi^d_{n_d}(x_d)\} \tag{19}$$

where $\{\lambda_j; \varphi_j\}$ and $\{\mu_n; \psi_n\}$ are as above.

(iv)

$$A = (-\Delta + \rho)^{1/2} \tag{20}$$

with eigenvalues

$$\lambda_n = (|n|^2 + \rho)^{1/2} \tag{21}$$

and the exponentials as eigenfunction s.

Observe that the sequence $\{(k+\rho)^{1/2} \mid k \in \mathbb{Z}_+\}$ consists of rationally independent numbers for typical ρ.

Coming back to (1), denote $\{\varphi_n, \mu_n\}$ the eigenfunction s and corresponding eigenvalues for A, i.e.

$$A\,\varphi_n = \mu_n\,\varphi_n.$$

Fix some specific indices $n_1, n_2, \ldots, n_b$ and denote $\lambda_j = \mu_{n_j}$ $(j = 1, \ldots, b)$.

Then

$$u_0(x,t) = \sum_{j=1}^{b} a_j\ e^{i\lambda_j t}\ \varphi_{n_j}(x) \tag{22}$$

yields a solution of the linear equation

$$iu^0_t + Au^0 = 0. \tag{23}$$

This solution is quasi-periodic and corresponds to a flow on a b-torus $\mathbb{T}^b = \mathcal{T}_0\{|a_j|\}$. The problem we study is which of the solutions (22) "persist" for the perturbed equation

$$iu_t + Au + \varepsilon\,\frac{\partial H}{\partial \bar u} = 0. \tag{24}$$

The meaning of persistency is as follows. One has a quasi-periodic solution

$$u_\varepsilon(x,t) = \sum_{n,k} \hat u_\varepsilon\,(n,k)\ e^{i\langle\lambda',k\rangle t}\ \varphi_n(x) \tag{25}$$

of (24), where k is a $\mathbb{Z}^b$-index

$$\hat u_\varepsilon(n_j, e_j) = a_j \qquad (e_j = j^{\text{th}} \text{ unit vector}) \tag{26}$$

$$\sum_{(n,k)\notin\mathcal{S}} \omega(n,k)\ |\hat{u}_\varepsilon(n,k)\,| < \sqrt{\varepsilon} \tag{27}$$

$$|\lambda' - \lambda| < C\,\varepsilon. \tag{28}$$

In (27),

$$\mathcal{S} = \{(n_j, e_j) \mid j = 1, \ldots, b\} \tag{29}$$

is the "resonant set".

We let ω be a weight function. For instance

$$\omega(n,k) = e^{c(|n|+|k|)} \tag{30}$$

corresponds to analytic solutions. We will mostly consider weights of the form

$$\omega(n,k) = e^{(|n|+|k|)^c}. \tag{31}$$

For some $c > 0$ (because they lead to some technical simplifications later on). There is dependence of the new frequency λ' on $\{a_j\}$ and the perturbation $\varepsilon\ \frac{\partial H}{\partial \bar{u}}$. In fact, assuming $\lambda'_1, \ldots, \lambda'_b$ rationally independent, a time shift permits to assume $a_j (1 \le j \le b)$ real and hence λ' depends only on $\{|a_j|\}$. The solution (25) corresponds by (27) to a perturbed torus $\mathcal{T}_\varepsilon\{|a_j|\}$ of $\mathcal{T}_0\{|a_j|\}$ in the phase space.

Consider $\lambda = (\lambda_1, \ldots, \lambda_b)$ as a parameter taken in a parameter set Λ. Persistency of quasi-periodic solutions (22) after the Hamiltonian perturbation as described above will occur for λ in a large subset of the frequency set. More precisely, the persistency will hold provided

$$\lambda \notin \Lambda'_\varepsilon(|a_j|\ (1 \le j \le b),\ \text{perturbation}) \tag{32}$$

where

$$\text{mes}\ \Lambda'_\varepsilon \overset{\varepsilon\to 0}{\longrightarrow} 0\,. \tag{33}$$

The preceding deals with perturbations of linear equations. In order to obtain families of quasi-periodic solutions with b frequencies, one needs to consider a parameter dependent linear equation with b-parameters $p = (p_1, \ldots, p_b)$, such that if $\{\lambda_n(p)\}$ are the eigenvalues of $A(p)$, then

$$\det\left(\frac{\partial \lambda_{n_j}}{\partial p_k}\right)_{1\le j,\,k\le b} \neq 0. \tag{34}$$

In the previous examples, such parameters are obtained via the Fourier multiplier M_σ or by letting V range in some b-manifold of potentials. Recall here that if

$$A(V) = -\ \frac{d^2}{dx^2} + V(x) \tag{35}$$

then from first order variation

$$\frac{\partial \lambda}{\partial V} = |\varphi|^2 \tag{36}$$

where φ is the eigenfunction with eigenvalue λ. The use of parameter dependent equations may be avoided in various ways.

First, writing

$$\lambda' = \lambda + \varepsilon(\lambda, a) \tag{37}$$

and assuming initial non-resonance conditions (1^e Melnikov condition)

$$\sum_{j=1}^{b} k_j\, \lambda_j + \mu_n \neq 0 \quad (n \notin \{n, \dots, n_b\}) \tag{38}$$

(expressing non-resonance of the normal frequencies μ_n with ($\mathbb{Z}$-combinations of) the tangential frequencies $\lambda_1, \dots, \lambda_b$), conditions appearing later on in the process may be ensured by restricting $|a|$ to an appropriate Cantor set. This requires the dependence (37) of λ' on $|a|$ to be sufficiently non-degenerate (*amplitude-frequency modulation*).

Secondly, one may sometimes extract parameters out of the nonlinearity by putting the Hamiltonian

$$H(\phi) = \langle A\phi, \phi\rangle + \varepsilon\, H_1(\phi) \tag{39}$$

in an appropriate *Birkhoff normal form.*

Assume that one may write (1) in the form

$$\begin{cases} i\dot{q}_{n_j} + \lambda_j q_{n_j} + \varepsilon\; q_{n_j}\; |q_{n_j}|^2 + \varepsilon\delta\; \frac{\partial H_2}{\partial \bar{q}_{n_j}} = 0 \quad (j = 1, \dots, b) \\ i\dot{q}_n + \mu_n q_n + \varepsilon\delta\; \frac{\partial H_2}{\partial \bar{q}_n} = 0 \quad (n \neq n_1, \dots, n_b) \end{cases} \tag{40}$$

where

$$u_n(x,t) = \sum\nolimits_n q_n(t)\, \varphi_n(x).$$

Define for $j = 1, \dots, b$

$$I_j = |q_{n_j}|^2 \tag{41}$$

(action variables). From (40)

$$\dot{I}_j = 2\,\mathrm{Re}\,\dot{q}_{n_j}\bar{q}_{n_j}, \quad i\dot{I}_j = \varepsilon\delta\, \left(\frac{\partial H_2}{\partial q_{n_j}}\, q_{n_j} - \frac{\partial H_2}{\partial \bar{q}_{n_j}}\, \bar{q}_{n_j}\right). \tag{42}$$

Fix some values $\{I_j^0\}_{j=1}^b$ and write

$$I_j = I_j^0 + \sqrt{\delta}\; I_j'. \tag{43}$$

¿From (40), (42), (43) we get the system

$$\begin{cases} i\dot{q}_{n_j} + (\lambda_j + \varepsilon\, I_j^0)\, q_{n_j} + \varepsilon\sqrt{\delta}\; I_j'\, q_{n_j} + \varepsilon\,\delta\, \frac{\partial H_2}{\partial \bar{q}_{n_j}} = 0 \quad (j = 1, \dots, b) \\ i\dot{I}_j - \varepsilon\sqrt{\delta}\, \left(\frac{\partial H_2}{\partial q_{n_j}}\, q_{n_j} - \frac{\partial H_2}{\partial \bar{q}_{n_j}}\, \bar{q}_{n_j}\right) = 0 \quad (j = 1, \dots, b) \\ i\dot{q}_n + \mu_n q_n + \varepsilon\delta\, \frac{\partial H_2}{\partial \bar{q}_n} = 0 \quad (n \neq n_1, \dots, n_b) \end{cases} \tag{44}$$

which appears as $(\varepsilon\sqrt{\delta})$-perturbation of the linear system

$$\begin{cases} i\dot{q}_{n_j} + (\lambda_j + \varepsilon I_j^0) q_{n_j} = 0 & (j = 1, \dots, b) \\ i\dot{I}_j = 0 & (j = 1, \dots, b) \\ i\dot{q}_n + \mu_n q_n = 0 \end{cases} \tag{45}$$

with a b-parameter set of frequencies $\{\lambda_j + \varepsilon\, I_j^0\}_{j=1}^b$ obtained by variation of $\{I_j^0\}$. Hence the problem is reduced to a perturbation of a linear system as above.

A similar approach is applied in the study of perturbations of integrable systems. For instance, perturbations of the KdV-equation

$$u_t + u_{xxx} + uu_x = 0 \tag{46}$$

of the form

$$u_t + u_{xxx} + uu_x + \varepsilon\; f(u)_x = 0 \tag{47}$$

($f(u)$ polynomial or real analytic). The process of extracting the normal form here is rather involved since it is based on the Riemann surface correspondence (cf. [K2]).

Coming back to (1), we deal with the problem of persistency of finite dimensional tori in an infinite dimensional phase space. This is a generalization of the more classical KAM (Kolmogorov–Arnold–Moser) setting of persistency of n tori in $2n$-dimensional phase space. As a first method to approach (1), one may however, apply the basic KAM scheme, which consists in removing the perturbation εH in (39) by composing with canonical transformations. Given a Hamiltonian flow

$$iq_t = \varepsilon\, \frac{\partial F}{\partial \bar{q}} \qquad F = F(q, \bar{q}) \quad \text{Hamiltonian} \tag{48}$$

denote $q \to q'$ the time-1 shift, which is a symplectic transformation of phase space. Then the system

$$iq_t = \frac{\partial H}{\partial \bar{q}} \tag{49}$$

is transformed into

$$iq_t' = \frac{\partial H'}{\partial \bar{q}'} \tag{50}$$

where $H(q) = H'(q')$. Denote in complex coordinates the Poisson bracket

$$\{F_1, F_2\} = \operatorname{Im} \sum\nolimits_n \frac{\partial F_1}{\partial q_n} \frac{\partial F_2}{\partial \bar{q}_n}. \tag{51}$$

Then

$$H' = H + \varepsilon \{H, F\} + \frac{\varepsilon^2}{2!} \{\{H, F\}, F\} + \cdots = H + \varepsilon \{H, F\} + 0(\varepsilon^2). \tag{52}$$

Hence, for H given by (39)

$$H(q, \bar{q}) = \sum \mu_n |q_n|^2 + \varepsilon H_1(q, \bar{q}) = H_0 + \varepsilon H_1 \tag{53}$$

we find

$$H'(q', \bar{q'}) = \sum \mu_n |q_n|^2 + \varepsilon H_1(q, \bar{q}) + \varepsilon \{H_0, F\} + 0(\varepsilon^2). \tag{54}$$

The main idea is to choose F in order to reduce $H_1 + \{H_0, F\}$ and iterate the process. The drawback of this method is that it requires the spectrum of A to satisfy non-resonance conditions of the form (2^{e} Melnikov conditions)

$$\sum_{j=1}^{b} k_j \lambda_j + \mu_n - \mu_{n'} \neq 0 \qquad (n \neq n' \notin \{n_1, \ldots, n_b\}) \tag{55}$$

besides (38). Condition (55) is violated in the case of multiplicities in the normal frequencies and impairs the use of KAM in this situation (which in the context of (1) appears in 1D under periodic boundary conditions and in higher dimension).

To illustrate the appearance of (55), assume F of the form

$$F(q, \bar{q}) = \langle Bq, q \rangle, \quad B = B^*. \tag{56}$$

Then, by (51)

$$\{H_0, F\} = \frac{1}{2i} \sum_{n,n'} (\mu_n - \mu_{n'})\, b_{nn'}\, \bar{q}_n\, q_{n'} \sim \langle [A, B] q, q \rangle \tag{57}$$

where $[A, B] = AB - BA$.

If one rewrites (1) in the Fourier version ($wrt\ e^{i\langle \lambda', k \rangle t} \varphi_n(x)$), one gets

$$(-\langle \lambda', k \rangle + \mu_n)\, \hat{u}\,(n, k) + \varepsilon\, \widehat{\frac{\partial H}{\partial \bar{u}}}\,(n, k) = 0. \tag{58}$$

Hence (38) is the only type of condition which appears if one tries to solve (58) by a direct perturbative approach. If one considers such a ε-series, considerable problems appear due to the small divisors

$$\mu_n - \langle \lambda', k \rangle. \tag{59}$$

They yield approximative solutions by truncation, introducing only divisors such that

$$|\mu_n - \langle \lambda', k\rangle| \gg \varepsilon. \tag{60}$$

To perform their actual summation is a very delicate issue (leading to re-normalization problems) and was even in finite dimensional phase space (i.e. n ranges in a finite set) only recently fully understood[(*)]. The method followed here is to solve (58) by a Newton iteration scheme, converging much faster (double exponentially fast) and therefore less affected by the presence of small divisors. On the other hand, it will require to control the inverse of the linearization of (58) which is a non-diagonal operator.

We first perform a Lyapunov-Schmidt type decomposition. Consider the b equations

$$(-\lambda_j' + \lambda_j)\, a_j + \varepsilon\, \widehat{\frac{\partial H}{\partial \bar u}}\, (n_j, e_j) = 0 \quad (j = 1, \dots, b) \tag{61}$$

obtained by taking $(n,k) \in \mathcal{S}$ of (26), (29).

They form the (finite) system of Q-equations. The remaining (infinite) system

$$(-\langle \lambda', k\rangle + \mu_n)\, \hat u(n,k) + \varepsilon\, \widehat{\frac{\partial H}{\partial \bar u}}\, (n,k) = 0 \quad (n,k) \notin \mathcal{S} \tag{62}$$

are called the P-equations. The general procedure is to determine $\hat u|_{(n,k)\notin \mathcal{S}}$ from (62) (depending on λ') and then substitute in (61) to obtain the new frequencies $\lambda' = (\lambda_1', \dots, \lambda_b')$. Now these frequencies $\lambda_1', \dots, \lambda_b'$ need to be real. Thus (61) expresses, in fact, $2b$ conditions with b parameters. The formal solvability is a consequence of the Hamiltonian nature of (1) (result due to Poincaré). Assume $H(u, \bar u)$ is a sum of monomial s $u^j \bar u^k$ with real coefficients. In proving the persistency result, we may assume $a_j \in \mathbb{R}$ $(j = 1, \dots, b)$, considering time shifts as observed earlier. Hence, the system $\hat u|_{(n,k)\notin \mathcal{S}}$ produced from solving (62) will be real and so will be $\widehat{\frac{\partial H}{\partial \bar u}}\,(n_j, e_j)$ in (61). Thus (61) yields a real solution in λ' and the formal solvability is clear in this case.

The system (62) cannot be treated by a standard implicit function theorem because of the appearance of small divisors. We denote

$$v = \bar u. \tag{63}$$

Assuming the eigenfunction basis for A in (1) given by exponentials, we have thus

$$\hat v(n,k) = \overline{\hat u(-n,-k)} \tag{64}$$

(*) cf. [El]

(if one would consider a real eigenfunction basis $\{\varphi_n\}$, then clearly $\hat{v}(n,k) = \overline{\hat{u}(n,-k)}$).

Duplicate the equations (62) considering the system

$$\begin{cases} (-\langle\lambda', k\rangle + \mu_n)\, \hat{u}(n,k) + \varepsilon\, \widehat{\frac{\partial H}{\partial v}}\,(n,k) = 0 & (n,k) \notin \mathcal{S} \\ (\langle\lambda', k\rangle + \mu_{-n})\, \hat{v}(n,k) + \varepsilon\, \widehat{\frac{\partial H}{\partial u}}\,(n,k) = 0 & (n,k) \notin -\mathcal{S}. \end{cases} \tag{65}$$

We solve this system in $\big(\hat{u}|_{(n,k)\notin\mathcal{S}},\, \hat{v}|_{(n,k)\notin\mathcal{S}}\big)$ by Newton's algorithm. The relation (64) will be preserved for the approximative solutions.

Recall the formal scheme. Consider the equation

$$F(y) = 0. \tag{66}$$

Starting from y_0 (here 0), the consecutive approximations are defined by

$$\Delta_{i+1} y \equiv y_{i+1} - y_i = -[F'(y_i)]^{-1} F(y_i) \tag{67}$$

and

$$\|F(y_{i+1})\| = 0\,(\|y_{i+1} - y_i\|^2) = 0\,\big(\|[F'(y_i)]^{-1}\, F(y_i)\|^2\big)\,. \tag{68}$$

In our case (65), the linearized operator T is given by

$$T = D + \varepsilon\mathcal{S} \tag{69}$$

where D is diagonal

$$D = \begin{pmatrix} D^+ & 0 \\ 0 & D^- \end{pmatrix} \qquad \begin{aligned} D^+_{n,k} &= -\langle\lambda', k\rangle + \mu_n \\ D^-_{n,k} &= \langle\lambda', k\rangle + \mu_{-n} \end{aligned} \tag{70}$$

and

$$\mathcal{S} = \begin{pmatrix} S_{\frac{\partial^2 H}{\partial u \partial v}} & S_{\frac{\partial^2 H}{\partial v^2}} \\ S_{\frac{\partial^2 H}{\partial u^2}} & S_{\frac{\partial^2 H}{\partial u \partial v}} \end{pmatrix}. \tag{71}$$

S_ϕ expresses ϕ-multiplication in Fourier, thus here

$$S_\phi\,((n,k)\;(n',k')) = \widehat{\phi}(n-n',\; k-k'). \tag{72}$$

The operator $\mathcal{S}$ is selfadjoint and depends on the given approximative solution (u,v). Along the construction, we satisfy an estimate on their Fourier coefficients of the form

$$|\hat{u}(n,k)| < C\, e^{-(|n|+|k|)^c} \tag{73}$$

for some fixed $c > 0$. This fact clearly yields a corresponding off-diagonal decay estimate for each of the 4 matrixes S_ϕ appearing in (71)

$$S_\phi(x, x') < C\,e^{-|x-x'|^c}. \tag{74}$$

The main difficulty consists in controlling the inverse T^{-1}, due to the fact that the diagonal elements may be arbitrarily small. Fixing some $\varepsilon \lll \delta < 1$, call a site (n, k) singular if

$$|D^+_{n,k}| = |-\langle\lambda', k\rangle + \mu_n| < \delta$$

or

$$|D^-_{n,k}| = |\langle\lambda', k\rangle + \mu_{-n}| < \delta. \tag{75}$$

The geometric structure of the singular sites plays an important role in the study of T^{-1} (in particular, one exploits essentially their separation properties). Assume that one succeeds to ensure for these operators T bounds

$$(T_N)^{-1} < B(N) \tag{76}$$

on the inverse of the restriction

$$T_N = T|_{|n|+|k|<N} \tag{77}$$

where, for instance, $B(N)$ grows at most polynomially

$$B(N) < N^C \tag{78}$$

or at least much slower than exponential.

Assume $H(u, v)$ is a polynomial expression. At the jth step of the Newton iteration, an approximative solution y is obtained and we assume that

$$\operatorname{supp} \hat{u}_j \text{ is contained in an } (n,k) - \text{ball of radius } M^j \tag{79}$$

for some constant M. Letting $y_j = (\hat{u}_j, \hat{v}_j)$ and $F(y)$ the left number of (65), assume

$$\|F(y_j)\| < \varepsilon_j. \tag{80}$$

Since F is expressed as a polynomial in $\hat{u}, \hat{v}$, we get

$$\operatorname{supp} \widehat{F(y_j)} \subset CM^j. \tag{81}$$

Put $N = M^{j+1}$ and define y_{j+1} by (cf. (67))

$$y_{j+1} - y_j = -T_N^{-1}\,(F(y_j)). \tag{82}$$

Hence, by (78), (80)

$$\|y_{j+1} - y_j\| < N^C \varepsilon_j \tag{83}$$

and by (82)

$$\|F(y_{j+1})\| \leq \| (T - T_N)\ (y_{j+1} - y_j)\| + 0\ (\|y_{j+1} - y_j\|^2). \tag{84}$$

From (80), (83) and off-diagonal decay of T and T_N^{-1} (cf. (74)), the first term in (84) may be bounded by

$$\| (T - T_N)\ (y_{j+1} - y_j)\ \| < e^{-\frac{1}{2} N^c} \varepsilon_j. \tag{85}$$

Hence, from (84)

$$\varepsilon_{j+1} < e^{-\frac{1}{2} (M^{j+1})^c} \varepsilon_j + M^{2C(j+1)} \varepsilon_j^2. \tag{86}$$

Letting $c > 0$ be sufficiently small, one may conclude that

$$\varepsilon_j < e^{-4(M^j)^c} \tag{87}$$

$$\|y_{j+1} - y_j\| < e^{-3(M^j)^c} < e^{-2(M^{j+1})^c} \tag{88}$$

which is compatible with (73), (79).

The main difficulty is to obtain the estimate (76) and the off-diagonal estimate, say

$$|T_N^{-1}(x, x')| < e^{-\frac{1}{2} |x-x'|^c} \text{ for } |x - x'| > \frac{N}{100}. \tag{89}$$

Coming back to (38), assume (except for $(n, k) \in \mathcal{S}$)

$$| - \langle k, \lambda\rangle + \mu_n| > (1 + |k|)^{-C} \tag{90}$$

for some constant C. Assuming λ' as in (28), it follows that also

$$| - \langle k, \lambda'\rangle + \mu_n| > \frac{1}{2}\ (1 + |k|)^{-C} \text{ for } |k| < \varepsilon^{-1/2C}. \tag{91}$$

Thus, for

$$N < \varepsilon^{-1/2C} \tag{92}$$

one may invert $T_N = D_N + \varepsilon\, S_N$ by a Neumann series, i.e.

$$T_N^{-1} = D_N^{-1} + \sum_{j=1}^{\infty} (-1)^j \varepsilon^j \left(S_N D_N^{-1}\right)^j \tag{93}$$

since $\|S_N\, D_N^{-1}\| < \sqrt{\varepsilon}$, and (78), (89) are clearly satisfied.

In order to fulfill (78), (89) at later scales, requires to impose further conditions on (λ, λ', a); $a = (a_1, \dots, a_b)$, considered as initial parameters in the Newton iteration process (recall that at each stage, T depends on previous approximative solution). The effect of these further conditions is to excise from the (λ, λ', a)-parameter set exceptional subsets which measure tends rapidly to zero. Since λ' will be determined by solving the Q equations (61) as

$$\lambda' = \lambda + \varepsilon(\lambda, a) \tag{94}$$

our aim is to remain with conditions on (λ, a) as in (32), (34); in particular in a model when a is fixed, λ will be restricted to a Cantor set of positive measure. Essentially speaking, these conditions on (λ, λ', a) consist in keeping certain expressions $E(\lambda, \lambda', a)$ away from zero, where $E(\lambda, \lambda', a)$ is differentiable with more sensitivity on λ' than λ, a.

In order for $\hat{u}|_{(n,k)\notin S}$ to solve the P-equations, previous conditions on $(\lambda, \lambda' a)$ need to be fulfilled. We assume however, that $\hat{u}|_{(n,k)\notin S}$ is smoothly defined on the entire parameter set (λ, λ', a). Substitution in (61) allows then to solve in λ' invoking the standard implicit function theorem. One gets (94) or more precisely (assuming $a_j \in \mathbb{R}$)

$$\lambda_j' = \lambda_j + \varepsilon \, \frac{1}{a_j} \left[\frac{\partial H}{\partial \bar{u}} \left(\sum_{j=1}^{b} a_j \, e^{i\theta_j}, \ \sum_{j=1}^{b} a_j \, e^{-i\theta_j} \right) \right]^{\wedge} (n_j, e_j) + 0(\varepsilon^2) \tag{95}$$

where the second term depends on $a = (a_1, \dots, a_b)$ only.

In carrying out this program, there is a distinction between the case of time periodic solutions ($b = 1$) and quasi-periodic solutions ($b > 1$). The case $b = 1$ turns out to be significantly easier.

The arithmetic properties of the sequence $\{\mu_n\}$, more precisely the structure of clusters of the form

$$\{n \in \mathbb{Z}^d \mid |\mu_n - \mu| < 1\} \tag{96}$$

when $\mu \to \infty$ plays a basic role. In the PDE-context, the sequence $\{\mu_n\}$ is infinite and its properties depend on the space dimension d.

It turns out however that conditions of the form (55) are unnecessary and appear as an artifact of the KAM approach.

Remarks

(i) The method described above is more flexible than KAM. In fact, in the main body of the analysis, which consists in solving the P-equations, the Hamiltonian structure plays essentially no role.
There are variants. For instance, one may use a truncated perturbation series to

obtain an approximative solution of the P-equation up to ε^K, for any power K, and then apply the more rapidly converging Newton method and get an actual solution, ε^K-close to the approximative one.
(The Hamiltonian counterpart of this consists in achieving a normal form with non-resonant part $0(\varepsilon^K)$).
The condition (38)

$$\sum_{j=1}^{b} k_j \lambda_j + \mu_n \neq 0 \quad (n \notin \{n_1, \ldots, n_b\}) \tag{97}$$

(1^e Melnikov condition) expresses non-resonance of the normal frequencies with the tangential ones. In the resonant case, one modifies the previous scheme as follows. Define now the resonant set $\widetilde{\mathcal{S}} \supset \mathcal{S}$ as

$$\widetilde{\mathcal{S}} = \{(n,k) \mid -\langle \lambda, k\rangle + \mu_n = 0\} \tag{98}$$

and consider for the Lyapunov–Schmidt decomposition

$$\begin{cases} (-\lambda'_j + \lambda_j)a_j + \varepsilon\, \widehat{\frac{\partial H}{\partial \bar u}}\,(n_j, e_j) = 0 & (j = 1, \ldots, b) \\ (-\langle \lambda', k\rangle + \mu_n)\; \hat u(n,k) + \varepsilon\, \widehat{\frac{\partial H}{\partial \bar u}}\,(n,k) = 0 & \text{for } (n,k) \in \widetilde{\mathcal{S}}\backslash\mathcal{S} \end{cases} \tag{99}$$

(Q-equations)
and

$$(-\langle \lambda', k\rangle + \mu_n)\; \hat u(n,k) + \varepsilon\, \widehat{\frac{\partial H}{\partial \bar u}}\,(n,k) = 0 \text{ for } (n,k) \notin \widetilde{\mathcal{S}} \tag{100}$$

(P-equations).
Fix

$$(\lambda'_j)_{j=1,\ldots,b},\ \{a_j\}_{j=1,\ldots,b} \text{ and } \hat u|_{(n,k)\in\widetilde{\mathcal{S}}\backslash\mathcal{S}}$$

and solve $\hat u|_{(n,k)\notin\widetilde{\mathcal{S}}}$ from (100). One then uses (99) for the determination of λ' and $\hat u|_{\widetilde{\mathcal{S}}\backslash\mathcal{S}}$. A typical example of this appears in Theorem 131 below.

(ii) The Lyapunov–Schmidt approach to (1) yields a new method to solve stability problems in finite dimensional phase space as well. Observe that the Hamiltonian system

$$\dot p_n = \frac{\partial H}{\partial q_n} \quad \dot q_n = -\frac{\partial H}{\partial p_n} \quad (n = 1, \ldots, N) \tag{101}$$

Hamiltonian

$$H(p,q) = H(p_1, \ldots, p_N, q_1, \ldots, q_N) \tag{102}$$

is equivalent to

$$i\dot u = 2\,\frac{\partial H}{\partial \bar u} \tag{103}$$

where $u = p + iq$. Thus one gets a new proof of the KAM theorem (stability of N tori is 2N dimensional phase space) and Melnikov's theorem (more generally, stability of tori of dimension $n \leq N$) along the lines of Lyapunov's theorem (the periodic case $n = 1$). For Lyapunov's result, the non-resonance condition is

$$\mu_n \notin \lambda_1 \mathbb{Z} \tag{104}$$

corresponding to the case $b = 1$ in (38). Hence, in the finite dimensional case, we prove a Melnikov theorem for (1) under the non-resonance assumption (38), without the need of the extra assumption (55). This result is new. From the previous remark, it appears that besides multiple normal frequencies one may investigate also the case of normal frequencies resonant with the tangential ones, thus when (104), (38) are partially violated. Presently there has been no systematic study of this.

(iii) Consider a NLW

$$\partial_{tt} - \Delta y + Vy + \varepsilon F'(y) = 0 \tag{105}$$

and denote

$$B = (-\Delta + V)^{1/2} \tag{106}$$

(assuming this makes sense).
Rewrite (105) in the following Hamiltonian form

$$\begin{cases} \dot{y} = Bz \\ \dot{z} = -By - \varepsilon B^{-1} F'(y) \end{cases} \tag{107}$$

considering instead of L^2 the Hilbert space $H^{1/2}$ with scalar product

$$\langle u, v \rangle_{\frac{1}{2}} = \langle u, Bv \rangle. \tag{108}$$

Denoting $u = y + iz$, (107) yields

$$i\dot{u} - Bu + \varepsilon B^{-1} F'(\operatorname{Re} u) = 0 \tag{109}$$

which is of the form (1) with A replaced by $-B$. Thus the spectrum in 1D under periodic (resp. Dirichlet) boundary conditions is given by $-\sqrt{\lambda_j}$ (resp. $-\sqrt{\mu_n}$) and behaves as

$$\sqrt{\lambda_j} = \frac{\pi j}{2} + 0\left(\frac{1}{j}\right), \quad \sqrt{\mu_n} = \pi_n + 0\left(\frac{1}{n}\right). \tag{110}$$

In this particular case, $V = \rho$ (constant), we get for the periodic spectrum

$$\mu_n = (|n|^2 + \rho)^{1/2} \tag{111}$$

(cf example (iv) above).

Next, we state some concrete results.

THEOREM 112. *Consider a* 1D NLS [K1], [B1]

$$iu_t - u_{xx} + Vu + \varepsilon \frac{\partial H}{\partial \bar{u}} = 0 \tag{113}$$

or

$$iu_t = u_{xx} + M_\sigma u + \varepsilon \frac{\partial H}{\partial \bar{u}} = 0 \tag{114}$$

where $H = H(u, \bar{u})$ *is polynomial or real analytic,* V *a real analytic periodic potential and* M_σ *a real Fourier multiplier (as discussed in the examples). Consider an unperturbed solution of (113), (114) with* $\varepsilon = 0$

$$u_0(x,t) = \sum_{j=1}^{b} a_j \; e^{i\lambda_j t} \; \varphi_{n_j}(x) \tag{115}$$

where $\lambda_j = \mu_{n_j}$ $(j = 1, \dots, b)$ *and* $\{\varphi_{n_j}\}$ *are the corresponding eigenfunction s. Consider* $\lambda = (\lambda_1, \dots, \lambda_b)$ *as a b-parameter (for equation (113), this may be achieved by appropriate variation of* V*, cf. (35)-(36)). Assume a non-resonance condition (38)*

$$|\mu_n + \langle k, \lambda\rangle \;| \; \geq c\,(1 + |k|)^{-C} \tag{116}$$

satisfied, for $|k| < N$ *sufficiently large (depending on* H*).*
Then, for $\lambda \in \Lambda_\varepsilon(|a|)$*, a subset of the parameter set of small complementary measure when* $\varepsilon \to 0$*, there is a perturbed solution* u_ε *of (113), (114) with perturbed frequency* λ'

$$u_\varepsilon(x,t) = \sum_{n,k} \hat{u}_\varepsilon(n,k) \; e^{i\langle\lambda', k\rangle t} \; \varphi_n(x) \tag{117}$$

satisfying (26)-(28). Thus

$$\hat{u}_\varepsilon(n_j, e_j) = a_j \quad (j = 1, \dots, b) \tag{118}$$

$$\sum_{(n,k)\notin \mathcal{S}} e^{(|n|+|k|)^c} \; |\hat{u}_\varepsilon(n,k)| < \sqrt{\varepsilon} \tag{119}$$

$$|\lambda' - \lambda| < C\varepsilon \tag{120}$$

for some $c > 0$*, and where* $\mathcal{S} = \{(n_j, e_j)\,(j = 1, \dots, b\}$.

In the case of (113), with Dirichlet boundary conditions, assume V even and $H(u, \bar{u})$ even (hence $\frac{\partial H}{\partial \bar{u}}$ odd), see discussion in example (ii) above.

THEOREM 121. *The analogue of Theorem 112 holds for* 2D *Schrödinger equations, but in (113) the potential* $V(x_1, x_2)$ *should be assumed of the form* $V_1(x_1)+V_2(x_2)$ *(see discussion in example (iii) above).* [B3]

In 1D, a result similar to Theorem 112 may be stated for NLW equations [K1], [W]

$$u_{tt} - u_{xx} + V(x)u + \varepsilon\, f'(u) = 0 \tag{122}$$

and even derivative NLW equations [B1]

$$u_{tt} - u_{xx} + V(x)u + \varepsilon\, Bf'(u) = 0 \quad B = \left(\frac{-d^2}{dx^2}\right)^{1/2}. \tag{123}$$

In the next theorem, the role of outer parameters in the equation is replaced by amplitude-frequency modulation.

THEOREM 124. *Consider a* 1D NLW

$$u_{tt} - u_{xx} + \rho u + (u^3 + \textit{higher order terms}) = 0 \tag{125}$$

and assume $\rho \geq 0$ *is a typical number in the sense of linear independence of the sequence*

$$\mu_n = (n^2 + \rho)^{1/2}. \tag{126}$$

Fix a sequence

$$0 < n_1 < n_2 < \cdots < n_b \in \mathbb{Z}_+ \tag{127}$$

and consider the solution

$$u_0(x,t) = \sum_{j=1}^{b} a_j \ \cos\, n_j x \cdot \cos\, \lambda_j t \qquad (\lambda_j = \mu_{n_j}) \tag{128}$$

of the linear equation

$$u_{tt} - u_{xx} + \rho u = 0. \tag{129}$$

There is a Cantor set $C \subset \{a = (a_j)_{j=1,\ldots,b} \mid a_j > 0\}$ *of positive measure, in fact, of asymptotically full measure when* $|a| \to 0$*, such that for* $a \in C$ *the solution (128) of (129) persists for (125)*

$$u(x,t) = \sum_{j=1}^{b} q_j \ \cos\, n_j x \cdot \cos\, \lambda_j' t + 0(|a|^3). \tag{130}$$

The persistency problem for higher dimensional wave equations seems more difficult, due to the behavior of the frequencies $|n| = (n_1^2 + \ldots + n_d^2)^{1/2}$ when $d \geq 2$. One may treat however, the special case of time periodic solutions in any dimension (this is also the case for NLS).

THEOREM 131. *Consider the periodic wave equation in dimension d* [B2]

$$u_{tt} - \Delta u + \rho u + (u^3 + \textit{higher order terms}) = 0 \tag{132}$$

where again $\rho > 0$ is a typical number. More precisely, we require ρ to satisfy a condition of the form

$$\left|\sum_{j=0}^{r} k_j \, \rho^j\right| > \left(\sum |k_j|\right)^{-C_r} \quad \textit{for all } \{k_j\} \in \mathbb{Z}^{r+1}\backslash\{0\}. \tag{133}$$

Fix $n_0 \in \mathbb{Z}^d\backslash\{0\}$. There is a Cantor set C of positive measure in an interval $[0, \delta]$ and for $p_0 \in C$ a solution of (132) of the form

$$u(x,t) = p_0 \cos\left(\langle n_0, x\rangle + \lambda t\right) + 0\left(p_0^3\right) \tag{134}$$

where

$$\lambda^2 = \lambda(p_0)^2 = |n_0|^2 + \rho + \left(\frac{3}{4} + 0(1)\right) p_0^2. \tag{135}$$

The next two results are normal form reductions, bringing the problem back to perturbations of a linear problem with parameters. (cf. the discussion (39)-(45)).

THEOREM 136. *Consider a* 1D NLS

$$iu_t - u_{xx} + mu + f(|u|^2)u = 0 \tag{137}$$

where f is a polynomial or real analytic and satisfying a non-degeneracy condition

$$f'(0) \neq 0. \tag{138}$$

Consider (137) with periodic boundary conditions say. Fix a sequence of positive integers

$$n_1 < n_2 < \ldots < n_b. \tag{139}$$

Then for $a = \{a_j\}_{j=1,\ldots,b}$ in a Cantor family C of positive measure, there is a quasi-periodic solution

$$u(x,t) = \sum_{j=1}^{b} q_j \; e^{i(n_j x + \lambda_j' t)} + 0(|a|^3) \tag{140}$$

with frequencies $\lambda'_1, \dots, \lambda'_b$, *where*

$$\lambda'_j = n_j^2 + m + 0(|a|^2) \quad (j = 1, \dots, b). \tag{141}$$

This result is due to S. Kuksin and J. Pöschel (under Dirichlet boundary conditions) [K-P].

THEOREM 142. *Consider the* 2D *cubic* NLS [B3]

$$iu_t - \Delta u + cu|u|^2 = 0 \quad (c \neq 0) \tag{143}$$

or, more generally, an 2D NLS

$$iu_t - \Delta u + f(|u|^2)u = 0 \tag{144}$$

with f *as above in Theorem 136, with periodic boundary conditions.*
For the modes $n_1, \dots, n_b \in \mathbb{Z}^2$, *we fix 2 lattice points* n_1, n_2 *on a same circle*

$$|n_j| = R \quad (j = 1, 2), n_1 \neq -n_2 \tag{145}$$

(more complicated structures involving more then 2 points may be treated as well but this is the simplest case). There is a Cantor family C *of positive measure such that for* $a = \{a_j\}_{j=1,2} \in C$ *(143) (144) has a quasi-periodic solution*

$$u(x,t) = \sum_{j=1}^{2} a_j \, e^{i(\langle n_j, x\rangle + \lambda'_j t)} + 0(|a|^3) \tag{146}$$

with frequencies $\lambda' = (\lambda'_1, \lambda'_2)$

$$\lambda'_j = |n_j|^2 + 0(|a|^2) \quad (j = 1, 2). \tag{147}$$

Further Comments.
It seems a natural program to extend the classical theory of smooth dynamical systems to the setting of infinite dimensional phase space, in particular in the context of Hamiltonian PDE's as discussed here. A subject closely related to persistency of invariant tori is that of Nekhoroshev stability [N]. This phenomenon may be roughly stated as follows. Consider in 2N-dimensional phase space a perturbed Hamiltonian

$$H(I_1, \dots, I_N, \varphi_1, \dots, \varphi_N) = H_0(I) + \varepsilon\, H_1(I, \varphi) \tag{148}$$

where (I, φ) are action-angle variables.

The unperturbed Hamiltonian is assumed to be of the form

$$H_0(I) = \sum_{j=1}^{N} \lambda_j I_j + c \sum I_j^2 + 0(I^3) \tag{149}$$

where either the linear part is non-resonant or $c \neq 0$ (i.e. H_0 is strictly convex). Denote $I(t), \varphi(t)$ the evolution of I, φ for the H-flow

$$\begin{cases} \dot{I}_j = -\varepsilon \, \frac{\partial H_1}{\partial \varphi_j} \\ \dot{\varphi}_j = \frac{\partial H_0}{\partial I_j} + \varepsilon \, \frac{\partial H_1}{\partial I_j} \end{cases} \qquad (j = 1, \dots, N). \tag{150}$$

Then there is stability

$$|I(t) - I(0)| < \varepsilon^a \tag{151}$$

for exponentially long time

$$|t| < \exp(\varepsilon^{-b}) \tag{152}$$

where the relation a, b depends on the dimension N. It is expected that typically for sufficiently long time, there will be a "drift" of the action variables, known as Arnold diffusion.

The question we address is to what extend such result may be formulated for Hamiltonian PDE's. There are 2 immediate problems arising here.

(i) The fact that the phase space is infinite dimensional (in the proof of the previous result, at least in the convex case, the role of the dimension is significant).

(ii) The choice of an appropriate topology on the phase space.

Some results may be proved in the non-resonant region [B4]. For instance, consider a 1D NL wave equation

$$u_{tt} - u_{xx} + \rho\, u + \varepsilon\, f'(u) = 0 \tag{153}$$

with typical ρ, i.e. $\left\{\sqrt{n^2 + \rho}\right\}$ have good Diophantine properties. Assume $f'(u)$ odd and ε small. Then, for odd and smooth initial data $u(0)$, $u_t(0)$, the corresponding solution to (153) will evolve close to a quasi-periodic function of time for times t, $|t| < T_\varepsilon$ where T_ε may be taken to be any power of ε^{-1}. The key facts on which the argument is based are the following.

(i) The possibility to perform a perturbative analysis up to any order and obtain an approximative solution u_0 of (153).

(ii) The fact that the equation linearized at u_0 has zero Lyapunov exponents.

Presently, there do not seem to be satisfactory PDE counterparts of the Nekhoroshev theorem under convexity assumption of the Hamiltonian, even endowing the phase space with the weak topology. On the other hand, let us make also followingwing qualitative comment. Global smooth solutions u to (153) may be shown (in general) to satisfy estimates on higher derivatives of the form

$$\|u(t)\|_{H^s} < C^{|t|} \tag{154}$$

hence with exponential time dependence. Recently, this crude estimate was improved to a power-like bound (see [B5])

$$\|u(t)\|_{H^s} < (1+|t|)^{A(s-1)} \quad (s>1) \tag{155}$$

yielding an upper bound on how fast energy may travel from low to high Fourier modes. Related examples show that this is essentially an optimal result, up to the value of the constant A. The approach used is fairly general and combines algebraic conservation properties with the local theory on the initial value problem. Similar results may be obtained for the nonlinear Schrödinger equation, for instance the 1D cubic NLS

$$iu_t + \nabla u - u|u|^2 = 0 \tag{156}$$

(known to be non-integrable) and in infinite volume, i.e. on the line instead of with periodic boundary conditions.

References (strictly for the purpose of this exposé)

[B1] J. Bourgain: Construction of quasi-periodic solutions of Hamiltonian perturbations of linear equations and applications to nonlinear PDE, *IMRN*, Vol.11 (1944).

[B2] J. Bourgain: Construction of time periodic solutions of nonlinear wave equations in higher dimension, preprint 1995, to appear in *Geometric and Functional Analysis* (GAFA).

[B3] J. Bourgain: Quasi-periodic solutions of Hamiltonian perturbations of 2D linear Schrödinger equations, preprint IHES/M/95/1.

[B4] J. Bourgain: Construction of approximative and almost periodic solutions of perturbed linear Schrödinger and wave equations, preprint.

[B5] J. Bourgain: On the behavior of higher Sobolev norms of smooth solutions of nonlinear Hamiltonian PDE, preprint.

[C-W] W. Craig, E. Wayne: Newton's method and periodic solutions of nonlinear wave equations, *Comm. Pure and Applied Math.* **46** (1993), 1409-1501.

[K1] S. Kuksin: Nearly Integrable Infinite-Dimensional Hamiltonian Systems, *LNM* **1556** (Springer).

[K2] S. Kuksin: Perturbation theory for quasi-periodic solutions of infinite dimensional Hamiltonian systems and its applications to the Korteweg-de Vries equation, *Math. USSR Sbornik* **64** (1989), 397-413.

[K-P] S. Kuksin, J. Pöschel: *Annals of Math.*, to appear.

[N] N. Nekhoroshev: An exponential estimate of the time of stability of nearly-integrable Hamiltonian systems, *Russian Math. Surv.* Vol.32, $N^0 6$ (1977), 1-66.

[W] E. Wayne: Periodic and quasi-periodic solutions of nonlinear wave equations via KAM theory, *Comm. Math. Phys.* **127** (1990), 479-528.

Scaling Limits for Lattice Gas Models

S.R.S. VARADHAN Courant Institute of Mathematical Sciences, New York University, New York, NY 10012 (varadhan@math.nyu.edu)

H.T. YAU Courant Institute of Mathematical Sciences, New York University, New York, NY 10012 (yau@math.nyu.edu)

1. INTRODUCTION

When we consider a large system of interacting particles evolving in time, one of the natural things is to provide a simplified description of the state of the system. Rather than provide a detailed "microscopic" picture, we describe the system by providing the values of certain "macroscopic" parameters. This is best illustrated by an example. The model considered here describes particles on sites in a d-dimensional periodic lattice. Let us take a small parameter ϵ and consider a periodic lattice in $\boldsymbol{R}^d$ of length $L = \epsilon^{-1}$. We have a certain number, n_ϵ, of particles distributed over the sites in the lattice. The important restriction is that there may be at most one particle per site. We shall scale the lattice by ϵ so that the lattice is embedded in the standard d-dimensional torus for every value of ϵ. We denote by x points of the embedded lattice $\mathbb{Z}_L^d$ as well as points in the d-torus T^d. For each lattice site x the variable $\eta(x)$ which can take the values of zero or one signifies the absence or presence of a particle at site x. Clearly

$$\sum_{x \in \mathbb{Z}_L^d} \eta(x) = n_\epsilon$$

As ϵ gets small we say that the particles are distributed according to density $\rho(x)$ where $\rho(\cdot)$ is a nonnegative measurable function on T^d with $0 \leq \rho(x) \leq 1$ for all x if for every bounded continuous test function $J(x)$ on T^d we have

$$\lim_{\epsilon \to 0} \epsilon^d \sum_x J(x)\eta(x) = \int_{T^d} J(x)\rho(x)\, dx$$

Clearly this implies that the total number of particles, n_ϵ, should be proportional to L^d as $\epsilon \to 0$, the constant of proportionality is the total integral of $\rho(x)$ on T^d.

Imagine now that the system of particle evolves in some stochastic manner. Particles may jump at some Poisson rates to nearby sites if any of them should be vacant. The Poisson rate could depend on the immediate environment of the current particle. This would of course happen simultaneously for all the particles. Although locally the system will be changing fast, the local "density" of particles will change rather slowly because no particles are created or destroyed . If we speed thing up then the "density" changes normally but locally the system changes its microscopic state very rapidly. Assuming that density is the only conserved quantity there will be an one parameter family of invariant measures for the evolution. Because of the rapid time scale, locally the system will be near equilibrium dictated by the local density. The local density then is a function of space and time. One of the goals is to write down a partial differential equation that controls the evolution of this density function $\rho(t, x)$.

2. THE MODEL

In order to describe the dynamics mathematically we proceed in the following manner. Let us denote by Ω the space of all configurations on the periodic lattice $\mathbb{Z}_L^d$. Ω is only a finite set even if it is a large one. Our evolution is a Markov process with Ω as state space. In order to describe our Markovian evolution on Ω we first define a probability measure μ on Ω and define a Dirichlet form relative to μ. The infinitesimal generator, $\mathcal{L}$, will be the operator corresponding to the Dirichlet form. Suppose b is a bond connecting two nearest neighbor sites on the lattice. For any configuration η we can define a new configuration η^b obtained by interchanging the values of η at the two ends of the bond. This corresponds to a particle moving from the occupied site at one end of the bond to the free site at the other end of the same bond. Of course if both ends are free or both are occupied then nothing happens and $\eta^b = \eta$. For any function f on Ω and any bond b we can define

$\nabla_b\,(f)(\eta) = [f(\eta^b) - f(\eta)]$. The Dirichlet form $D_L(f)$ then takes the form

$$D_L(f) = \int_\Omega \sum_b [\nabla_b(f)(\eta)]^2 \, d\mu$$

The measure μ on the space Ω is specified as a Gibbs measure which means that the probabilities for the various points η in Ω are defined by

$$\mu(\eta) = \exp\,[-\sum_{x\in \boldsymbol{Z}_L^d} F_x(\eta)]/Z$$

Here Z is a normalization constant, $F(\eta)$ is a local function of η depending on $\{\eta(y)\}$ for a few y near the origin and F_x is the translation of F by x so that $F_x(\eta)$ is the same function but around the site x. The sum in the exponent is to be thought of as the energy so that $\mu(\eta) = exp[-H(\eta)]$. The total energy H is the sum of local energies F_x. If F depends only on $\eta(0)$ then μ is a product measure. Otherwise μ has some dependence built into it. A simple class of examples are the Ising measures.

Given a μ through the F and given the Dirichlet form we have a Markovian evolution. In this evolution the total number of particles is conserved. If we denote by $\Omega_{L,n}$ those configuration on our lattice of length L which have n particles, this subset will be an irreducible invariant subset of our Markov process so that the restriction $\mu_{L,n}$ of μ to this set, correctly normalized, will be the invariant ergodic measures and one should really think of the evolution taking place on any one of these irreducible components. One final comment is that we need to speed up the evolution by a factor of ϵ^{-2} and this means that the Dirichlet form gets a factor of ϵ^{-2} in front.

3. HYDRODYNAMIC SCALING

We make a qualitative assumption concerning the energy function F. We assume that it is such that it possesses a unique infinite volume Gibbs measure and more over certain stronger conditions known as the mixing conditions of Dobrushin and Shlosman are satisfied. This is the case if F is small in some sense. According to the work of [LY, Y], such a condition implies a uniform estimate on the spectral gaps for our(speeded up) Markov generators on various $\Omega_{L,n}$. Let us start with some initial configuration η that has asymptotically $\rho_0(x)$ as macroscopic density, i.e.

$$\lim_{\epsilon\to 0} \sum_x J(x)\eta(x) = \int_{T^d} J(x)\rho_0(x)\,dx$$

Then with respect to the measure P_ϵ, representing our Markovian evolution, the following limit holds uniformly on any finite interval in probability:

$$\lim_{\epsilon\to 0} \sum_x J(x)\eta(t,x) = \int_{T^d} J(x)\rho(t,x)\,dx$$

Here the function $\rho(t,x)$ is given as the unique solution of the following nonlinear diffusion equation.

$$\frac{\partial\rho}{\partial t} = \frac{1}{2}\nabla D(\rho)\nabla(\rho),$$
$$\rho(0,x) = \rho_0(x)$$

The diffusion matrix D is to be determined. It can be specified by a Green-Kubo formula or, as we do here, through a variational formula which makes it easier to work with. In the definition of $\mu(\eta)$ we can add to the exponent a term of the form $\lambda\sum_x \eta(x)$ so that the formula in a finite volume reads

$$\mu(\eta) = \exp[\lambda\sum_x \eta(x) - \sum_{x\in \mathbb{Z}_L^d} F_x(\eta)]/Z$$

In the infinite volume limit this will produce a stationary process with density ρ which is a function of the real number λ. As λ varies from $-\infty$ to ∞, $\rho(\lambda)$ will go from zero to one and $\lambda(\rho)$ is the inverse function. An infinite volume Gibbs measure μ can be specified by its density ρ or by its chemical potential λ.

4. FORMULA FOR THE DIFFUSION MATRIX

For simplicity let us do the calculation in one dimension. We want to calculate the change in time of an object like $\epsilon\sum_x J(x)\eta(t,x)$ where $\eta(t,.)$ is the configuration at time t. Computing the change in time gives us two terms. One of them is a martigale term which is seen to be negligible for small ϵ by a mean square calculation. The other term is obtained by applying the infinitesimal generator to the quantity of interest and is of the form

$$-\sum_x J'(x)W_{x,x+1}(\eta(t,.)).$$

Here $W_{x,x+1}$ is the current from x to $x+1$. The system is called a non-gradient system because $W_{x,x+1}$ is not the difference $V_{x+1} - V_x$ for some local functional V of the configuration. Here the functions $W_{x,x+1}$ and V_x are basically the same function translated by shift to the point x on the lattice. In a gradient system we can do a summation by parts and reduce the above quantity to

$$\epsilon\sum_x J''(x)V_x$$

We can calculate the expected value $v(\rho)$ of V_x in the Gibbs state with density ρ and this leads to the diffusion equation

$$\frac{\partial\rho}{\partial t} = \frac{\partial^2 v(\rho(t,x))}{\partial x^2}$$

Our case is non-gradient and we have to replace $W_{x,x+1}$ by a term of the form $D(\eta(x+1)-\eta(x))$ for a suitable constant D depending on the local density ρ. This leads to the diffusion equation

$$\frac{\partial \rho(t,x)}{\partial t} = \frac{\partial}{\partial x} D(\rho(t,x)) \frac{\partial \rho(t,x)}{\partial x}$$

The function $D(\rho)$ as a function of ρ is determined in the following manner. We start in equilibrium with the Gibbs measure with density ρ in infinite lattice. If f is a local function then $g = \mathcal{L}f$, $W_{0,1}$,$(\eta(1)-\eta(0))$ are all mean zero functions. Their integrals over time has a central limit theorem. Let us denote by

$$\sigma^2(f,D,\rho) = \lim_{l\to\infty} \frac{1}{2l+1} \lim_{t\to\infty} \frac{1}{t} \times$$
$$Var\left(\int_0^t \sum_{-l\le x\le l} [W_{x,x+1}(\eta(s,\cdot)) - g_x(\eta(s,\cdot)) - D(\eta(x+1)-\eta(x))]\right) ds$$

Then $D(\rho)$ is determined so that

$$\inf_f \sigma^2(f, D = D(\rho), \rho) = 0 \, .$$

There is a multi-dimensional analog for replacing the current by $D(\rho)$ times $\nabla\rho$ and this determines the matrix $D(\rho)$.

REFERENCES

[LY] Lu, S. L. and Yau, H.-T.: Spectral gap and Logarithmic Sobolev Inequality for Kawasaki and Glauber dynamics, Commun. Math. Phys. **156**, 399-433, 1993.

[Y] Yau, H.-T.: Logarithmic Sobolev inequality for lattice gases with mixing conditions, preprint.

Multivariate Distributions with Gaussian Conditional Structure

BARRY C. ARNOLD Department of Statistics, University of California, Riverside, California

JACEK WESOLOWSKI Mathematical Institute, Warsaw University of Technology, Warsaw, Poland

Key words: *quasi-Gaussian distributions, classical normal distribution, normal conditionals distribution, elliptical contours, linear regression, mixtures, Kagan class.*

ABSTRACT

Multivariate distributions exhibiting some features of the conditional structure associated with the classical normal model are investigated. Features considered include conditional distributions of subvectors and conditional moments. Our understanding of the classical normal model is enhanced by the study of such quasi-Gaussian distributions together with investigation of additional assumptions required to characterize the classical normal model. Special attention is paid to the class of distributions exhibiting Gaussian conditional structure of the second order, i.e. those in which the conditional moments of orders one and two match the Gaussian model.

1 THE CLASSICAL MULTIVARIATE NORMAL DISTRIBUTION

A random vector $\underline{X} = (X_1, X_2, \ldots, X_k)$ is said to have a classical multivariate normal distribution if it admits a representation of the form

$$\underline{X} = \underline{\mu} + \Sigma^{1/2}\underline{Z}$$

where $Z_1, Z_2, \ldots, Z_k$ are i.i.d. standard univariate normal random variables. In such a case we write $\underline{X} \sim N^{(k)}(\underline{\mu}, \Sigma)$. Here $\underline{\mu}\epsilon\mathbf{R}^k$ and Σ is a non-negative definite $k \times k$ matrix. Such random variables have remarkable properties. For example:

1. All one dimensional marginals are normal.
2. All ℓ dimensional marginals, $\ell < k$, are ℓ-variate normal.
3. All linear combinations are normal. In fact for any $\ell \times k$ matrix B we have
$$B\underline{X} \sim N^{(\ell)}(B\underline{\mu}, B\Sigma B') \ .$$
4. All conditionals are normal. Thus if we partition $\underline{X} = (\dot{\underline{X}}, \ddot{\underline{X}})$ then the conditional distribution of $\dot{\underline{X}}$ given $\ddot{\underline{X}} = \ddot{\underline{x}}$ is multivariate normal.
5. All regressions are linear. Thus for any i and any $j_1, j_2, \ldots, j_\ell (\neq i)$ $E(X_i | X_{j_1}, \ldots, X_{j_\ell})$ is a linear function of $X_{j_1}, X_{j_2}, \ldots, X_{j_\ell}$.
6. All conditional variances are constant. Thus $var(X_i | X_{j_1}, \ldots, X_{j_\ell})$ is nonrandom for any i, and any $j_1, j_2, \ldots, j_\ell (\neq i)$.
7. If Σ is positive definite, the joint density of $\underline{X}$ is elliptically contoured.
8. $\underline{X}$ has linear structure, i.e. $\underline{X}$ admits a representation of the form
$$\underline{X} = \underline{a}_0 + A\underline{Z}$$
where the Z_i's are independent random variables.

Most of these properties, taken individually, fail to characterize the classical multivariate normal distribution. Combinations of these properties can be used to characterize the classical model . Condition 3 does characterize the classical model. Condition 4 also will characterize the classical model provided $k > 2$. None of the others alone will do it. Conditions 7 and 8, together, will characterize the classical distribution.

The present paper will focus mainly on two issues: the possibility of weakening the assumption of property 4 and still preserving a k-variate normal characterization (Section 2), and a discussion of models which, though not classical normal, mimic the conditional moment structure of the classical models (Section 3 and 4). Additional conditions for such structures leading to multivariate normality are outlined in Section 5.

Some useful notational conventions follow. Suppose $\underline{X}$ denotes a k-dimensional random vector and $\underline{x}\epsilon\mathbf{R}^k$. A partition of $\underline{X}$ into two subvectors of dimension $\dot{k}$ and $\ddot{k}$ with $\dot{k}+\ddot{k}=k$ will be denoted by $(\dot{\underline{X}},\ddot{\underline{X}})$. The corresponding partition of $\underline{x}$ will be denoted $(\dot{\underline{x}},\ddot{\underline{x}})$. X_i will denote the ith coordinate of $\underline{X}$. $\underline{X}_{(i)}$ is the $k-1$ dimensional vector obtained from $\underline{X}$ by deleting X_i. $\underline{X}_{(i,j)}$ is obtained from $\underline{X}$ by deleting X_i and X_j. Analogously real vectors $\underline{x}_{(i)}$ and $\underline{x}_{(i,j)}$ are defined.

2 CONDITIONAL CHARACTERIZATIONS OF THE CLASSICAL NORMAL MODEL

Suppose that for each i and for each $\underline{x}_{(i)}\epsilon\mathbf{R}^{k-1}$ the conditional distribution of X_i given $\underline{X}_{(i)}=\underline{x}_{(i)}$ is normal with a mean and variance that may depend on $\underline{x}_{(i)}$, i.e.

$$X_i|\underline{X}_{(i)}=\underline{x}_{(i)}\sim N(\mu_i(\underline{x}_{(i)}),\sigma_i^2(\underline{x}_{(i)}))\ . \tag{1}$$

In this case, generalizing the early results of Bhattacharya (1943) and solving an appropriate set of functional equations, one may verify that $\underline{X}$ must have what we may call a k-variate normal conditionals distribution with density of the form:

$$f_{\underline{X}}(\underline{x})=\exp[-\frac{1}{2}G(\underline{x})] \tag{2}$$

where

$$G(\underline{x})=\sum_{i_1=0}^{2}\sum_{i_2=0}^{2}\cdots\sum_{i_k=0}^{2}\gamma_{i_1i_2,\ldots,i_k}(\prod_{j=1}^{k}x_j^{i_j})\ . \tag{3}$$

There are necessary restrictions on the ranges of the γ's in (2.3) in order to ensure integrability and to ensure that all expressions for conditional variances are uniformly positive. Of course $\gamma_{00\ldots0}$ is not really a parameter, it is a normalizing factor that is a function of the remaining γ's chosen to ensure that the integral of the joint density is 1. If $\underline{X}$, of dimension k, has a density of the form (2.2) we will write

$$\underline{X}\sim NC^{(k)}(\underline{\gamma})\ .$$

See Arnold, Castillo and Sarabia (1992) for a more detailed introduction to the normal conditionals model.

The classical k-variate normal distribution is of course a special case of the normal conditionals model (2.2), since obviously it satisfies the required condition (2.1). It can be recognized by the fact that for such a distribution all coefficients $\gamma_{\underline{i}}$ for which $\sum_{j=1}^{k}i_j>2$ must be zero since, in order for (2.2) to represent a classical normal model, $G(\underline{x})$ must be a quadratic form.

Many characterization programs may be viewed as beginning with conditional normal requirements leading to the model (2.2), or some related submodel, and then imposing additional conditions to ensure vanishing of the "unwanted" coefficients (i.e. $\gamma_{\underline{i}}$'s with $\sum_{j=1}^{k}i_j>2$).

To begin with, we may recall that the classical normal distribution actually has far more conditional normal distributions associated with it than those alluded to in (2.1). In fact, if $\underline{X} \sim N^{(k)}(\underline{\mu}, \Sigma)$, then for any partition of $\underline{X}$ into subvectors $\dot{\underline{X}}$ and $\ddot{\underline{X}}$ of dimensions $\dot{k}$ and $\ddot{k}$ with $k = \dot{k} + \ddot{k}$ we have

$$\dot{\underline{X}}|\ddot{\underline{X}} = \ddot{\underline{x}} \sim N^{(k)}(\dot{\underline{\mu}}(\ddot{\underline{x}}), \dot{\Sigma}(\ddot{\underline{x}})) \ . \tag{4}$$

Since all subvectors of $\underline{X}$ are again classical normal, even more conditional distributions, analogous to those in (2.4) but now based on partitioning subvectors of $\underline{X}$, are again guaranteed to be normal.

Assumption (2.1) is not enough to guarantee the classical model. Assumption (2.4) is more than enough (provided $k > 2$, otherwise (2.1) and (2.4) coincide and fail to characterize the classical normal model). In fact one may prove (see Arnold, Castillo and Sarabia (1994)) that, for $k > 2$, a sufficient condition to guarantee a classical multivariate normal model is an assumption that for each i, j and each $\underline{x}_{(i,j)} \epsilon \mathbf{R}^{k-2}$

$$(X_i, X_j)|\underline{X}_{(i,j)} \sim N^{(2)}(\underline{\mu}_{(ij)}(\underline{x}_{(i,j)}), \Sigma_{ij}(\underline{x}_{(i,j)})) \ . \tag{5}$$

The key observation is that (2.5) implies that for each i, $X_i|\underline{X}_{(i,j)} = \underline{x}_{(i,j)}$ is normal, since the classical bivariate normal has normal marginals. Consequently (2.5) is enough to guarantee that

$$\underline{X} \sim NC^{(k)}(\underline{\gamma}) \tag{6}$$

and for each i,

$$\underline{X}_{(i)} \sim NC^{(k-1)}(\underline{\gamma}^{(i)}) \ . \tag{7}$$

However marginals of a normal conditionals distribution (2.2) can only be of the normal conditionals form if certain $\underline{\gamma}$'s are zero. In fact (2.7) guarantees that all the "unwanted" $\underline{\gamma}$'s are zero, and the fact that $\underline{X}$ must have a classical normal distribution is a consequence.

Of course the conditional mean functions and conditional variance functions which are encountered in the normal conditionals model (2.2) are not the familiar linear regressions and constant conditional variances associated with the classical model. If we are willing to assume, in addition to the assumption that each X_i given $\underline{X}_{(i)}$ is normal, that the conditional variances are constant, i.e. that

$$X_i|\underline{X}_{(i)} = \underline{x}_{(i)} \sim N(\mu_i(\underline{x}_{(i)}, \sigma_i^2) \ , \tag{8}$$

then the unwanted $\underline{\gamma}$'s in (2.2) are forced to be zero and we must have $\underline{X}, \sim N^{(k)}(\underline{\mu}, \Sigma)$, i.e. classical normal. An analogous alternative sufficient prescription is the requirement that in (2.1) each $\mu_i(\underline{x}_{(i)})$ be a linear function of $\underline{x}_{(i)}$.

It is indeed well known that for a classical k-variate normal random vector $\underline{X}$ we can explicitly write the parameters in the conditional distribution of $\dot{\underline{X}}$ given $\ddot{\underline{X}}$ in terms of the original parameters of the distribution of $\underline{X}$. Thus with $\underline{X} = (\dot{\underline{X}}, \ddot{\underline{X}})$ and $\underline{\mu} = (\dot{\underline{\mu}}, \ddot{\underline{\mu}})$ we have

$$\dot{\underline{X}}|\ddot{\underline{X}} = \ddot{\underline{x}} \sim N^{(\dot{k})}(\dot{\underline{\mu}} + \Sigma_{12}\Sigma_{22}^{-1}(\ddot{\underline{x}} - \ddot{\underline{\mu}}), \Sigma_{11.2}) \tag{9}$$

where

$$\Sigma = \begin{pmatrix} \Sigma_{11} & \Sigma_{12} \\ \Sigma_{21} & \Sigma_{22} \end{pmatrix}$$

and

$$\Sigma_{11.2} = \Sigma_{11} - \Sigma_{12}\Sigma_{22}^{-1}\Sigma_{21} .$$

The linear regressions and constant conditional variances are explicitly displayed in (2.9). Linear regressions are not that unusual in multivariate distributions. Constant conditional variances are unexpected. In some ways they are even counterintuitive. Taken together, the requirements of linear regressions and constant conditional variances seem potentially so restrictive as to possibly, alone, suffice to characterize the classical normal model. They don't. But verifying that they don't and asking what additional requirements will lead to characterizations is an interesting exercise that enriches our understanding of the real nature of the curious classical multivariate normal model. The topic will be addressed in the next section.

Before leaving the study of conditional normality assumptions to focus on conditional moment assumptions, it is worth returning to the list of 8 properties of the classical model listed in section 1. Which of these in addition to (2.1) (i.e. X_i given $\underline{X}_{(i)} = \underline{x}_{(i)}$ is normal $\forall i, \forall \underline{x}_{(i)}$) will guarantee classical multivariate normality. We have already considered properties 5 and 6. Property 1 has potential, since marginals of normal conditionals models are typically not of the normal conditionals form and a fortiori not (classical) normal. In fact, if all one dimensional marginals of $\underline{X}$ are normal and (2.1) holds then the unwanted $\underline{\gamma}$'s in (2.3) must disappear and the classical normal model is obtained. Turning to condition 2, it is a condition that subsumes 1 and consequently can be used to characterize the classical model. Actually far less is needed. For example, if in addition to (2.1), each $\underline{X}_{(i)}$ is classical $(k-1)$-variate normal, then $\underline{X}$ must be classical k-variate normal. Indeed any marginal normality statement sufficient to guarantee one dimensional normal marginals will obviously suffice. Turn next to condition 3. If k linearly independent linear combinations of the coordinates of $\underline{X}$ are normally distributed and if (2.1) holds then, by a suitable linear transformation, we have $\underline{Y} = B\underline{X}$ with normal conditionals (i.e. (2.1)) and normal one dimensional marginals. Then $\underline{Y}$ and consequently also $\underline{X}$ is a classical normal random vector. Next turn to condition 7, elliptical contours. This is easily dealt with. The contours of the normal conditionals density are determined by $G(\underline{x})$ in (2.3). Their form will be elliptical only if the unwanted $\underline{\gamma}$'s are all zero; i.e. only in the classical normal case

Finally consider condition 8. The assumption of linear structure turns out to be particularly fruitful in conjunction with certain conditional moment assumptions, as we shall see in the next section. In conjunction with the normal conditionals assumption, i.e. (2.1), the role of linear structure is less evident. Assumption 8 does however imply the existence of a linear transformation of $\underline{X}$ (with density (2.2)) that has a density which can be factored. This does imply that the unwanted $\underline{\gamma}$'s in (2.3) must be zero and does indeed guarantee classical multivariate normality.

3 GAUSSIAN CONDITIONAL STRUCTURE

What can we say about collections of random variables which exhibit linear regression functions and constant conditional variances? We will, following Wesolowski (1991), call this property Gaussian conditional structure of the second order.

Formally we will say that a random element (or indexed collection of random variables) $X = \{X_\alpha : \alpha \epsilon A\}$ exhibits Gaussian conditional structure of the second order and write $X \epsilon GCS_2(A)$ if for any $n = 2, 3, \dots$ and any $\alpha_1, \alpha_2, \dots, \alpha_n \epsilon A$,

(i) $E(X_{\alpha_1}|X_{\alpha_2}, \dots, X_{\alpha_n})$ is a linear function of $X_{\alpha_2}, X_{\alpha_3}, \dots, X_{\alpha_n}$ and

(ii) $var(X_{\alpha_1}|X_{\alpha_2}, \dots, X_{\alpha_n})$ is non-random.

At times it is convenient to use the term Gaussian conditional structure of the second order to refer to the distributions or probability measures associated with the random element rather than with the random element per se; we will do this at times without explanation and without fear of confusion. To avoid trivial examples we will usually implicitly assume that the X_α's are linearly independent and not uncorrelated. Collections of independent random variables could otherwise provide uninteresting examples of Gaussian conditional structure of the second order.

Observe that A could correspond to the natural numbers or the reals or positive reals. Consequently time series will be subsumed in the class of random elements under consideration. Spatial processes can be viewed as being random elements associated with a set A that is a subset of $\mathbf{R}^k$. Any normal process or, more generally, any Gaussian random element, will obviously exhibit Gaussian conditional structure of the second order. Our main focus will be however on random vectors of dimension k; i.e. on random elements where $A = \{1, 2, \dots, k\}$. If $X = (X_1, \dots, X_k)$ exhibits Gaussian conditional structure of the second order we will write $X \epsilon GCS_2(k)$.

A remark is in order about the subscript 2 that appears in our definition of Gaussian conditional structure of the second order. One could obviously ask that the random element mimic the conditional moment structure of a Gaussian element with regard to more than the first two conditional moments. One could ask for the first j conditional moments to behave as they do for Gaussian elements. The class of random elements exhibiting such behavior would be denoted by $GCS_j(A)$ instead of $GCS_2(A)$. Our focus will be on $GCS_2(A)$. Only once will we briefly mention how we might construct non-Gaussian members of $GCS_j(A)$, for $j > 2$.

If $\underline{X} \epsilon GCS_2(k)$, it is natural to ask whether $\underline{X}$ must necessarily be Gaussian. The question is already meaningful and reasonably challenging when $k = 2$; i.e. in the case of bivariate distributions. Kagan, Linnik and Rao (1973) provide the following lemma indicating the nature of characteristic functions associated with $GCS_2(2)$ distributions.

Lemma 3.1: In order that the two-dimensional random vector (X, Y) satisfy (i) $E(Y|X) =$

$\alpha + \beta X$ and (ii) $var(Y|X) = \sigma^2$ (a constant), it is necessary and sufficient that the characteristic function of (X, Y) satisfies

$$\frac{\partial}{\partial t_2}\phi(t_1, t_2)|_{t_2=0} = i\alpha\phi(t_1, 0) + \beta\frac{d}{dt_1}\phi(t_1, 0) \tag{1}$$

and

$$\frac{\partial^2}{\partial t_2^2}\phi(t_1, t_2)|_{t_2=0} \equiv -(\sigma^2 + \alpha^2)\phi(t_1, 0) + 2i\alpha\beta\frac{d}{dt_1}\phi(t_1, 0) + \beta^2\frac{d^2}{dt_1^2}\phi(t_1, 0) \,. \tag{2}$$

If one, as do Kagan, Linnik, and Rao, then assumes that (X, Y) has linear structure (i.e., satisfies condition 8, of section 1), then we may verify that indeed (X, Y) must have a classical bivariate normal structure.

Examples of non-Gaussian characteristic functions satisfying the conditions of Lemma 3.1 are not that easy to visualize. It is in fact probably an inappropriate approach to the problem of verifying that there do exist random vectors with GCS_2 that are not classical normal random vectors. It is probably more fruitful to seek non-Gaussian density functions that will exhibit the required conditional properties (and a fortiori will have characteristic functions satisfying the conditions in the Lemma). The first example of this genre was provided by Kwapien sometime prior to 1985. It was first reported in Bryc and Plucinska (1985). It was in fact presented in terms of the joint characteristic function. He considers a random vector (X, Y) whose joint characteristic function is given by

$$\phi_{X,Y}(s, t) = p\,\cos(s + t) + (1 - p)cos(s - t) \,. \tag{3}$$

where $p \in (0, 1)$ and, to avoid independence, $p \neq 1/2$. It is obvious that (3.3) does not correspond to a Gaussian random vector and it is not hard to verify that conditions (3.1) and (3.2) hold, as do the parallel conditions corresponding to interchanging the roles of X and Y. Consequently (X, Y) with characteristic function (3.3) does exhibit Gaussian conditional structure of order 2, i.e. $(X, Y) \in GCS_2(2)$.

Where did (3.3) come from? And, why does it work? The picture is clearer if we look at the following joint discrete density of a random vector (X, Y)

$$f_{X,Y}(x, y): \quad \begin{array}{cc|cc|} & x & & \\ y & & -1 & 1 \\ & & & \\ & -1 & \frac{1-p}{2} & \frac{p}{2} \\ & 1 & \frac{p}{2} & \frac{1-p}{2} \end{array} \tag{4}$$

where $p \in (0, 1)$ and, to avoid independence, $p \neq 1/2$. It is readily verified that this is indeed Kwapien's example (the corresponding characteristic function is given by (3.3)). But the joint distribution (3.4) has marginals with only two possible values. This gives linear regression functions by default (any function with a two point domain is linear!). Constant conditional variances are a consequence of the fact that $p(1 - p) = (1 - p)p$.

The elegant simplicity of the Kwapien example would suggest ready extension to higher dimensions. However, only recently (Nguyen, Rempala and Wesolowski (1994)), have any other (other than relabeled versions of the Kwapien example) non-Gaussian examples been described in either two or more dimensions. Indeed there were some disturbing indications that Gaussian conditional structure of the second order might be more restrictive than one would initially imagine. Bryc and Plucinska (1985) showed that if we consider a random element where $A = \{1, 2, \ldots\}$ then under mild regularity conditions, Gaussian conditional structure of order 2 i.e. $GCS_2(1, 2, \ldots)$ is sufficient to guarantee that X must be a normal process. Earlier, in a series of papers (Plucinska (1983), Wesolowski (1984) and Bryc (1985)), an analogous result was obtained in case on which $A = \mathbf{R}^+$.

The first step to finding non-Gaussian random vectors of dimension greater than 2 with Gaussian conditional structure of order 2, would focus on 3 dimensional examples and, following Kwapien's lead, would focus on simple discrete examples in which the conditional moment conditions will transform into relatively simple equations in the unknown cell probabilities. Thus for example we might seek a 3-dimensional discrete distribution whose second order conditional structure will match that of a 3 dimensional classical normal distribution of the form

$$\begin{pmatrix} X_1 \\ X_2 \\ X_3 \end{pmatrix} \sim N^{(3)} \left(\begin{pmatrix} 0 \\ 0 \\ 0 \end{pmatrix}, \begin{pmatrix} 1 & \frac{1}{2} & \frac{1}{2} \\ \frac{1}{2} & 1 & \frac{1}{2} \\ \frac{1}{2} & \frac{1}{2} & 1 \end{pmatrix} \right). \tag{5}$$

For such a distribution we will have $E(X_i) = 0$ and $var(X_i) = 1, i = 1, 2, 3$. The conditional moments will be given by

$$\begin{aligned} E(X_i|X_j, X_k) &= \frac{1}{3}(X_j + X_k) \\ var(X_i|X_j, X_k) &= \frac{2}{3} \\ E(X_i|X_j) &= X_j/2 \end{aligned} \tag{6}$$

and

$$var(X_i|X_j) = \frac{3}{4}$$

for all choices of $i \neq j \neq k$ (since the distribution is clearly symmetric). Following the lead of the Kwapien example we would seek a 3 dimensional random vector (Y_1, Y_2, Y_3) with a discrete distribution with possible values for each Y_i being $-m, m-1, \ldots, 1, 1, \ldots, m$ and with probabilities $p_{ijk} = P(Y_1 = i, Y_2 = j, Y_3 = k)$. The joint distribution should be exchangeable. The marginal means and variances should be 0's and 1's. The conditional means and variances of the Y_i's should agree with those in (3.6). Our task is then to solve for the unknown values of $\{p_{ijk} : -m \leq i \leq j \leq k \leq m\}$ subject to the given constraints. If $m \leq 3$, there are more constraints than variables. For $m = 4$, there are 120 variables (p_{ijk}'s) which must be non-negative and satisfy 91 linear constraints. A promising situation, although a solution is not guaranteed. Unfortunately efforts to solve such a system of equations have not proved successful. The search for a solution continues since a simple discrete example may shed additional light on the nature of the class of distributions with Gaussian conditional structure. However, the problem of constructing a non-Gaussian multivariate distributions (of finite dimension > 2) with Gaussian conditional structure was recently resolved by Nguyen, Rempala and Wesolowski (1994). The solution is ingenious but, retrospectively, obvious. Inspired by their examples, the following simple construction is possible.

Take $f_0(\underline{x})$ to be the joint density of a classical k-variate normal distribution with mean vector $\underline{\mu}$ and variance-covariance matrix Σ. We now construct a k-dimensional density which has the same conditional means and variances as does $f_0(\underline{x})$. Pick two distinct bounded densities g_1 and g_2 each supported on the interval $(-1, 1)$ and each having mean 0 and variance 1. There are of course a plethora of such densities. Now consider the new k-dimensional density defined by

$$f^*(\underline{x}) = f_0(\underline{x}) + c\prod_{i=1}^{k}[g_1(x_i) - g_2(x_i)] \tag{7}$$

where c is chosen small enough to guarantee that the expression in (3.7) is uniformly positive (possible since the g_i's are bounded densities). Obviously $f^*(\underline{x})$ is non-Gaussian but all of its marginals are Gaussian and it is readily verified that all of its first and second conditional moments match those of $f_0(\underline{x})$. The density $f^*(\underline{x})$ thus belongs to $GCS_2(k)$; and in fact $GCS_2(k)$ is not just non-empty but contains an enormous variety of distributions constructed in a fashion analogous to that used to define (3.7). It is indeed possible, by putting additional higher moment conditions on the g_i's (used in the construction of f^*), to find k-variate non-Gaussian distributions whose conditional moments up to the m'th order ($m > 2$) match those of a classical normal k-variate distribution.

4 THE STRUCTURE OF THE CLASS $GCS_2(k)$

From the discussion in Section 3, we are aware that the class $GCS_2(k)$ is quite extensive. Our goal in the present section is to identify characteristic properties of the class and to identify conditions sufficient to guarantee that a member of the class indeed is a classical Gaussian distribution. For notational simplicity, some of the discussion is restricted to the bivariate case (i.e. $k = 2$).

Suppose that $\underline{X} \epsilon GCS_2(k)$. Obviously any vectors of the form $\underline{Y} = (c_1X_1 + b_1, c_2X_2 + b_2, \ldots, c_kX_k + b_k)$ for $c_1, \ldots, c_k > 0$ and $\underline{b} \epsilon \mathbf{R}^k$ will again belong to $GCS_2(k)$. Consequently there is no loss in generality if we focus on standardized members of $GCS_2(k)$. These are random vectors $\underline{Z} \epsilon GCS_2(k)$ with the property that $E(Z_i) = 0$ and $var(Z_i) = 1$; $i = 1, 2, \ldots, k$. Throughout this section we will adopt the convention that if we use the notation $\underline{X}$, we are dealing with a general member of the class $GCS_2(k)$ while, if we use the notation $\underline{Z}$, we are referring to a standardized random vector in $GCS_2(k)$.

Thus we are concerned with random vectors $\underline{X}$ such that, with $Z_i = (X_i - E(X_i))/\sqrt{var X_i}$, $\underline{Z}$ satisfies: for any $i, j_1, \ldots, j_\ell$ ($\ell \le k-1$)

$$\text{(i)} \qquad E(Z_i|Z_{j_1}, \ldots, Z_{j_\ell}) = \sum_{m=1}^{\ell} \delta_{\underline{j},i,m} Z_{j_m} \tag{1}$$

and

$$\text{(ii)} \qquad var(Z_i|Z_{j_1}, \ldots, Z_{j_k}) = \sigma^2_{\underline{j},i} \tag{2}$$

for constants $\delta_{\underline{j},i,m} \epsilon \mathbf{R}$ and $\sigma^2_{\underline{j},i} \epsilon \mathbf{R}^+$.

A random vector $\underline{Z}$ satisfying (4.1) and (4.2) will have a corresponding variance-covariance matrix $\Sigma = R$ (with unit entries in the diagonal). To avoid trivial cases we assume R is not a diagonal matrix. Clearly there are quite complicated inter-relationships that must hold among the coefficients appearing in (4.1) and (4.2) since they must be consistent with some diagonal variance-covariance matrix R. Of course, for a given R, there are many $GCS_2(k)$ distributions. It is convenient to introduce the notation $GCS_2(k, \Sigma)$ to denote all random vectors $\underline{X}$ with Gaussian conditional structure of the second order with a given k-dimensional variance-covariance matrix Σ. Analogously if we write $\underline{Z} \in GCS_2(k, R)$ we mean that $\underline{Z}$ is a standardized vector with Gaussian conditional structure of the second order and correlation matrix R.

If (4.1) and (4.2) hold, the joint characteristic function of $\underline{Z}$ is severely constrained. Conditions analogous to those displayed in equations (3.1) and (3.2) must hold for various first and second partial derivatives of the joint characteristic function. In the bivariate case, we have $\underline{Z} \in GCS_2(2)$ iff

$$\text{(i)} \qquad E(Z_1|Z_2) = \rho Z_2, E(Z_2|Z_1) = \rho Z_1 \tag{3}$$

and

$$\text{(ii)} \qquad var(Z_1|Z_2) = var(Z_2|Z_1) = 1 - \rho^2 \tag{4}$$

where

$$\rho = cov(Z_1, Z_2)(\in (-1, 1)) \ .$$

Conditions (3.1) and (3.2) may be rewritten for such standardized variables as follows.

<u>Lemma 4.1</u>: $\underline{Z} \in GCS_2(2)$ iff for some $\rho \in (-1, 1)$ its joint characteristic function $\phi(t_1, t_2)$ satisfies

$$\frac{\partial}{\partial t_1}\phi(t_1, t_2)|_{t_1=0} = \rho\frac{d}{dt_2}\phi(0, t_2) \tag{5}$$

$$\frac{\partial}{\partial t_2}\phi(t_1, t_2)|_{t_2=0} = \rho\frac{d}{dt_1}\phi(t_1, 0) \tag{6}$$

$$\frac{\partial^2}{\partial t_1^2}\phi(t_1, t_2)|_{t_1=0} = (\rho^2 - 1)\phi(0, t_2) + \rho^2\frac{d^2}{dt_2^2}\phi(0, t_2) \tag{7}$$

and

$$\frac{\partial^2}{\partial t_2^2}\phi(t_1, t_2)|_{t_2=0} = (\rho^2 - 1)\phi(t_1, 0) + \rho^2\frac{d^2}{dt_1^2}\phi(t_1, 0) \ . \tag{8}$$

It is not hard to verify that a classical bivariate normal random vector with unit variances and correlation coefficient ρ, has a joint characteristics function which satisfies (4.5) - (4.8). Similarly the joint characteristic function of the Kwapien distribution (3.3) clearly satisfies (4.5) - (4.8) with $\rho = 2p - 1$.

The class $GCS_2(k)$ contains Gaussian distributions, non-Gaussian densities as in (3.7) and, when $k = 2$, even discrete distributions. The common features of all the members can be expressed in terms of properties of conditional moments or of derivatives of the joint characteristic function. The class is however diverse. Some closure properties are however available for the class $GCS_2(k)$. For example each subclass $GCS_2(k,\Sigma)$, for fixed Z, is closed under mixtures.

Theorem 4.2: Suppose $\{\underline{X}_\alpha : \alpha \in \Lambda\}$ is an indexed collection of random vectors with $X_\alpha \in GCS_2(k,\Sigma)$ for every α. If we define Z to be a random vector with distribution function

$$F_{\underline{Z}}(\underline{z}) = \int_\Lambda F_{\underline{X}_\alpha}(\underline{z})dH(\alpha)$$

for any probability distribution H on Λ, then $\underline{Z} \in GCS_2(k,\Sigma)$.

Proof: The bivariate case ($k=2$) with Λ of cardinality 2 was reported by Bryc (1985). The general result is straightforward if we write the joint characteristic function as a mixture

$$\phi_{\underline{Z}}(t) = \int_\Lambda \phi_{\underline{X}_\alpha}(t)dH(\alpha)$$

and observe that the conditions (4.5) - (4.8) (and their k-dimensional analogs) are preserved by mixtures since the covariance structures (and hence the coefficients in (4.5) - (4.8)) are the same for every α.

Linear combinations of independent random vectors in $GCS_2(k,\Sigma)$ will yield random vectors in $GCS_2(k)$ but with a different covariance matrix. Specifically we have

Theorem 4.3: Suppose that $\underline{X}^{(1)}$ and $\underline{X}^{(2)}$ are independent members (not necessarily identically distributed) of $GCS_2(k,\Sigma)$ then for $(a,b,)\neq(0,0)$, $a\underline{X}^{(1)}+b\underline{X}^{(2)} \in GCS_2(k,(a^2+b^2)\Sigma)$. In particular if $a^2+b^2=1$, then $a\underline{X}^{(1)}+b\underline{X}^{(2)} \in GCS_2(k,\Sigma)$.

Proof: We provide a proof in the bivariate case. More extensive equations analogous to (4.5) - (4.8) must be verified in higher dimensional cases.

For simplicity and without loss of generality we assume that $\underline{X}^{(1)}$ and $\underline{X}^2$ have been standardized and we will denote them by $\underline{Z}^{(1)}$ and $Z^{(2)}$. By assumption $\underline{Z}^{(1)}$ and $Z^{(2)}$ have common correlation ratio ρ and their joint characteristic functions satisfy (4.5) - (4.8). Denote the joint characteristic functions of $\underline{Z}^{(1)}, \underline{Z}^{(2)}$ and $a\underline{Z}^{(1)}+b\underline{Z}^{(2)}$ by f_1, f_2 and f_3. Because $\underline{Z}^{(1)}$ and $\underline{Z}^{(2)}$ are independent we have

$$f_3(t_1,t_2) = f_1(at_1,at_2)f_2(bt_1,bt_2)\ .$$

Consequently, using (4.5) for f_1 and f_2

$$\begin{aligned}\frac{\partial}{\partial t_1}f_3(t_1,t_2)|_{t_1=0} &= a\frac{\partial}{\partial t_1}f_1(at_1,at_2)f_2(bt_1,bt_2)\\ &+f_1(at_1,at_2)b\frac{\partial}{\partial t_1}f_2(bt_1,bt_2)|_{t_1=0}\\ &= a\rho\frac{d}{dt_2}f_1(0,at_2)f_2(0,bt_2)\end{aligned}$$

$$+b\rho f_1(0, at_2)\frac{d}{dt_2}f_2(0, bt_2)$$

$$= \rho\frac{d}{dt_2}f_3(0, t_2) .$$

Thus (4.5) holds for f_3. Similarly (4.6) may be verified. Differentiating twice and using (4.7) for f_1 and f_2 we find

$$\frac{\partial^2}{\partial t_1^2}f_3(t_1, t_2)|_{t_1=0} = (a^2 + b^2)(\rho^2 - 1)f_3(0, t_2)$$

$$+\rho^2\frac{d^2}{dt_2^2}f_3(0, t_2) .$$

When $a^2 + b^2 = 1$ this implies that (4.7) continues to hold for f_3. In parallel fashion, since (4.8) holds for f_1 and f_2 then, when $a^2 + b^2 = 1$, it continues to hold for f_3. Since conditions (4.5) - (4.8) are sufficient for membership in $GSC_2(2)$ the conclusion of the theorem follows.

Naturally we can extend Theorem 4.3 to deal with sums of more than 2 independent members of $GCS_2(\Sigma)$. Indeed we can consider infinite convolutions since clearly the class $GCS_2(k, \Sigma)$ is closed under weak convergence (i.e. if $\underline{X}^{(n)} \epsilon GCS_2(k, \Sigma), n = 1, 2, \dots$ and $\underline{X}^{(n)} \xrightarrow{d} \underline{X}^{(\infty)}$ then $\underline{X}^{(\infty)} \epsilon GCS_2(k, \Sigma)$). Thus we may state

Theorem 4.4: Suppose $\underline{X}^{(1)}\underline{X}^{(2)}\cdots$ are independent random vectors each being a member of $GCS_2(k, \Sigma)$ (the same Σ for every $\underline{X}^{(i)}$). Define $\underline{Y} = \sum_{i=1}^{\infty} a_i \underline{X}^{(i)}$ where $\sum_{i=1}^{\infty} a_i^2 = 1$. It follows that $\underline{Y} \epsilon GCS_2(k, \Sigma)$.

Example: (Uniform and Cantor marginals) Suppose that $\underline{X}^{(i)}, i = 1, 2, \dots$ are i.i.d. Kwapien random vectors (with characteristic function (3.3) and joint density (3.4)). Consider a random vector $\underline{Y} = \sum_{i=1}^{\infty} a_i \underline{X}^{(i)}$ where $\sum_{i=1}^{\infty} a_i^2 < \infty$. Since each $\underline{X}^{(i)} \epsilon GCS_2(2)$ with correlation $2p-1$, it follows that $\underline{Y} \epsilon GCS_2(2)$ with the same correlation, $2p-1$. Particular choices for the a_i's yield interesting examples. If we choose $a_i = 1/2^i, i = 1, 2, \dots, \underline{Y}$ will have a continuous bivariate distribution with uniform $(-1, 1)$ marginals (and Gaussian conditional structure). We conjecture but are unable to prove that this joint distribution is singular (unless $p = 1/2$, the uninteresting case of independent marginals). If we choose $a_i = 2/3^i, i = 1, 2, \dots$ then $\underline{Y}$ will have a singular joint distribution with Cantor (and thus clearly singular) marginals. Thus we have a singular continuous example with Gaussian conditional structure. It is well known that sums of independent Cantor-like singular random variables can have non-singular (indeed uniform) distributions. Our present construction (using $a_i = 2/3^i$) allows us to give an example of dependent Cantor-like random variables whose sum is uniform. To do this, consider the special case $\underline{Y} = \sum_{i=1}^{\infty} \frac{2}{3^i}\underline{X}^{(i)}$ where the $\underline{X}^{(i)}$'s are Kwapien random vectors with $p = 2/3$. Here Y_1 and Y_2 are singular (Cantor) distributed on $(-1, 1)$ but $(Y_1 + Y_2)/2$ is uniform on $(-1, 1)$ (as is easily proved by looking at the convergent infinite product representation of its characteristic function obtained using the expression for the Kwapien characteristic function given in (3.3)).

5 FROM $GCS_2(k)$ TO CLASSICAL NORMAL

The examples of section 4 clearly indicate that additional conditions, besides appropriate behavior of conditional moments up to order 2 will be required to characterize the classical normal model. In this section we survey some known and some new results in this area. First a result due to Szablowski (1989).

Theorem 5.1: If $\underline{X} \in GCS_2(k)$ and if $\underline{X}$ is elliptically contoured then $\underline{X} \sim N^{(k)}(\underline{\mu}, \Sigma)$.

Next, we consider the generalized independence models described by Kagan (1988) classes.

Definition 5.2: A k-dimensional random vector $\underline{X}$ belongs to the Kagan class $D_{k,j}(loc), j = 1, 2, \ldots, k,\ k = 1, 2, \ldots,$ if its characteristic function $\phi_{\underline{X}}$, in some neighborhood, V, of the origin in $\mathbf{R}^k$ has the form

$$\phi_{\underline{X}}(\underline{t}) = \prod_{1 \le i_1 < \cdots < i_j \le k} R_{i_1,\ldots,i_j}(t_{i_1} t_{i_2}, \ldots, t_{i_j})$$

where each $R_{\underline{i}}$ is a continuous complex function with $R_{\underline{i}}(\underline{0}) = 1,\ \forall \underline{i}$.

It is plausible that any $\underline{X}$ in a Kagan class $D_{k,j}(loc)$ that exhibits Gaussian conditional structure of the second order might be classical k-variate normal. Some progress towards proving this result is provided in the following result due to Wesolowski (1991).

Theorem 5.3: If $\underline{X} \in GCS_2(k)$ for some $k > 2$ and if $\underline{X} \in D_{k,2}(loc)$ then $\underline{X} \sim N^{(k)}(\underline{\mu}, \Sigma)$.

If we assume that $\underline{X}$, in addition to having Gaussian conditional classical structure of the second order, is infinitely divisible, then it must be classical normal. This result is due to Wesolowski (1993). We refer the reader to the original paper for a general proof. Here we provide a simple illuminating proof for the bivariate case only.

Theorem 5.4: If $\underline{X} \in GCS_2(k)$ and if $\underline{X}$ is infinitely divisible then $\underline{X} \sim N^{(k)}(\underline{\mu}, \Sigma)$.

Proof: (in the bivariate case, i.e. $k = 2$, the case $k > 2$ was proved by another approach in Wesolowski (1993)). As usual, without loss of generality we assume zero means and unit variances. Since $\underline{X} = (X_1, X_2)$ is infinitely divisible, the logarithm of its joint characteristic function is of the form

$$\begin{aligned} \psi(\underline{t}) = \log \phi(\underline{t}) &= -\frac{1}{2}(t_1^2 + 2\rho t_1 t_2 + t_2^2) \\ &+ \int_{\mathbf{R}^2} [e^{i(t_1 x + t_2 y)} - 1 - i(t_1 x + t_2 y)] \frac{dK(x,y)}{x^2 + y^2} \end{aligned}$$

for some measure K. It then follows that

$$\frac{\partial^2}{\partial t_1^2} \psi(\underline{t})|_{t_1=0} = -1 - \int_{\mathbf{R}^2} e^{it_2 y} \frac{x^2}{x^2 + y^2} dK(x,y)$$

and

$$\frac{d^2}{dt_2^2}\psi(0,t_2) = -1 - \int_{\mathbf{R}^2} e^{it_{2y}} \frac{y^2}{x^2+y^2} dK(x,y) \ .$$

However since $\underline{X} \in GCS_2(2)$, we know that (4.8) holds. Consequently we have

$$\int_{\mathbf{R}^2} \frac{x^2}{x^2+y^2} dK(x,y) = \rho^2 \int \frac{y^2}{x^2+y^2} dK(x,y) \qquad (1)$$

(where $\rho^2 < 1$). Analogously, by considering $\frac{\partial^2}{\partial t_2^2}\psi(\underline{t})|_{t_2=0}$ and $\frac{d^2}{dt_1^2}\psi(t_1,0)$ we find that (5.1) again holds with the roles of x and y interchanged. Summing we conclude that for $\rho^2 < 1$,

$$\int_{\mathbf{R}^2} dK(x,y) = \rho^2 \int_{\mathbf{R}^2} dK(x,y) \ ,$$

i.e. $dK \equiv 0$. Consequently $\underline{X}$ must be classical bivariate normal.

6 REMARKS

Progress towards understanding the class of distributions with Gaussian conditional structure is accelerating. Many interesting questions remain open. Perhaps the most frustrating lacuna in the current inventory of examples involves the absence of any discrete example with Gaussian conditional structure of dimension greater than 2 (as discussed in Section 3). Theorem 4.3 together with the celebrated Kwapien example permits construction of a plethora of two dimensional discrete distributions with Gaussian conditional structure. The elusive 3 dimensional examples should appear soon.

REFERENCES

1. Arnold, B.C., Castillo, E. and Sarabia, J.M., Conditionally Specified Distributions, *Lecture Notes in Statistics*, Vol. 73, Springer-Verlag, Berlin, (1992).

2. Arnold, B.C., Castillo, E. and Sarabia, J.M., A conditional characterization of the multivariate normal distribution, *Statistics and Probability Letters*, 19, 313-315, (1994).

3. Bhattacharya, A., On some sets of sufficient conditions leading to the normal bivariate distribution, *Sankhya*, 6, 399-406, (1943).

4. Bryc, W., Some remarks on random vectors with nice enough behavior of conditional moments, *Bull. Polish Acad. Sci. Math.*, 33, 677-684, (1985).

5. Bryc, W. and Plucinska, A., A characterization of infinite Gaussian sequences by conditional moments, *Sankhya*, A47, 166-173, (1985).

6. Kagan, A.M., New classes of dependent random variables and a generalization of the Darmois-Skitovich to several forms, *Theory of Probability and Applications*, 33, 286-295, (1988).

7. Kagan, A.M., Linnik, J.V. and Rao, C.R., *Characterization Problems of Mathematical Statistics,* Wiley, New York, (1973).

8. Nguyen, T.T., Rempala, G. and Wesolowski, J., Non-Gaussian measures with Gaussian structure, to appear in *Probability and Mathematical Statistics*, (1994).

9. Plucinska, A., On a stochastic process determined by the conditional expectation and the conditional variance, *Stochastics*, 10, 115-129, (1983).

10. Szablowski, P.J., Can the first two conditional moments identify a mean square differentiable process?, *Comput. Math. Appl.*, 18, 329-348, (1989).

11. Wesolowski, J., A characterization of the Gaussian process based on properties of conditional moments, *Demonstratio Math.*, 18, 795-808, (1984).

12. Wesolowski, J., Gaussian conditional structure of the second order and the Kagan classification of multivariate distributions, *Journal of Multivariate Analysis*, 39, 79-86, (1991).

13. Wesolowski, J., Multivariate infinitely divisible distributions with the Gaussian conditional structure of the second order. In *Stability Problems for Stochastic Models* (Kalashnikov, V.V. and Zolotarev, V.M. eds). *Lecture Notes in Mathematics*, Vol. 1546, 180-183, Springer-Verlag, Berlin, (1993).

The Minimal Projection from L^1 onto π_n

BRUCE L. CHALMERS Department of Mathematics, University of California, Riverside, CA 92521

FREDERIC T. METCALF Department of Mathematics, University of California, Riverside, CA 92521

ABSTRACT

Simple equations are presented which are shown to be necessary and sufficient for the (unique) projection from $L^1[a,b]$ onto a finite-dimensional Haar subspace to be minimal. In particular these equations provide the minimal $L^1[-1,1]$ projection onto the algebraic polynomials of degree n, for which numerical solutions are given for $n \le 5$.

1. Introduction and Preliminaries

In [1] are derived sufficient and necessary (assuming the subspace is "smooth") equations for finite rank L^1 projections to be minimal (Theorem A below). (The existence of a minimal projection in this setting is proved in [5].) As an application of these conditions in [1] is obtained a sufficient condition, labeled "Prescription," for determining a minimal projection from $L^1[-1,1]$ onto a finite-dimensional subspace $\mathcal{V}$. In this paper we show that the Prescription is in fact also necessary whenever $\mathcal{V}$ is a Haar space, the projection (identity) action on V is generalized to an arbitrary non-singular action on $\mathcal{V}$, and $L^1[-1,1]$ is generalized to $L^1([-1,1],\nu)$, where ν is an arbitrary finite nonatomic Borel measure on $[-1,1]$. (Recall that $\mathcal{V}$ is an n-dimensional Haar space on $[a,b]$ means that the elements of $\mathcal{V}$ are continuous functions on $[a,b]$ and any nonzero $v \in \mathcal{V}$ has no more than $n-1$ zeros in $[a,b]$, or, equivalently, that any n distinct point-evaluation functionals (supported in $[a,b]$) are independent on $\mathcal{V}$.) The proof is based on an application of the classical Hobby-Rice theorem, a theorem of fundamental importance in the the theory of best approximation in the L^1-norm (cf [2]). As applications, we will use the Prescription to find numerical solutions for the minimal projections from $L^1[-1,1]$ onto $\mathcal{V} = \pi_{n-1}$, the space of $(n-1)$-degree algebraic polynomials, $n-1 \leq 5$.

More generally, for (T,Σ,ν) a complete measure space, let $P = \Sigma u_i \otimes v_i$ be a linear operator from $L^1(T,\nu))$ onto $\mathcal{V}$ with $u_i \in L^\infty$ and $v_i \in \mathcal{V}$, satisfying

$$\int_T v_i(t)u_j(t)\,d\nu(t) = a_{ij} \qquad (i,j=1,\ldots,n),$$

with the matrix $A = (a_{ij})$ fixed, but non-zero.

Additional notation which will be used in the following is

$$\vec{y} = (y_1,\ldots,y_n), \quad \vec{y}\cdot\vec{z} = y_1z_1 + \ldots + y_nz_n, \quad |\vec{y}| = (\vec{y}\cdot\vec{y})^{1/2}, \quad \vec{u}\odot\vec{v} = \sum u_i \otimes v_i,$$

and

$$V_i(t) = \int_T v_i(s)\operatorname{sgn} k(t,s)d\nu(s) \quad \text{where} \quad k(t,s) = \sum_{j=1}^{n} u_j(t)v_j(s).$$

Also, the Lebesgue function $L(t)$ for P is given by

$$L(t) = \int_T |k(t,s)|\,d\nu(s) = \vec{u}\,(t)\cdot\vec{V}(t), \quad t\in T;$$

note that $\|P\|$ = ess sup $L(t)$; see [3, Lemma 2]. Throughout this paper the notation $t \in T' \subseteq T$ will mean for almost all t out of T' relative to the measure ν.

In the following all the statements and results will refer to the operator P (and the associated "action" matrix A). Note that if P is a projection then A is the identity matrix.

The following theorem proved in [1] provides necessary and sufficient (equality) conditions for the operator P to be minimal.

Theorem A ([1]). *Let (T, Σ, ν) be a complete measure space for which ν is strictly localizable. Let $\mathcal{V}$ be a finite-dimensional subspace of L^1, and let $P = \vec{u} \odot \vec{v} = \sum_{i=1}^{n} u_i \otimes v_i$ be an operator mapping L^1 into $\mathcal{V}$ with $u_i \in L^\infty$ $(i = 1, \ldots, n)$, $\vec{v} = (v_1, \ldots, v_n)$ a fixed basis for $\mathcal{V}$, and the matrix $A = \int_T v_i(t) u_j(t)\, d\nu(t) = a_{ij}$ $(i, j = 1, \ldots, n)$ fixed. In order that P be minimal, the following equality conditions are necessary and sufficient:*
There exists a non-zero $n \times n$ matrix M such that
(a) the Lebesgue function $L(t) = \|P\|$ on $T' = \operatorname{supp}(M\vec{v})$, and
(b) there exists a positive function ϕ such that

$$\phi(t)\,\vec{V}(t) = M\,\vec{v}\,(t), \qquad t \in T'. \tag{1}$$

(In fact, $\phi = \vec{u} \cdot M\vec{v}\,/\|P\|$.)

Notation. We will denote by $P_{\min}$ a minimal operator given by Theorem A.

Recall that, in an L^1 space, the subspace $\mathcal{V}$ is said to be *smooth* if and only if each member of $\mathcal{V}\backslash 0$ is almost everywhere different from 0.

Corollary A ([1]). *If $\mathcal{V}$ is smooth, then the Lebesgue function for $P_{\min}$ in Theorem A is constant on T.*

Theorem B ([1]). *$P_{\min}$ is unique if $\mathcal{V}$ is smooth and $\vec{u}\,(t) \cdot \vec{v}$ is determined up to a scalar factor by its roots.*

2. Main Theorem

Lemma 1 (Hobby-Rice Theorem [4]). *Let $\mathcal{V}$ be an n-dimensional subspace of $L^1([-1,1], \nu)$, where ν is a finite nonatomic Borel measure. Then there exist points $-1 = t_0 \leq t_1 \leq \cdots \leq t_{n+1} = 1$ such that*

$$\sum_{i=0}^{n} (-1)^i \int_{t_i}^{t_{i+1}} v(t)\,d\nu(t) = 0, \quad \forall\, v \in \mathcal{V}. \tag{2}$$

Theorem 1 (Prescription). *Let $\mathcal{V}$ be an n-dimensional Haar subspace of $L^1([-1,1], \nu)$, where ν is a finite nonatomic Borel measure. Then $P_{\min} = \vec{u} \odot \vec{v} = \sum_{i=1}^{n} u_i \otimes v_i$ from $L^1([-1,1], \nu)$ into $\mathcal{V}$ with respect to the fixed action A is given uniquely by the following prescription:*

$$\begin{pmatrix} u_1(t) \\ u_2(t) \\ \cdot \\ \cdot \\ \cdot \\ u_n(t) \end{pmatrix} = \begin{pmatrix} \hat{V}_1(\vec{x}(t)) & \hat{V}_2(\vec{x}(t)) & \cdots & \hat{V}_n(\vec{x}(t)) \\ v_1(x_1(t)) & v_2(x_1(t)) & \cdots & v_n(x_1(t)) \\ \cdot & \cdot & \cdots & \cdot \\ \cdot & \cdot & \cdots & \cdot \\ \cdot & \cdot & \cdots & \cdot \\ v_1(x_{n-1}(t)) & v_2(x_{n-1}(t)) & \cdots & v_n(x_{n-1}(t)) \end{pmatrix}^{-1} \begin{pmatrix} \lambda\sigma(t) \\ 0 \\ \cdot \\ \cdot \\ \cdot \\ 0 \end{pmatrix}, \tag{3}$$

where

$$\hat{V}_i(\vec{x}) := \left(\int_{-1}^{x_1} - \int_{x_1}^{x_2} + \cdots + (-1)^{n-1}\int_{x_{n-1}}^{1}\right) v_i(s)\, d\nu(s), \quad i = 1, \ldots, n, \tag{4}$$

and $\vec{x}(t)$ *are solutions to*

$$m_{i+1}(t)\hat{V}_i(\vec{x}(t)) - m_i(t)\hat{V}_{i+1}(\vec{x}(t)) = 0, \quad i = 1, \ldots, n-1, \tag{5}$$

where $m_i(t) := \vec{m}_i \cdot \vec{v}(t), \quad i = 1, \ldots, n,$ *for some* $n \times n$ *matrix*

$$M := \begin{pmatrix} \vec{m}_1 \\ \cdot \\ \cdot \\ \cdot \\ \vec{m}_n \end{pmatrix}, \tag{6}$$

with

$$\sigma(t) := \operatorname{sgn}\left[\frac{\widehat{V}_n(\vec{x}(t))}{\vec{m}_n \cdot \vec{v}(t)}\right] \tag{7}$$

and

$$\lambda := \|P\|. \tag{8}$$

Note: The n^2 parameters m_{ij} (normalized so that one of $m_{ij} = 1$) and the norm parameter λ are determined from the n^2 orthonormalization conditions

$$\int_{-1}^{+1} v_i(t)u_j(t)\, d\nu(t) = a_{ij} \qquad (i, j = 1, \ldots, n). \tag{9}$$

Proof. Keeping in mind equation (b) of Theorem A, for a given matrix M with non-zero rows $\vec{m}_i$ $(i = 1, \ldots, n)$, we would like to solve

$$\frac{\widehat{V}_i(\vec{x}(t))}{\vec{m}_i \cdot \vec{v}(t)} = \frac{\widehat{V}_n(\vec{x}(t))}{\vec{m}_n \cdot \vec{v}(t)}, \quad i = 1, \ldots, n-1, \tag{10}$$

for $\vec{x}(t) = (x_1(t), \ldots, x_{n-1}(t))$, with $-1 < x_1(t) < \cdots < x_{n-1}(t) < 1$ for each $t \in (-1, 1)$.

Note that (10) can be rewritten as

$$\vec{m}_{i+1} \cdot \vec{v}(t)\widehat{V}_i(\vec{x}(t)) - \vec{m}_i \cdot \vec{v}(t)\widehat{V}_{i+1}(\vec{x}(t))$$

$$= \left(\int_{-1}^{x_1} - \int_{x_1}^{x_2} + \cdots + (-1)^{n-1}\int_{x_{n-1}}^{1}\right) [m_{i+1}(t)v_i(s) - m_i(t)v_{i+1}(s)]\, d\nu(s), \quad i = 1, \ldots, n,$$

where $m_i^t := \vec{m}_i \cdot \vec{v}(t)$, $i = 1, \ldots, n$. The existence of $x_i = x_i(t)$, $i = 1, \ldots, n-1$, now follows directly from Lemma 1 (the Hobby-Rice Theorem).

Next note that the invertibility of the $n \times n$ matrix in (3) is equivalent to the $n-1$ point-evaluations δ_{x_i} $(i = 1, ..., n-1)$ and the functional

$$\beta(\cdot) := \left(\int_{-1}^{x_1} - \int_{x_1}^{x_2} + \cdots + (-1)^{n-1} \int_{x_{n-1}}^{1} \right) (\cdot)(s)\, d\nu(s)$$

being independent on $\mathcal{V}$. But, indeed if $0 \neq v \in \mathcal{V}$ such that $v(x_i) = 0$, $i = 1, ..., n-1$, then, by the Haar assumption, for $x_0 := -1$ and $x_n := 1$, we have $\mathrm{sgn}(v) = \epsilon\,(-1)^i$ on (x_{i-1}, x_i), $i = 1, ..., n$, where $\epsilon = -\mathrm{sgn}(v(-1))$, and thus $|\beta(v)| > 0$, establishing the independence.

Note that $\sigma(t)\hat{V}_i(\vec{x}\,(t))$ will be $V_i(t)$ of Theorem A, $i = 1, \ldots, n$. The homogeneity of the system (10) allows one of the m_{ij} to be normalized to be 1, leaving exactly n^2 equations in n^2 unknowns. Finally, by the Haar condition, for each $t \in (-1, +1)$, $k(t, s) = \vec{u}(t) \cdot \vec{v}(s)$, as constructed according to the above prescription, changes sign only at $s = x_i(t)$ ($i = 1, \ldots, n-1$) and thus $\lambda = \vec{u}(t) \cdot \vec{V}(\vec{x}(t)) = \int_T |k(t, s)|\, d\nu(s) = L(t) > 0$, where $L(t)$ is the Lebesgue function for P. Thus Theorem A guarantees that we have $P_{\min}$.

Finally, by Theorem B, $P_{\min}$ is unique. ■

REMARK. System (3) is equivalent to

$$\begin{pmatrix} v_1(x_1) & \cdots & v_{n-1}(x_1) \\ \vdots & & \vdots \\ v_1(x_{n-1}) & \cdots & v_{n-1}(x_{n-1}) \end{pmatrix} \begin{pmatrix} \psi_1 \\ \vdots \\ \psi_{n-1} \end{pmatrix} = \begin{pmatrix} -v_n(x_1) \\ \vdots \\ -v_n(x_{n-1}) \end{pmatrix}, \tag{3a}$$

where $\psi_i = u_i/u_n$ ($i = 1, \ldots, n-1$), and

$$u_n = \frac{\lambda\sigma}{\widehat{V}_1(\vec{x})\psi_1 + \cdots + \widehat{V}_{n-1}(\vec{x})\psi_{n-1} + \widehat{V}_n(\vec{x})}. \tag{3b}$$

Thus, if $\mathcal{V}_{n-1} := [v_1, ..., v_{n-1}]$ is also Haar, then the $(n-1) \times (n-1)$ matrix in (3a) above is invertible. In particular, in the algebraic case, $\mathcal{V}_{n-1} := [1, t, \ldots, t^{n-2}]$ and the $(n-1) \times (n-1)$ matrix above is in fact the classical (invertible) Vandermonde matrix.

3. Applications

All the applications in this section are directed towards the determination of the minimal projection from $L^1[-1,1]$ onto π_{n-1} (i.e., the action $A = I$, the measure ν is standard Lebesgue measure, and $\mathcal{V} = [1, t, \ldots, t^{n-1}]$).

The first two applications are repeated from [1] for the sake of completeness and as an aid to the reader to identify the various parts of the Prescription.

In Application 1 (n=2), for each M, the single function $x(t)$, being the root of a quadratic, is determined explicitly, and then M :=diag$(1, m)$ and λ are determined to meet the (two) remaining (after symmetry) orthonormality conditions (via a numerical method).

In Application 2 (n=3), for each M, the two functions $x_i(t)$, $i = 1, 2$, are also determined explicitly (in terms of the solution of a quartic), and then M and λ are determined to meet the (five) remaining (after symmetry) orthonormality conditions (via a numerical method.

In Applications n-1 ($4 \leq n \leq 6$), for each M, determining the $n-1$ functions $x_i(t)$, $i = 1, ..., n-1$, involves solving (for each t) an $(n-1) \times (n-1)$ system of polynomial equations of degree n and thus must be determined entirely numerically (e.g. by Newton's method). M and λ are then determined to meet the remaining (after symmetry) orthonormality conditions (via a numerical method). In all cases we give the M matrix (up to 4 decimal places) and the projection norm λ (up to 5 decimal places).

APPLICATION 1. $\mathcal{V} = [1, t]_{[-1,1]}$ (Franchetti-Cheney [3]).

Consider the process described by the above prescription. In this case $n = 2$, $\widehat{V}_1(x) = 2x$ and $\widehat{V}_2(x) = x^2 - 1$. Using symmetry considerations, equation (5) becomes

$$\frac{2x(t)}{1} = \frac{x^2(t) - 1}{mt}, \tag{1.5}$$

from which the admissible solution is $x(t) = mt - \operatorname{sgn}(mt)\sqrt{m^2t^2 + 1}$. Equations (3a), (3b) and (7) are then

$$\psi(t) = -x(t) \qquad\qquad (= u_1(t)/u_2(t)\,) \tag{1.3a}$$

$$u_2(t) = \frac{\lambda\sigma(t)}{2x(t)\psi(t) + x^2(t) - 1} \tag{1.3b}$$

$$\sigma(t) = \operatorname{sgn} x(t), \tag{1.7}$$

which give

$$u_1(t) = \frac{\lambda|x(t)|}{1 + x^2(t)}$$

$$u_2(t) = -\frac{\lambda \operatorname{sgn} x(t)}{1 + x^2(t)}.$$

Using the symmetry of u_1 and u_2, equations (9) become

$$\int_0^1 u_1(t)\,dt = \int_0^1 tu_2(t)\,dt = \frac{1}{2}, \tag{1.9}$$

which are identical to equations (8) and (10) of [3], and result in the equation

$$2\psi(1)[1 - \psi^2(1) + \psi(1)]\log|\psi(1)| + 1 - \psi^2(1) = 0 \tag{12}$$

for $\psi(1)$, and hence m. It then follows that $\lambda > 0$ and Theorem 1 guarantees we have $P_{\min}$. In this case

$$M = \begin{pmatrix} 1 & 0 \\ 0 & 1.3605 \end{pmatrix}$$

and

$$P_{\min} = \lambda = 1.22040\dots .$$

APPLICATION 2. $\mathcal{V} = [1, t, t^2]_{[-1,1]}$ (Chalmers-Metcalf [1]).

In this case $n = 3$ and

$$\widehat{V}_1(x_1, x_2) = 2(x_1 - x_2 + 1), \quad \widehat{V}_2(x_1, x_2) = x_1^2 - x_2^2, \quad \widehat{V}_3(x_1, x_2) = \frac{2}{3}(1 + x_1^3 - x_2^3).$$

Using symmetry considerations, equations (5) become

$$\frac{2[x_1(t) - x_2(t) + 1]}{m_{11} + m_{13}t^2} = \frac{x_1^2(t) - x_2^2(t)}{t} = \frac{\frac{2}{3}[x_1^3(t) - x_2^3(t) + 1]}{m_{31} + m_{33}t^2}. \tag{2.5}$$

Letting

$$\gamma_1 = \frac{m_{11} + m_{13}t^2}{2t} \quad \text{and} \quad \gamma_3 = \frac{m_{31} + m_{33}t^2}{\frac{2}{3}t},$$

equations (2.5) may be rewritten

$$x_1 - x_2 + 1 = \gamma_1(x_1^2 - x_2^2), \qquad x_1^3 - x_2^3 + 1 = \gamma_3(x_1^2 - x_2^2). \tag{2.5$'$}$$

Introducing the variables $y_1 = x_1 - x_2$ and $y_2 = x_1 + x_2$ leads to

$$y_1 + 1 = \gamma_1 y_1 y_2 \qquad \text{and} \qquad y_1 \frac{3y_2^2 + y_1^2}{4} + 1 = \gamma_3 y_1 y_2,$$

which reduce to the single quartic equation for y_2:

$$(\gamma_1 y_2 - 1)^2[3y_2^2 + 4(\gamma_1 - \gamma_3)y_2 - 4] + 1 = 0.$$

This equation is then solved yielding admissible x_1 and x_2 $(-1 \le x_1(t) < x_2(t) \le 1)$. The function $\sigma(t)$ (in (7)) is -1 for $|t| < t_0$ and $+1$ for $t_0 < |t|$, where $\pm t_0$ are points where the admissible solutions of the quartic equation switch from one pair of roots to another. The values of λ, m_{11}, m_{13}, m_{31}, and m_{33} are determined from the five non-trivial orthonormality conditions

$$\begin{aligned} 1 &= \int_{-1}^{1} u_1(t)\,dt = \int_{-1}^{1} t^2 u_3(t)\,dt = \int_{-1}^{1} t u_2(t)\,dt \\ 0 &= \int_{-1}^{1} t^2 u_1(t)\,dt = \int_{-1}^{1} u_3(t)\,dt. \end{aligned} \tag{2.9}$$

The solution of these equations (for example, by the iteration method of §3 in [1]) yields $t_0 = 0.45710\ldots$ and the values of λ and M given below. Hence, the $x_i(t)$ are specified, and $P_{\min} = \vec{u} \odot \vec{v}$, where $\vec{u}(t)$ is given by

$$\begin{pmatrix} \hat{V}_1(\vec{x}(t)) & \hat{V}_2(\vec{x}(t)) & \hat{V}_3(\vec{x}(t)) \\ 1 & x_1(t) & x_1^2(t) \\ 1 & x_2(t) & x_2^2(t) \end{pmatrix} \begin{pmatrix} u_1(t) \\ u_2(t) \\ u_3(t) \end{pmatrix} = \begin{pmatrix} \lambda\sigma(t) \\ 0 \\ 0 \end{pmatrix},$$

or

$$\begin{pmatrix} u_1(t) \\ u_2(t) \\ u_3(t) \end{pmatrix} = \frac{3\lambda \operatorname{sgn}(ty_2)}{y_1^3 - 2 - 3(y_1^2 - y_2^2)/2} \begin{pmatrix} (y_2^2 - y_1^2)/4 \\ -y_2 \\ 1 \end{pmatrix},$$

and $\|P_{\min}\| = \lambda$.

$$M = \begin{pmatrix} 1 & 0 & -.6675 \\ 0 & .9336 & 0 \\ -.1552 & 0 & 1.2711 \end{pmatrix}$$

and

$$P_{\min} = \lambda = 1.35948\ldots .$$

APPLICATION 3. $\mathcal{V} = [1, t, t^2, t^3]_{[-1,1]}$.

$$M = \begin{pmatrix} 1 & 0 & -.4726 & 0 \\ 0 & 1.4760 & 0 & -1.1095 \\ -.0797 & 0 & 1.0520 & 0 \\ 0 & -.2648 & 0 & 1.3017 \end{pmatrix}$$

and

$$P_{\min} = \lambda = 1.46184\ldots .$$

APPLICATION 4. $\mathcal{V} = [1, t, t^2, t^3, t^4]_{[-1,1]}$.

$$M = \begin{pmatrix} 1 & 0 & .2767 & 0 & -.9250 \\ 0 & 1.2925 & 0 & -.7046 & 0 \\ -.1590 & 0 & 2.3605 & 0 & -1.8019 \\ 0 & -.1523 & 0 & 1.1257 & 0 \\ -0.0126 & 0 & -.1178 & 0 & 1.0642 \end{pmatrix}$$

and

$$P_{\min} = \lambda = 1.54874\ldots .$$

APPLICATION 5. $\mathcal{V} = [1, t, t^2, t^3, t^4, t^5]_{[-1,1]}$.

$$M = \begin{pmatrix} 1 & 0 & -.0466 & 0 & -.4955 & 0 \\ 0 & .8893 & 0 & .9039 & 0 & -1.4609 \\ -.0781 & 0 & 1.6151 & 0 & -.9900 & 0 \\ 0 & -.5593 & 0 & 3.5525 & 0 & -2.6180 \\ -.0743 & 0 & -.0675 & 0 & .9124 & 0 \\ 0 & -.1769 & 0 & .6439 & 0 & .2685 \end{pmatrix}$$

and

$$P_{\text{min}} = \lambda = 1.61031\ldots .$$

REFERENCES

[1] Chalmers, B. L. and F. T. Metcalf, *The determination of minimal projections and extensions in L^1*, Trans. Amer. Math. Soc. **329**(1992), 289-305.

[2] Cheney, E. W., *Applications of fixed-point theorems to approximation theory*, Theory of Approx. with Appl., Academic Press Inc., New York, (1976), 1-8.

[3] Franchetti, C. and E. W. Cheney, *Minimal projections in L^1-space*, Duke Math. J. **43**(1976), 501-510.

[4] Hobby, Charles R. and John R. Rice, *A moment problem in L^1 approximation*, Proc. Amer. Math. Soc. **16**(1965), 665-670.

[5] Morris, P. D. and Cheney, E. W., *On the existence and characterization of minimal projections*, J. Reine Angew. Math., **270**(1974), 61-76.

"Proofs" and Proofs of the Eckart–Young Theorem

JOHN S. CHIPMAN Department of Economics, University of Minnesota, Minneapolis, Minnesota 55455

INTRODUCTION

In 1936 Eckart and Young formulated the problem of approximating an $n \times k$ matrix X of rank k by an $n \times k$ matrix of rank $r < k$. This has come to be known as the Eckart–Young theorem. It has important applications to factor analysis in psychometrics (for which it was originally developed by Eckart and Young), to clustering and aggregation in econometrics (cf. Fisher, 1962, 1969), to quantum chemistry (cf. Goldstein and Levy, 1991; Aiken, Erdos and Goldstein, 1980), as well as to the theory of biased estimation (cf. Marquardt, 1970) in statistics. Marquardt showed that if in the regression model

$$y = X\beta + \varepsilon, \quad \mathcal{E}\varepsilon = 0, \quad \mathcal{E}\varepsilon\varepsilon' = \sigma^2 I,$$

Work supported by a Humboldt Research Award for Senior U. S. Scientists. I wish to thank John Eagon, Joel Roberts, and Paul Garrett of the University of Minnesota's School of Mathematics for their help. In particular, Lemma 2 and Theorem 1 were supplied by Roberts and the idea for Theorem 3 by Garrett, with whom I had many valuable discussions. Both of them declined coauthorship, but they deserve most of the credit for the results. Upon presentation of this paper at the Delhi Workshop on Generalized Inverses, 14 December 1992, George Styan drew my attention to an unpublished technical report by Rao and Styan (1976), some of the results of which were reported by Rao (1979, 1980); this raised some of the same issues as the present paper, and presented alternative proofs. I was also privileged to read some unpublished notes by Styan (1976). A relevant unpublished paper by Sondermann (1980) should also be mentioned. Finally, I wish to thank Renate Meyer for bringing by attention to the paper by Mirsky (1960)—see also Schmidt (1907), von Neumann (1937), Stewart & Sun (1990), and Meyer (1993, p. 67)—and to Jerome Goldstein for stimulating conversations.

where rank $X = k$, the square of the normalized length of β (i.e., $\beta'\beta/\sigma^2$) is less than the sum of the reciprocals of the $k - r$ smallest eigenvalues of $X'X$, then the estimator $\hat{\beta}_{(r)} = X^{\dagger}_{(r)}y$ —where $X_{(r)}$ is the best approximation of X by a matrix of rank $r < k$ and $X^{\dagger}_{(r)}$ is the Moore-Penrose generalized inverse of $X_{(r)}$—has lower mean-square error $\mathcal{E}(\hat{\beta}_{(r)} - \beta)'(\hat{\beta}_{(r)} - \beta)$ than the ordinary least-squares estimator $\hat{\beta} = X^{\dagger}y$ of β. This result has been extensively applied to aggregation problems in econometrics by the present writer (Chipman, 1978, 1983, 1985).

Eckart and Young (1936) stated their result without proof, although they presented a heuristic argument. A somewhat more elaborate argument appeared in Householder and Young (1938), but still with important gaps. Golub and Kahan (1965, p. 220) presented an alternative proof, but it contains a serious *lacuna* analyzed below. A detailed proof is presented in Stewart (1973, pp. 322–3) but here again some of the crucial steps are omitted. It was subsequently pointed out by Stewart and Sun (1990, pp. 208–10) that the result had been proved by Mirsky (1960) for complex matrices and arbitrary unitarily invariant norms, and earlier still by Schmidt (1907) for integral operators and the Hilbert–Schmidt norm. They therefore call the result the Schmidt–Mirsky theorem. Mirsky's proof is based on von Neumann's (1937) theory of symmetric gauge functions. In this paper some simple proofs will be provided for the Frobenius norm that do not require this apparatus. While they will be stated for real matrices X and orthogonal matrices P and Q, the proofs go through without change for complex matrices X and unitary matrices P and Q.

A basic problem is that the set of $n \times k$ matrices of rank $r < k$ is not closed. Such a set is defined by the conditions that all minors of order greater than r vanish, and at least one minor of order r be nonvanishing. One wishes to find in this set a matrix that is closest to X in the Frobenius norm; but since the set is obviously not closed, the existence of such a matrix is not at all obvious. Stewart (1973, pp. 322–3) recognizes that there is an existence problem, but limits himself to stating: "We prove the theorem under the assumption that the minimum ... actually exists (this assumption can easily be established by analytical considerations)." Implicitly what is done is to deal with the set of matrices of rank $\leqq r$; this is a closed set, and can be compactified, hence a matrix in this set exists that is closest to X; but on the face of it, it might have rank $< r$. It has to be shown that it has rank exactly r.

FORMULATION AND SOLUTION OF THE PROBLEM

We denote by $\mathcal{X}$ the set of all real $n \times k$ matrices X. For $X, Y \in \mathcal{X}$ with entries x_{ij}, y_{ij} we define the inner product

$$\langle X, Y\rangle = \operatorname{trace}(Y'X) = \sum_{i,j} x_{ij}y_{ij}.$$

Write $\|X\| = \langle X, X\rangle^{\frac{1}{2}}$; this is the Frobenius norm. The function $d(X, Y) = \|X - Y\|$ defines the Frobenius distance between X and Y, and is a metric on $\mathcal{X}$. We shall denote by $\mathcal{X}_r$ the subset of $n \times k$ matrices of rank $\leqq r$.

Lemma 1 to follow shows that the set of $n \times k$ matrices $\tilde{X}$ of rank $\leqq r$ is closed, and that a suitable subset of this set consisting of matrices close to the given X is compact. Hence X can be best approximated by a matrix of rank $\leqq r$. Lemma 2 is used to prove Theorem 1, which states that the matrix closest to X actually has rank r. As far as existence of a closest matrix of rank r is concerned, this is the end of the matter. However, the Eckart–Young theorem states that the best approximation of X by a matrix of rank r can be obtained by replacing all but the the r largest singular values of X by zeros. It must still be shown that this procedure provides the correct result, and that the resulting matrix actually has rank r. This is done in Theorem 2, which is based on Stewart (1973). Theorem 3 provides an alternative, streamlined, proof. Finally, an extremely simple proof furnished to the author by Heinz Neudecker is presented in the Appendix.

LEMMA 1 Let X be a given $n \times k$ matrix of rank $> r$, where $r < m = \min(n, k)$. Then within the class $\mathcal{X}_r$ of $n \times k$ matrices $\tilde{X}$ of rank $\leqq r$, there exists a matrix $\hat{X}$ closest to X in the Frobenius norm, i.e., such that

$$\|X - \hat{X}\| = \min_{\tilde{X} \in \mathcal{X}_r} \|X - \tilde{X}\|. \tag{1}$$

Proof: The set $\mathcal{X}_r$ of $n \times k$ matrices of rank $\leqq r$ is defined by the condition that all minors of order $r + 1$ of such matrices are equal to zero. Since these minors are polynomials in the elements of the matrices $\tilde{X}$, these equations define a closed set in the nk-dimensional space of matrices X. Let B be the ball of radius $\|X\|$ centered at X. Then $B \cap \mathcal{X}_r$ is compact, and is nonempty since it contains at least the zero matrix 0. Therefore the continuous function $f(\tilde{X}) = \|X - \tilde{X}\|$ has a minimum, $\hat{X}$, on $B \cap \mathcal{X}_r$, which is clearly the minimum on $\mathcal{X}_r$. ∎

LEMMA 2 Let E_{ij} be the $n \times k$ matrix with 1 in the i,jth position and 0s elsewhere, and let A be any $n \times k$ matrix. Then there exists a real number $\lambda \neq 0$ such that the nk matrices

$$\lambda E_{ij} - A \quad (i = 1, 2, \ldots, n;\ j = 1, 2, \ldots, k) \tag{2}$$

form a basis in nk-dimensional space.

Proof: If the set of matrices (2) is linearly dependent, then the $nk \times nk$ matrix whose columns are the column vectors of successive columns of (2) is a matrix of the form $\lambda I_{nk} - M$, where M has all its nk columns equal to to the column vector of columns of A, and λ is an eigenvalue of M. But the eigenvalues of M are 0 with a multiplicity of $nk - 1$, and just one other real number (namely, the sum of the elements of A). For any real number other than this one or 0, the set (2) is therefore linearly independent. ∎

THEOREM 1 Let X be an $n \times k$ matrix of rank $\geqq r$, and let $\hat{X}$ be a matrix of rank $\leqq r$ which is closest to X (in the Frobenius norm) among all matrices in the set $\mathcal{X}_r$ of $n \times k$ matrices of rank $\leqq r$. Then rank $\hat{X} = r$.

Proof: Suppose rank $\hat{X} < r$, and let λ be such as to satisfy Lemma 2. Since rank $\lambda E_{ij} = 1$ for $\lambda \neq 0$, we have, since multiplication of a matrix by a non-zero scalar does not affect its rank,

$$\operatorname{rank}[(1 - t)\hat{X} + t\lambda E_{ij}] \leqq \operatorname{rank} \hat{X} + \operatorname{rank} \lambda E_{ij} \leqq r \tag{3}$$

for any real number t, since the rank of the sum of two matrices is less than or equal to the sum of their ranks (because the column space of the sum of two matrices is contained in the sum of the column spaces of the two matrices, as is easily verified), and (3) holds for $t = 0$ and $t = 1$. Thus, of all points on the line $(1 - t)\,\hat{X} + t\lambda E_{ij}$, the closest to X will be $\hat{X}$, since $\hat{X}$ has been assumed to be a closest point to X of all $\tilde{X}$ with $\tilde{X} \in \mathcal{X}_r$. Thus, the matrix $X - \hat{X}$ is perpendicular to the matrix $\hat{X} - \lambda E_{ij}$, since the shortest distance from a point to a line is along the perpendicular. But then $X - \hat{X}$ is perpendicular to all of nk-dimensional space, since $\hat{X} - \lambda E_{ij}$ is a basis for this space. Thus, $X - \hat{X} = 0$, i.e., $X = \hat{X}$. ■

We now present our first proof of the Eckart–Young theorem. The proof presented here follows that of Stewart (1973, pp. 322–2), which is in turn based on that of Householder and Young (1938). In particular, the crucial Step 3a has been added, showing that the relevant matrix $\bar{D}_{11}$ has rank r. Steps 4 and 5 provide some additional needed details.

THEOREM 2 (Eckart and Young) Let X be a given $n \times k$ matrix of rank $p > r$, where $p \leqq m = \min(n, k)$. Then a matrix that minimizes $\|X - \tilde{X}\|$ over the set $\mathcal{X}_r = \{\tilde{X} : \text{rank } \tilde{X} \leqq r\}$ is given by

$$\hat{X} = QD_{(r)}P', \tag{4}$$

where

$$X = QDP' \tag{5}$$

is a singular-value decomposition of X, and the $n \times k$ matrix $D_{(r)}$ is obtained from D by replacing all but a set of its r largest diagonal elements by 0s. Further, $\hat{X}$ is a minimizer only if it is obtained in this way.

Proof: From Lemma 1, there exists a matrix $\hat{X}$ of rank $\leqq r$ closest to X in the Frobenius norm, i.e., such that (1) holds, and since rank $X > r$, this matrix $\hat{X}$ has rank r, by Theorem 1.

Let a singular-value decomposition of $\hat{X}$ be denoted

$$\hat{X} = \hat{Q}\hat{D}\hat{P}' \tag{6}$$

where $\hat{D}$ is an $n \times k$ diagonal matrix of the form

$$\hat{D} = \begin{bmatrix} \hat{S} & 0 \\ 0 & 0 \end{bmatrix}, \tag{7}$$

and $\hat{S}$ is an $r \times r$ diagonal matrix $\text{diag}\{\hat{s}_1, \hat{s}_2, \ldots, \hat{s}_r\}$ with $\hat{s}_1 \geqq \hat{s}_2 \geqq \ldots \geqq \hat{s}_r > 0$. The main task of the proof is to show that $\hat{D} = D_{(r)}$ where the latter is obtained from D in the manner described in the statement of the theorem.

Define

$$\bar{D} = \hat{Q}'X\hat{P}, \tag{8}$$

and denote

$$\bar{D} = \begin{bmatrix} \bar{D}_{11} & \bar{D}_{12} \\ \bar{D}_{21} & \bar{D}_{22} \end{bmatrix} \tag{9}$$

where $\bar{D}_{11}$ is of order $r \times r$. $\bar{D}$ has rank p. Owing to the orthogonal invariance of the Frobenius norm, it follows from (6) and (8) that (1) is equivalent to

$$\|\bar{D} - \hat{D}\| = \min_{\text{rank } \tilde{D} \leqq r} \|\bar{D} - \tilde{D}\|. \tag{10}$$

First we show that the matrix $\bar{D}$ of (8) and (9) must be of the form

$$\bar{D} = \begin{bmatrix} \hat{S} & 0 \\ 0 & \bar{D}_{22} \end{bmatrix}. \tag{11}$$

This will be done in a series of three steps.

Step 1. We show that $\bar{D}_{21} = 0$. Suppose not; then define

$$\tilde{D} = \begin{bmatrix} \hat{S} & 0 \\ \bar{D}_{21} & 0 \end{bmatrix}. \tag{12}$$

This has the same rank as $\hat{S}$, which is r; hence the matrix

$$\tilde{X} = \hat{Q}\tilde{D}\hat{P}'$$

has rank r, and by the orthogonal invariance of the Frobenius norm,

$$\|X - \tilde{X}\| = \|X - \hat{Q}\tilde{D}\hat{P}'\| = \|\hat{Q}'X\hat{P} - \tilde{D}\| = \|\bar{D} - \tilde{D}\|.$$

But from (9) and (12),

$$\|\bar{D} - \tilde{D}\| = \left\| \begin{bmatrix} \bar{D}_{11} - \hat{S} & \bar{D}_{12} \\ 0 & \bar{D}_{22} \end{bmatrix} \right\| < \left\| \begin{bmatrix} \bar{D}_{11} - \hat{S} & \bar{D}_{12} \\ \bar{D}_{21} & \bar{D}_{22} \end{bmatrix} \right\| = \|\bar{D} - \hat{D}\|,$$

and from the orthogonal invariance of the Frobenius norm, $\|\bar{D} - \hat{D}\| = \|X - \hat{X}\|$. Therefore, $\tilde{X}$, which has rank r, is closer to X than $\hat{X}$, in contradiction to (10). This contradiction proves that $\bar{D}_{21} = 0$.

Step 2. A completely similar proof shows that $\bar{D}_{12} = 0$.

Step 3a. We show that rank $\bar{D}_{11} = r$. Suppose not; then since $\bar{D}$ has rank $p > r$, and $\bar{D}_{21}$ and $\bar{D}_{12}$ have been shown to be zero, we can find a partition

$$\bar{D}_{22} = \begin{bmatrix} \bar{D}_{22,11} & \bar{D}_{22,12} \\ \bar{D}_{22,21} & \bar{D}_{22,22} \end{bmatrix}$$

of $\bar{D}_{22}$ such that $\bar{D}_{22,11} \neq 0$ and the $n \times k$ matrix

$$\tilde{D} = \begin{bmatrix} \bar{D}_{11} & 0 & 0 \\ 0 & \bar{D}_{22,11} & 0 \\ 0 & 0 & 0 \end{bmatrix}$$

has rank r. Accordingly,

$$\|\bar{D} - \tilde{D}\| = \left\| \begin{bmatrix} 0 & 0 & 0 \\ 0 & 0 & \bar{D}_{22,12} \\ 0 & \bar{D}_{22,21} & \bar{D}_{22,22} \end{bmatrix} \right\| < \left\| \begin{bmatrix} \bar{D}_{11} - \hat{S} & 0 & 0 \\ 0 & \bar{D}_{22,11} & \bar{D}_{22,12} \\ 0 & \bar{D}_{22,21} & \bar{D}_{22,22} \end{bmatrix} \right\| = \|\bar{D} - \hat{D}\|.$$

This contradicts (10) as before; hence rank $\bar{D}_{11} = r$.

Step 3b. We show finally that $\bar{D}_{11} = \hat{S}$. Suppose not; then, defining

$$\tilde{D} = \begin{bmatrix} \bar{D}_{11} & 0 \\ 0 & 0 \end{bmatrix},$$

this matrix has rank r, and

$$\|\bar{D} - \tilde{D}\| = \left\| \begin{bmatrix} 0 & 0 \\ 0 & \bar{D}_{22} \end{bmatrix} \right\| < \left\| \begin{bmatrix} \bar{D}_{11} - \hat{S} & 0 \\ 0 & \bar{D}_{22} \end{bmatrix} \right\| = \|\bar{D} - \hat{D}\|,$$

leading to a contradiction, as before. Therefore $\bar{D}$ must be of the form (11).

Step 4. Now let

$$\bar{D}_{22} = Q_{22} R P'_{22} \tag{13}$$

be a singular-value decomposition of the $(n-r) \times (k-r)$ matrix $\bar{D}_{22}$, where Q_{22} and P_{22} are, respectively, $(n-r) \times (n-r)$ and $(k-r) \times (k-r)$ orthogonal matrices, and R is an $(n-r) \times (k-r)$ diagonal matrix of singular values of $\bar{D}_{22}$. Define further the partitions $\hat{P} = [\hat{P}_1, \hat{P}_2]$ and $\hat{Q} = [\hat{Q}_1, \hat{Q}_2]$ of $\hat{P}$ and $\hat{Q}$ into their first r and last $k-r$ and $n-r$ columns, respectively. Finally, define the rectangular $n \times k$ diagonal matrix

$$D = \begin{bmatrix} \hat{S} & 0 \\ 0 & R \end{bmatrix} \tag{14}$$

and the $k \times k$ and $n \times n$ matrices

$$P = [\hat{P}_1, \hat{P}_2] \begin{bmatrix} I_r & 0 \\ 0 & P_{22} \end{bmatrix}, \quad Q = [\hat{Q}_1, \hat{Q}_2] \begin{bmatrix} I_r & 0 \\ 0 & Q_{22} \end{bmatrix}, \tag{15}$$

which are readily verified to be orthogonal. Then we verify from (15), (14), (13), (11), and (8) that

$$QDP' = \hat{Q}\bar{D}\hat{P}' = X; \tag{16}$$

thus, QDP' is a singular-value decomposition of X, in accordance with (5); and from the orthogonal invariance of the Frobenius norm,

$$\|D\| = \|\bar{D}\| = \|X\|.$$

On the other hand, it is clear from (15), (7), and (6) that

$$Q\hat{D}P' = \hat{Q}\hat{D}\hat{P}' = \hat{X}, \tag{17}$$

so that $Q\hat{D}P'$ is a singular-value decomposition of $\hat{X}$. It remains to show that $\hat{D} = D_{(r)}$, establishing (4).

From (16), (14), and (13) we have

$$\|X\|^2 = \|\hat{S}\|^2 + \|R\|^2 = \|\hat{S}\|^2 + \|\bar{D}_{22}\|^2, \tag{18}$$

so that X has the diagonal elements of $\hat{S}$ as r of its singular values, and the sum of squares of its remaining $m-r$ singular values is equal to $\|\bar{D}_{22}\|^2$. From (16), (17), (14), (7), and (13) we have

$$\|X - \hat{X}\| = \|D - \hat{D}\| = \|\bar{D}_{22}\| = \|R\|. \tag{19}$$

Since by hypothesis, (19) is a minimum (satisfying (1)), this can only be the case if, in (18), the diagonal elements of $\hat{S}$ are the r largest singular values of X, and those of R are the $m-r$ smallest (with possible ties). It follows that, if the singular values of X are ordered as $s_1 \geqq s_2 \geqq \ldots \geqq s_r \geqq s_{r+1} \geqq \ldots \geqq s_m$, $\hat{S}$ must contain $s_1, s_2, \ldots, s_r$, and R must contain $s_{r+1}, \ldots, s_m$. (If $s_r = s_{r+1}$, $\hat{X}$ is not unique.) Applying this requirement to (7) and (14) we have $\hat{D} = D_{(r)}$ and the main part of the theorem is proved.

Step 5. Finally, let $X = \check{Q}\check{D}\check{P}'$ be any other singular-value decomposition of X and let $\check{D}_{(r)}$ be obtained from $\check{D}$ by replacing all but a set of its r largest singular values by 0s. Define $\check{X} = \check{Q}\check{D}_{(r)}\check{P}'$. Then by the orthogonal invariance of the Frobenius norm we have

$$\|X - \check{X}\| = \|\check{D} - \check{D}_{(r)}\| = \|D - D_{(r)}\| = \|X - \hat{X}\|. \blacksquare$$

This may be compared with the theorem as presented by Golub and Kahan (1965, p. 220), who proceed as follows[1] (where I have substituted the notation of the present paper):

"THEOREM" Let X be an $n \times k$ matrix of rank $p \leqq k < n$ and let its singular-value decomposition be given by (5), where D is an $n \times k$ diagonal matrix of the form

$$D = \begin{bmatrix} S & 0 \\ 0 & 0 \end{bmatrix},$$

and S is a $p \times p$ diagonal matrix of singular values of X in descending order $s_1 \geqq s_2 \geqq \ldots \geqq s_p$. Let $\tilde{\mathcal{X}}_r$ be the set of all $n \times k$ matrices of rank $r < p$. Let $\hat{D}$ be the $n \times k$ diagonal matrix obtained from D by setting all but its r largest singular values equal to zero, and define $\hat{X} = Q\hat{D}P'$. Then

$$\|X - \hat{X}\| \leqq \|X - \tilde{X}\| \quad \text{for all } \tilde{X} \in \tilde{\mathcal{X}}_r.$$

"Proof": From the orthogonal invariance of the Frobenius norm,

$$\|X - \tilde{X}\| = \|D - Q'\tilde{X}P\|.$$

Denote $\tilde{D} = Q'\tilde{X}P$. Then

$$\|D - \tilde{D}\|^2 = \sum_{j=1}^{k}(s_j - \tilde{d}_{jj})^2 + \sum_{i \neq j} \tilde{d}_{ij}^2 \geqq \sum_{j=1}^{k}(s_j - \tilde{d}_{jj})^2.$$

Since $\|X - \hat{X}\|^2 = \|D - \hat{D}\|^2 = \sum_{j=r+1}^{k} s_j^2$, it follows that $\|X - \tilde{X}\|$ is minimized when $\tilde{d}_{jj} = s_j$ for $j = 1, 2, \ldots, r$ and $\tilde{d}_{ij} = 0$ otherwise. ■

As noted above, the set $\tilde{\mathcal{X}}_r$ is not closed; and no use in the proof is made of the hypothesis that $\tilde{X}$, hence $\tilde{D}$, has rank r. However, the problem is simpler: the last sentence asserts in effect that

$$\sum_{j-1}^{k}(s_j - \tilde{d}_{jj})^2 \geqq \sum_{j=r+1}^{k} s_j^2 \quad \text{for all } \tilde{D} = Q'\tilde{X}P \text{ such that } \tilde{X} \in \tilde{\mathcal{X}}_r.$$

Suppose $k = p = 3$ and $r = 2$, and let $s_1 = 3, s_2 = 2$, and $s_3 = 1$. Then the matrix

$$\tilde{D} = \begin{bmatrix} 3 & 3 & 3 \\ 2 & 2 & 2 \\ 1 & 1 & .5 \end{bmatrix}$$

provides a counterexample to the statement. (By setting all elements in the bottom row of $\tilde{D}$ equal to .5, one would obtain the same result but violate the rank condition.) Thus, nondiagonal $\tilde{D}$ would have to be disposed of by a separate argument (cf. Styan, 1976).

THEOREM 3 Let X be a fixed element of $\mathcal{X}$ and have rank $p \leqq m = \min(n, k)$, and for any $r < p$ let $\hat{X}$ be any matrix $\in \mathcal{X}_r$ that is closest to X. Then there exists a singular-value

[1]See also Problem 10 in section 1f.3 of Rao (1965, p. 56; 1973, p. 70), and section 21 of Chapter 6 of Ben-Israel & Greville (1974, pp. 246–9), as well as the revision in Ben-Israel & Greville (1980, pp. 246–9), where a few more references will also be found.

decomposition $X = QDP'$ of X, where

$$D = \begin{bmatrix} d_1 & & & & & \\ & \ddots & & & 0 & \\ & & d_p & & & \\ & & & 0 & & \\ & 0 & & & \ddots & \\ & & & & & 0 \end{bmatrix}$$

and $d_1 \geqq d_2 \geqq \ldots \geqq d_p > 0$, such that, defining a matrix $\hat{D}$ by

$$\hat{D} = \begin{bmatrix} d_1 & & & & & \\ & \ddots & & & 0 & \\ & & d_r & & & \\ & & & 0 & & \\ & 0 & & & \ddots & \\ & & & & & 0 \end{bmatrix},$$

where $d_r > 0$, we have $\hat{X} = Q\hat{D}P'$. Thus, $\hat{X}$ has rank r.

Proof: Let $\Delta \in \mathcal{X}$ be a matrix having one of the three patterns (block decompositions)

$$\begin{bmatrix} 0 & * \\ 0 & 0 \end{bmatrix}, \begin{bmatrix} 0 & 0 \\ * & 0 \end{bmatrix}, \begin{bmatrix} * & 0 \\ 0 & 0 \end{bmatrix},$$

where the northwest block is $r \times r$ and the southeast block is $(n-r) \times (k-r)$. Let $\hat{X} \in \mathcal{X}_r$ be a matrix closest to X (which exists by Lemma 1), and let $\hat{X} = \hat{Q}\hat{D}\hat{P}'$ be a singular-value decomposition of $\hat{X}$ such that

$$\hat{D} = \begin{bmatrix} D_1 & 0 \\ 0 & 0 \end{bmatrix}$$

where D_1 has diagonal entries $d_1 \geqq d_2 \geqq \ldots \geqq d_r \geqq 0$. Define, for real t, and for Δ of one of the three patterns above,

$$\hat{X}(t) = \hat{X} + t\hat{Q}\Delta\hat{P}'.$$

Then $\hat{X}(t)$ is certainly still in $\mathcal{X}_r$.

Since $\hat{X} \in \mathcal{X}_r$ minimizes $\|X - \hat{X}\|^2$, it follows that

$$\frac{\partial}{\partial t}\left(\|X - \hat{X}(t)\|^2\right)\Big|_{t=0} = 0$$

for Δ of all three patterns above. Multiplying this out and taking the derivative, we obtain

$$\langle X - \hat{X}, \hat{Q}\Delta\hat{P}'\rangle = 0.$$

Rearranging, this is

$$\langle \hat{Q}'X\hat{P} - \hat{D}, \Delta\rangle = 0.$$

Since this holds for Δ of all three patterns above, it must be that $\hat{Q}'X\hat{P} - \hat{D}$ is of the form

$$\hat{Q}'X\hat{P} - \hat{D} = \begin{bmatrix} 0 & 0 \\ 0 & Z_{22} \end{bmatrix}$$

where Z_{22} is some $(n-r) \times (k-r)$ matrix. It follows then that

$$\hat{Q}'X\hat{P} = \begin{bmatrix} D_1 & 0 \\ 0 & Z_{22} \end{bmatrix}.$$

Let A_{22} and B_{22} be orthogonal matrices of orders $n-r$ and $k-r$ respectively, such that Z_{22} has singular-value decomposition $Z_{22} = A_{22}D_2B_{22}'$, where

$$D_2 = \begin{bmatrix} d_{r+1} & & 0 \\ & \ddots & \\ 0 & & d_m \end{bmatrix},$$

with $d_{r+1} \geqq d_{r+2} \geqq \ldots \geqq d_m \geqq 0$. Define

$$A = \begin{bmatrix} I_r & 0 \\ 0 & A_{22} \end{bmatrix}, \; B = \begin{bmatrix} I_r & 0 \\ 0 & B_{22} \end{bmatrix}, \text{ and } D = \begin{bmatrix} D_1 & 0 \\ 0 & D_2 \end{bmatrix}.$$

Then

$$A'\hat{Q}'X\hat{P}B = D.$$

But $\hat{Q}A$ and $\hat{P}B$ are orthogonal matrices, so the singular values of X are the diagonal elements of D_1 and D_2. But the singular values of $\hat{X}$ are the diagonal elements of D_1 plus adjoining zeros. Thus, defining $P = \hat{P}B$ and $Q = \hat{Q}A$, we have

$$Q'(X - \hat{X})P = \begin{bmatrix} 0 & 0 \\ 0 & D_2 \end{bmatrix}.$$

Therefore, the distance between X and $\hat{X}$ is

$$\|X - \hat{X}\| = (d_{r+1}^2 + \ldots + d_m^2)^{\frac{1}{2}}.$$

The permutations of the singular values which minimize this distance are obviously those for which the $m-r$ singular values $d_{r+1}, \ldots, d_m$ are the $m-r$ smallest and the r singular values $d_1, \ldots, d_r$ the r largest. Since $\hat{X}$ has been defined as a matrix closest to X in $\mathcal{X}_r$, D_1 must contain the r largest singular values; and the singular-value decomposition $\hat{X} = \hat{Q}\hat{D}\hat{P}'$ was already chosen so that the singular values of $\hat{X}$, which by the above are the r largest singular values of X, are in descending order. Likewise the singular-value decomposition $Z_{22} = A_{22}D_2B_{22}'$ was chosen so that the singular values $d_{r+1}, \ldots, d_m$ are in descending order. Hence, QXP' arranges all the singular values of X in descending order. Since $r < p$ and $d_p > 0$, clearly $d_r > 0$ and $\hat{X}$ has rank exactly r. ■

We conclude with an extremely simple proof of necessity of the Eckart–Young condition furnished to me by Heinz Neudecker, which is contained in the Appendix following.[2]

[2]For the methodology followed see Magnus and Neudecker (1991), pp. 358ff.

APPENDIX: A Proof of the Eckart–Young Theorem

HEINZ NEUDECKER, University of Amsterdam, Amsterdam, The Netherlands

Let $\hat{X}$ be closest to the given $n \times k$ matrix X in the Frobenius norm. Its rows may be expressed without loss of generality as linear combinations of r $1 \times k$ orthonormal vectors, i.e.,

$$\hat{X} = AB', \quad B'B = I_r$$

where A is $n \times r$ and B is $k \times r$. We therefore wish to find A and B that solve the problem

$$\text{Minimize tr } (X - AB')'(X - AB') \quad \text{subject to} \quad B'B = I,$$

or equivalently,

$$\text{Maximize } \psi \equiv 2 \text{ tr } BA'X - \text{tr } A'A \quad \text{subject to} \quad B'B = I.$$

Setting up the Lagrangean expression

$$\varphi = 2 \text{ tr } BA'X - \text{tr } A'A - \text{tr } L(B'B - I),$$

we see without difficulty that since $B'B$ is symmetric, without loss of generality the Lagrangean multiplier matrix L may be taken to be symmetric. Using this symmetry we obtain for variations in A and B

$$d\varphi = 2 \text{ tr } (B'X' - A')dA + 2 \text{ tr } (A'X - LB')dB.$$

Setting $d\varphi = 0$ for arbitrary dA and dB yields, with the given constraint,

(i) $XB = A$

(ii) $A'X = LB'$

(iii) $B'B = I.$

From these three equations we obtain

$$A'A = A'XB = LB'B = L$$

whence L is also positive definite. From the first two equations and the symmetry of L we obtain

$$X'XB = X'A = BL' = BL.$$

From these it follows that

$$\psi = 2 \text{ tr } BA'X - \text{tr } A'A = 2 \text{ tr } BLB' - \text{tr } L = \text{tr } L,$$

which is to be a maximum.

Write

$$L = T\Lambda T'$$

where T is orthogonal and Λ is diagonal, and define

$$\tilde{A} = AT, \quad \tilde{B} = BT.$$

Then

$$\tilde{A}'\tilde{A} = T'A'AT = T'LT = \Lambda$$

and

$$\tilde{B}'\tilde{B} = T'B'BT = T'T = I.$$

Equations (i) to (iii) above then become

(i′) $X\tilde{B} = \tilde{A}$

(ii′) $\tilde{A}'X = T'LB' = T'LT\tilde{B}' = \Lambda\tilde{B}'$

(iii′) $\tilde{B}'\tilde{B} = I.$

From these equations it follows that

$$X'X\tilde{B} = \tilde{B}\Lambda \quad \text{and} \quad \tilde{B}'\tilde{B} = I.$$

Thus, Λ, whose trace is to be maximized (being equal to the trace of L), is a diagonal matrix of r eigenvalues of $X'X$, and $\tilde{B}$ is the matrix whose r columns constitute an associated orthonormal set of r eigenvectors of $X'X$. Λ is maximized when these r eigenvalues are a set of r largest eigenvalues of $X'X$. ■

REFERENCES

Aiken, John G., John A. Erdos and Jerome A. Goldstein (1980). Unitary Approximation of Positive Operators, *Illinois Journal of Mathematics, 24* (Spring): 61–72.

Ben-Israel, Adi, and Thomas N. E. Greville (1974). *Generalized Inverses: Theory and Applications.* New York: John Wiley & Sons. Reprint edition with corrections, Huntington, New York: Robert E. Krieger Publishing Company, 1980.

Chipman, John S. (1978). Towards the Construction of an Optimal Aggregative Model of International Trade: West Germany, 1963–1975, *Annals of Economic and Social Measurement, 6* (Winter-Spring): 535–554.

Chipman, John S. (1983). Dynamic Adjustment of Internal Prices to External Price Changes, Federal Republic of Germany, 1958–1979: An Application of Rank-Reduced Distributed-Lag Estimation by Spline Functions, *Quantitative Studies on Production and Prices* (Wolfgang Eichhorn, Rudolf Henn, Klaus Neumann, and Ronald W. Shephard, eds.), Würzburg: Physica-Verlag, Rudolf Liebing GmbH, pp. 195–230.

Chipman, John S. (1985). Testing for Reduction of Mean-Square Error by Aggregation in Dynamic Econometric Models, *Multivariate Analysis – VI. Proceedings of the Sixth International Symposium on Multivariate Analysis* (Paruchuri R. Krishnaiah, ed.), Amsterdam: North-Holland Publishing Company, pp. 97–119.

Eckart, Carl, and Gale Young (1936). The Approximation of One Matrix by Another of Lower Rank, *Psychometrika, 1* (September): 211–218.

Eckart, Carl, and Gale Young (1939). A Principal Axis Transformation for Non-Hermitian Matrices, *Bulletin of the American Mathematical Society, 45* (February): 118–121.

Fisher, Walter D. (1962). Optimal Aggregation in Multi-Equation Prediction Models, *Econometrica, 30* (October): 774–769.

Fisher, Walter D. (1969). *Clustering and Aggregation in Economics.* Baltimore: The Johns Hopkins Press.

Goldstein, Jerome A., and Mel Levy (1991). Linear Algebra and Quantum Chemistry, *American Mathematical Monthly, 98* (October): 710–718.

Golub, G., and W. Kahan (1965). Calculating the Singular Values and Pseudo-Inverse of a Matrix, *Journal of the Society for Industrial and Applied Mathematics*, Series B, *Numerical Analysis, 2* (No. 2): 205–224.

Householder, A. S., and Gale Young (1938). Matrix Approximations and Latent Roots, *American Mathematical Monthly, 45* (March): 165–171.

Magnus, Jan R., and Heinz Neudecker (1991). *Matrix Differential Calculus with Applications in Statistics and Econometrics.* Chichester and New York: John Wiley & Sons. Reprinted 1994.

Marquardt, Donald W. (1970). Generalized Inverses, Ridge Regression, Biased Linear Estimation, and Nonlinear Estimation, *Technometrics, 12* (August): 591–612.

Meyer, Renate (1993). *Matrix-Approximation in der multivariaten Statistik.* Aachen: Verlag der Augustinus Buchhandlung.

Mirsky, L. (1960). Symmetric Gauge Functions and Unitarily Invariant Norms, *Quarterly Journal of Mathematics*, Oxford Second Series, *11* (March): 50–59.

von Neumann, John (1937). Some Matrix-Inequalities and Metrization of Matric-Space, *Tomsk Univ. Rev., 1:* 286–300.

Rao, C. Radhakrishna (1965). *Linear Statistical Inference and Its Applications.* New York: John Wiley & Sons. 2nd edition, 1973.

Rao, C. Radhakrishna (1979). Separation Theorems for Singular Values of Matrices and Their Applications in Multivariate Analysis, *Journal of Multivariate Analysis, 9:* 362–377.

Rao, C. Radhakrishna (1980). Matrix Approximations and Reduction of Dimensionality in Multivariate Statistical Analysis, *Multivariate Analysis – V. Proceedings of the Fifth International Symposium on Multivariate Analysis* (Paruchuri R. Krishnaiah, ed.), Amsterdam: North-Holland Publishing Company, pp. 3–22.

Rao, C. Radhakrishna, and George P. H. Styan (1976). Notes on a Matrix Approximation Problem and Some Related Matrix Inequalities, Indian Statistical Institute, Delhi Campus, Discussion Paper No. 137, March.

Schmidt, Erhard (1907). Zur Theorie der linearen und nichtlinearen Integralgleichungen. I. Theil: Entwicklung willkürlicher Funktionen nach Systemen vorgeschriebener, *Mathematische Annalen, 63:* 433–476.

Sondermann, Dieter (1980). Best Approximate Solutions to Matrix Equations under Rank Restrictions. Report No. 23/80, Institute for Advanced Studies, The Hebrew University,

Mount Scopus, Jerusalem, Israel (August).

Stewart, G. W. (1973). *Introduction to Matrix Computations.* New York: Academic Press.

Stewart, G. W. and Ji-guang Sun (1990). *Matrix Perturbation Theory.* San Diego: Academic Press, Inc.

Styan, George P. H. (1976). "The Berlin Notes" (MS).

An Analytic Semigroup Associated to a Degenerate Evolution Equation

ANGELO FAVINI* Dipartimento di Matematica, Università di Bologna, Piazza di Porta S. Donato, 5, 40126 Bologna (Italy)

JEROME A. GOLDSTEIN** Department of Mathematical Sciences, University of Memphis, Memphis, TN 38152

SILVIA ROMANELLI * Dipartimento di Matematica, Universita' di Bari, via E. Orabona, 4, 70125 Bari (Italy)

1. Introduction It is well known that an important diffusion process is described with the help of the differential operator

$$Au(x) := x(1-x)u''(x) \qquad x \in (0,1)$$

whose domain $D(A)$ includes the so-called Wentzell boundary conditions, i.e.

$$\lim_{x \to 0^+, x \to 1^-} Au(x) = 0.$$

the corresponding semigroup has been studied by many authors since Feller's work [9]. It arises in many ways in the applications, for instance, in a diffusion approximation limit

*Supported by M.U.R.S.T. 60% and 40% and by G.N.A.F.A. of C.N.R.

**Partially supported by a USNSF grant.

for a sequence of Markov chains related to the Wright - Fisher model in genetics (see [8], Chapter 10). From the point of view of the generation problem, the results of Clément and Timmermans in [6], assure that A with domain

$$D(A) := \{u \in C[0,1] \cap C^2(0,1) | \lim_{x \to 0^+, x \to 1^-} Au(x) = 0\}$$

is the generator of a C_o-contraction semigroup on $C[0,1]$ equipped with the sup-norm $\|.\|_\infty$ and many interesting consequences are derived in approximation theory as shown in the monograph [1].

A subsequent direct approach to the study of existence and uniqueness results concerning Cauchy problems associated to the partial differential equation

$$\alpha(x)\frac{\partial^2 u}{\partial x^2}(x,t) - \frac{\partial u}{\partial t}(x,t) = 0, \qquad 0 < x < 1, \quad t > 0,$$

with boundary conditions

$$u(0,t) = u(1,t) = 0,$$

where $\alpha(x) := p(x)x(1-x)$ with $p \in C[0,1]$ and $p(x) > 0$ for all $x \in [0,1]$, was given in the space $H_o^1(0,1)$ by Fichera in [10], highlighting also other properties concerning, in particular, the eigenvalues of A. Hence, in a natural way the question arose if analyticity holds for the semigroup generated by $(A, D(A))$ in some of the above considered spaces.

Stimulated by these investigations, in [2] Attalienti and Romanelli examined the more general problem of analyticity for C_o-semigroups generated by differential operators of the type $A_\alpha u = \alpha u''$ on $C[0,1]$ with Wentzell boundary conditions, provided that $\alpha \in C[0,1]$ and $\alpha(x) > 0$, for $x \in (0,1)$ and $\alpha(0) = 0 = \alpha(1)$. Unfortunately, assumptions on α leading to analyticity of the semigroup gave rise to some restrictions of A_α included the condition that

$$\int_{(0,1)} \frac{1}{\sqrt{\alpha(x)}}\, dx = +\infty,$$

which obviously fails when $\alpha(x) := x(1-x)$.

Recently, analyticity of the semigroups generated by operators of the type A_α, with or without Wentzell boundary conditions, in L^p-weighted spaces $(1 < p < \infty)$, was studied in [3]. In particular, for $\alpha(x) := x(1-x)$, it follows that if $D(A_\alpha)$ is defined as the completion of $C_o^\infty(0,1)$ in the norm

$$\|u\|_{2,\alpha} := (\|u\|^2_{L^2_{\frac{1}{\alpha}}} + \|u'\|^2_{L^2} + \|x(1-x)u''\|^2_{L^2_{\frac{1}{\alpha}}})^{\frac{1}{2}},$$

then $(A_\alpha, D(A_\alpha))$ generates an analytic semigroup on

$$L^2_{\frac{1}{\alpha}}(0,1) := \{u \in L^2_{loc}(0,1) | \int_{(0,1)} \frac{|u(x)|^2}{\alpha(x)}\, dx < +\infty\}.$$

Our purpose, here, is to give an explicit description of the domain of A in $H_o^1(0,1)$, which allows us to obtain the analyticity of the associated semigroup. Interesting consequences are derived in connection with the adjoint problem (see [5]). It is also shown that the operator A (with suitable domain) generates a holomorphic semigroup on $W^{1,p}(0,1)$, for $1 < p < \infty$.

This work was completed during the visit of J.A. Goldstein at the Universities of Bari and Bologna, in May 1996. The authors are grateful to G.N.A.F.A. of C.N.R. for having supported this invitation and J.A. Goldstein is most grateful for the exceptional hospitality of his two coauthors and Enrico Obrecht during this visit and a previous visit to Bologna and Bari in 1994, when preliminary insight into this research was initiated.

2. Main results in Hilbert spaces Let us introduce the operator A on $H_o^1(0,1)$ given by

$$D(A) := \{u \in H_o^1(0,1) \,|\, u'' \text{ exists (in the sense of distributions) with } x(1-x)u'' \in H_o^1(0,1)\}$$

and

$$Au := x(1-x)u'', \quad \text{for } u \in D(A).$$

We have

Theorem 1. *$(A, D(A))$ generates a uniformly bounded semigroup analytic in the right half plane on $H_o^1(0,1)$.*

Proof. First of all, let us consider $H_o^1(0,1)$ endowed with the inner product

$$< u, v > := \int_{(0,1)} u'(x)\overline{v'(x)}\, dx \qquad u, v \in H_o^1(0,1),$$

that is equivalent to the usual inner product

$$\int_{(0,1)} u(x)\overline{v}(x)\, dx + \int_{(0,1)} u'(x)\overline{v'(x)}\, dx$$

in view of the Poincaré inequality. To motivate our choice of the space, we observe that if $Re\,\lambda > 0$ with

$$\lambda u - x(1-x)u'' = f \in H_o^1(0,1), \quad u \in D(A), \tag{1}$$

then

$$\lambda \frac{u}{x(1-x)} - u'' = \frac{f}{x(1-x)} \tag{1'}$$

implies necessarily

$$\int_{(0,1)} u''(x)\overline{u}(x)\, dx = -\int_{(0,1)} |u'(x)|^2\, dx, \tag{2}$$

that is, $[u'(x)\overline{u}(x)]_{x=0}^{x=1}$ vanishes.

Indeed,

$$\int_{(0,1)} u''(x)\overline{u}(x)\,dx = [u'(x)\overline{u}(x)]_{x=0}^{x=1} - \int_{(0,1)} |u'(x)|^2\,dx. \tag{3}$$

Moreover,

$$\lambda\int_{(0,1)} \frac{|u(x)|^2}{x(1-x)}\,dx - \int_{(0,1)} u''(x)\overline{u}(x)\,dx = \int_{(0,1)} \frac{f(x)\overline{u}(x)}{x(1-x)}\,dx.$$

Now $f \in H_o^1(0,1)$ gives

$$\begin{aligned}
\int_{(0,1)} \frac{|f(x)|^2}{x(1-x)}\,dx &= \int_{(0,\frac{1}{2})} \frac{1}{x(1-x)}|\int_0^x f'(t)\,dt|^2\,dx + \int_{(\frac{1}{2},1)} \frac{1}{x(1-x)}|\int_x^1 f'(t)\,dt|^2\,dx \\
&\le \int_{(0,\frac{1}{2})} \frac{1}{x(1-x)}(\int_0^x |f'(t)|\,dt)^2\,dx + \int_{(\frac{1}{2},1)} \frac{1}{x(1-x)}(\int_x^1 |f'(t)|\,dt)^2\,dx \\
&\le \int_{(0,\frac{1}{2})} \frac{1}{x(1-x)}(\int_0^x 1^2\,dt)(\int_0^x |f'(t)|^2\,dt)\,dx + \\
&+ \int_{(\frac{1}{2},1)} \frac{1}{x(1-x)}(\int_x^1 1^2\,dt)(\int_x^1 |f'(t)|^2\,dt)\,dx \\
&\le 2\|f\|_{H_o^1}, \qquad (4)
\end{aligned}$$

and therefore $\int_{(0,1)} \frac{f(x)\overline{u}(x)}{x(1-x)}\,dx$ converges since $u \in H_o^1(0,1)$. This implies that $u''\overline{u}$ is summable on $(0,1)$ and (3) is verified.

On the other hand, this also implies that both limits

$$\lim_{x\to 0+} u'(x)\overline{u}(x), \qquad \lim_{x\to 1^-} u'(x)\overline{u}(x)$$

exist and belong to $\mathbf{C}$.

In order to show that they vanish, we prove that, for all $u \in D(A)$, the limits

$$\lim_{x\to 0+} u'(x), \qquad \lim_{x\to 1^-} u'(x)$$

exist and are in $\mathbf{C}$.

To see this, observe that $x(1-x)u'' = g \in H_o^1(0,1)$, so that $g(x) = \int_0^x g'(t)\,dt$ yields

$$|g(x)| \le \sqrt{x}(\int_0^x |g'(t)|^2\,dt)^{\frac{1}{2}} \le C\sqrt{x}$$

for a suitable constant C depending on u.

Analogously,

$$g(x) = -\int_x^1 g'(t)\,dt, \qquad x \in (0,1),$$

gives

$$|g(x)| \leq C\sqrt{1-x}, \quad x \in (0,1).$$

Hence, $0 \leq y \leq x \leq \frac{1}{2}$ implies

$$\begin{aligned}
|u'(x) - u'(y)| &= |\int_y^x u''(t)\,dt| - |\int_y^x \frac{g(t)}{t(1-t)}\,dt| \\
&\leq \int_y^x \frac{|g(t)|}{t(1-t)}\,dt \\
&\leq 2C \int_y^x \frac{dt}{\sqrt{t}} \\
&= 4C(\sqrt{x} - \sqrt{y}) \to 0
\end{aligned}$$

as $x, y \to 0^+$. The same argument assures that any function u satisfying our equation admits $\lim_{x\to 1^-} u'(x) \in \mathbf{C}$. Then (2) is proved. Multiplying (1') by $\overline{u}(x)$, integrating on $(0,1)$ and taking real and imaginary parts, we deduce that

$$Re\,\lambda \int_{(0,1)} \frac{|u(x)|^2}{x(1-x)}\,dx + \|u'\|_{L^2}^2 = Re \int_{(0,1)} \frac{f(x)\overline{u}(x)}{x(1-x)}\,dx$$

$$|Im\,\lambda| \int_{(0,1)} \frac{|u(x)|^2}{x(1-x)}\,dx = |Im \int_{(0,1)} \frac{f(x)\overline{u}(x)}{x(1-x)}\,dx|.$$

This yields

$$(Re\,\lambda + |Im\,\lambda|) \int_{(0,1)} \frac{|u(x)|^2}{x(1-x)}\,dx + \|u'\|_{L^2}^2 \leq 2(\int_{(0,1)} \frac{|f(x)|^2}{x(1-x)}\,dx)^{\frac{1}{2}} (\int_{(0,1)} \frac{|u(x)|^2}{x(1-x)}\,dx)^{\frac{1}{2}}.$$

Consequently

$$|\lambda| (\int_{(0,1)} \frac{|u(x)|^2}{x(1-x)}\,dx)^{\frac{1}{2}} \leq 2(\int_{(0,1)} \frac{|f(x)|^2}{x(1-x)}\,dx)^{\frac{1}{2}} \leq 4\|f\|_{H_0^1(0,1)}.$$

On the other hand, multiplying (1) by $-\overline{u}''(x)$ and integrating on $(0,1)$ we have

$$-\lambda \int_{(0,1)} u(x)\overline{u}''(x)\,dx + \int_{(0,1)} x(1-x)|u''(x)|^2\,dx = -\int_{(0,1)} f(x)\overline{u}(x)\,dx;$$

in view of (3), this reads

$$\lambda \int_{(0,1)} |u'(x)|^2\,dx + \int_{(0,1)} x(1-x)|u''(x)|^2\,dx = -\int_{(0,1)} f(x)\overline{u}(x)\,dx. \tag{5}$$

Taking real and imaginary parts in (5), we easily obtain

$$\begin{aligned}(Re\,\lambda + |Im\,\lambda|)\|u'\|^2_{L^2} + \int_{(0,1)} x(1-x)|u''(x)|^2\,dx &\le \\ &\le 2|\int_{(0,1)} f(x)\overline{u}''(x)\,dx| \\ &= 2|\int_{(0,1)} \frac{f(x)}{\sqrt{x(1-x)}}\sqrt{x(1-x)}|u''(x)|^2\,dx| \\ &\le 4\|f\|_{H^1_o(0,1)}(\int_{(0,1)} x(1-x)|u''(x)|^2\,dx)^{\frac{1}{2}}.\end{aligned}$$

Notice that $x(1-x)u'' \in H^1_o(0,1)$ implies

$$\int_{(0,1)} \frac{1}{x(1-x)}|x(1-x)u''(x)|^2\,dx = \int_{(0,1)} x(1-x)|u''(x)|^2\,dx < +\infty,$$

by the above remark. Hence

$$(\int_{(0,1)} x(1-x)|u''(x)|^2\,dx)^{\frac{1}{2}} \le 4\|f\|_{H^1_o(0,1)}.$$

Moreover, since $\int_{(0,1)} f(x)\overline{u}''(x)\,dx$ is convergent by

$$\int_{(0,1)} f(x)\overline{u}''(x)\,dx = \int_{(0,1)} \frac{f(x)}{\sqrt{x(1-x)}}\sqrt{x(1-x)}\overline{u}''(x)\,dx,$$

and the Cauchy-Schwarz inequality, we deduce that

$$\int_{(0,1)} f(x)\overline{u}''(x)\,dx = [f(x)\overline{u}'(x)]_{x=0}^{x=1} - \int_{(0,1)} f'(x)\overline{u}'(x)\,dx$$

and then both limits

$$\lim_{x\to 0^+} f(x)\overline{u}'(x), \quad \lim_{x\to 1^-} f(x)\overline{u}'(x)$$

exist. Since $f \in H^1_o(0,1)$ and both limits $\lim_{x\to 0^+} u'(x)$ and $\lim_{x\to 1^-} u'(x)$ belong to $\mathbf{C}$, we conclude that for all u verifying (1) we have

$$\int_{(0,1)} f(x)\overline{u}''(x)\,dx = -\int_{(0,1)} f'(x)\overline{u}'(x)\,dx.$$

Therefore, from (4), rewritten as

$$\lambda\,\|u'\|^2_{L^2} + \int_{(0,1)} x(1-x)|u''(x)|^2\,dx = \int_{(0,1)} f'(x)\overline{u}'(x)\,dx,$$

we deduce the a priori bound

$$|\lambda| \|u\|_{H^1_o} \leq c\|f\|_{H^1_o} = c\|(\lambda - A)u\|_{H^1_o},$$

for some absolute constant c. The preceding arguments show that A is symmetric on $H^1_o(0,1)$ too, since

$$\begin{aligned}
< Au, v > &= \int_{(0,1)} (x(1-x)u'')'\overline{v}'(x)\,dx \\
&= -\int_{(0,1)} x(1-x)u''(x)\overline{v}''(x)\,dx \\
&= \int_{(0,1)} u'(x)(x(1-x)\overline{v}'')'(x)\,dx \\
&=< u, Av >
\end{aligned}$$

for all $u, v \in D(A)$. Moreover

$$< Au, u >= -\int_{(0,1)} x(1-x)|u''(x)|^2\,dx \leq 0,$$

so that A is nonpositive.

On the other hand, we observe that, for all $u, v \in D(A)$,

$$< (I - A)u, v >_{H^1_o} = \int_{(0,1)} u'(x)\overline{v}'(x)\,dx + \int_{(0,1)} x(1-x)u''(x)\overline{v}''(x)\,dx.$$

Let us introduce the Hilbert space V defined by

$$V := \{u \in H^1_o(0,1) | \int_{(0,1)} x(1-x)|u''(x)|^2\,dx < \infty\}.$$

It coincides with the completion of $C_c^\infty(0,1)$ with respect to the norm

$$\|u\|_V^2 := \int_{(0,1)} |u'(x)|^2\,dx + \int_{(0,1)} x(1-x)|u''(x)|^2\,dx.$$

To see this, note that for $u \in V$,

$$\begin{aligned}
\int_{(0,1)} |u'(x)|^2\,dx &= Re \int_{(0,1)} -\overline{u}(x)u''(x)\,dx \\
&= Re[\int_{(0,1)} \frac{-\overline{u}(x)}{\sqrt{x(1-x)}} u''(x)\sqrt{x(1-x)}\,dx] \\
&\leq [\int_{(0,1)} \frac{|u(x)|^2}{x(1-x)}\,dx. \int_{(0,1)} |u''(x)|^2 x(1-x)\,dx]^{\frac{1}{2}} < \infty,
\end{aligned}$$

by (4) and Cauchy - Schwarz inequality. The sesquilinear form

$$a(u,v) := \int_{(0,1)} u'(x)\overline{v}'(x)\,dx + \int_{(0,1)} x(1-x)u''(x)\overline{v}''(x)\,dx$$

is continuous on $V \times V$ and coercive, because $\|u\|_V^2 = a(u,u)$. It follows that the operator $\widetilde{B}$ associated to $a(.,.)$ (see [13], Theorems 2.22 and 2.23, pp.28-29) is an isomorphism from V to its dual V^* and the part B of $\widetilde{B}$ in $H_o^1(0,1)$ is positive definite and self-adjoint. Since

$$D(B) = \{u \in V |\, \widetilde{B}u \in H_o^1(0,1)\},$$

the operator B is precisely $I - A$, so that $B = I - A$ is onto $H_o^1(0,1)$.

□

Corollary 1. *The operator W defined by*

$$D(W) := \{u \in H^1(0,1) |\, u'' \textit{ exists in the sense of distributions and} x(1-x)u'' \in H_o^1(0,1)\},$$
$$Wu := x(1-x)u'', \; \textit{for}\, u \in D(W),$$

generates an analytic semigroup on $H^1(0,1)$.

Proof. In order to solve the equation

$$\lambda w - x(1-x)w = f \in H^1(0,1), \tag{6}$$

with $w \in D(W)$, we notice that $f \in C[0,1]$ and, hence, we can introduce

$$h(x) := f(x) - (1-x)f(0) - xf(1), \quad x \in [0,1]$$

which, obviously, belongs to $H_o^1(0,1)$. As a consequence of Theorem 1, we can affirm that, for all λ, with $Re\,\lambda + |Im\,\lambda| \geq \epsilon_o > 0$, there exists a unique $u \in D(A)$ such that

$$\lambda u(x) - x(1-x)u''(x) = h(x). \tag{7}$$

This means that $u \in H_o^1(0,1)$ and $x(1-x)u'' \in H_o^1(0,1)$. But (7) can be rewritten as

$$\lambda(u(x) + \frac{1-x}{\lambda}f(0) + \frac{x}{\lambda}f(1)) - x(1-x)(u + \frac{1-x}{\lambda}f(0) + \frac{x}{\lambda}f(1))''(x) = f(x)$$

and, thus, $w := u + \frac{1-x}{\lambda}f(0) + \frac{x}{\lambda}f(1) \in H^1(0,1)$ solves precisely (6) with Wentzell boundary conditions. Moreover

$$\|w\|_{H^1} \leq \|u\|_{H^1} + \|\frac{1-x}{\lambda}f(0) + \frac{x}{\lambda}f(1)\|_{H^1}$$

is estimated (via Sobolev imbedding) as follows

$$\|w\|_{H^1} \leq C|\lambda|^{-1}\|f\|_{H^1} + |\lambda|^{-1}\|f\|_{H^1} \leq C'|\lambda|^{-1}\|f\|_{H^1}.$$

Since uniqueness easily follows from Theorem 1 too, this concludes the proof.

□

Remark 1.

As a by- product, we derive the following regularity property for a related degenerate differential operator with Neumann boundary conditions. We could restrict ourselves here to $\alpha(x) := x(1-x)$, but all subsequent arguments work in the case that $\alpha \in C^1[0,1]$, $\alpha > 0$ on $(0,1)$ and $\alpha(0) = \alpha(1) = 0$; so we assume these conditions on α in what follows.

Actually in [5] a detailed study is given for more general operators. Here we merely give a simple, direct approach.

Let us introduce the operator $(B, D(B))$ on $L^2(0,1)$ given by

$$D(B) := \{u \in L^2(0,1) \,|\, u \,\text{is locally absolutely continuous in}\,(0,1),\ \alpha u' \in H_o^1(0,1)\}$$

$$Bu := \frac{d}{dx}(\alpha \frac{du}{dx}).$$

Then, B is a closed densely defined operator on $L^2(0,1)$. From Corollary 1 we deduce that for all $f \in H^1(0,1)$ there exists a unique $u \in H^1(0,1)$ such that $\alpha u'' \in H_o^1(0,1)$ and

$$\lambda u(x) - \alpha(x)u''(x) = f(x), \quad x \in (0,1);$$

it follows that

$$\lambda u' - (\alpha u'')' = f' \in L^2(0,1).$$

Then, $v(x) := u'(x) \in L^2(0,1)$ satisfies

$$\lambda v(x) - (\alpha v')'(x) = f'(x),$$

where $\alpha v' \in H_o^1(0,1)$ and is the unique solution to

$$\lambda v - Bv - f'.$$

Thus, let $g \in L^2(0,1)$, and observe that $f(x) := \int_0^x g(t)\,dt \in H^1(0,1)$ and all functions $h \in H^1(0,1)$ with $g = h'$ reduce to $h = f + c$, with an arbitrary constant c.

Hence, if

$$\lambda v - (\alpha v')' = g = (f+c)', \quad v \in D(B)$$

then, necessarily $v = (u + \frac{c}{\lambda})' = u'$, where $u + \frac{c}{\lambda}$ satisfies for all $c, Re\,\lambda > 0$,

$$\lambda(u + \frac{c}{\lambda}) - \alpha(u + \frac{c}{\lambda})'' = f + c,$$

$$\alpha(u(x) + \frac{c}{\lambda})'' \to 0.$$

for $x \to 0^+$, $x \to 1^-$.

Therefore, taking into account that

$$|f(x)| \leq \sqrt{x}\,\|g\|_{L^2}, \qquad x \in (0,1)$$

and, hence

$$\|f\|_{L^2}^2 \leq \frac{1}{2}\|g\|_{L^2}^2,$$

we conclude that

$$\|v\|_{L^2} \leq \frac{M'}{|\lambda|}\|g\|_{L^2}.$$

This proves the following

Corollary 2. *The operator $(B, D(B))$, defined as above, generates a uniformly bounded C_o-semigroup on $L^2(0,1)$, analytic in the right half plane.*

3. Analytic semigroups in $W^{1,p}(0,1)\,(1 < p < \infty)$ and differentiable semigroups in $C^1[0,1]$.

The techniques of passing from $Au = (\alpha u')'$ (where $\alpha \in C^1[0,1]$), with generalized Neumann boundary conditions of the form

$$\alpha(x)u'(x) \longrightarrow 0 \,(\text{as}\, x \to 0,1),$$

to $Bu = \alpha u''$ with Wentzell boundary conditions by replacing $u(x)$ by $v(x) := \int_0^x u(y)\,dy$ enable us to translate properties of $A(= A_p)$ on $L^p(0,1)$ investigated in [5] to corresponding properties of $B(= B_p)$ on $W^{1,p}(0,1)$.

The following result is a particular case of [5], Theorem 2.9, reestablished here in a direct way.

Proposition 1. *Let α be in $C^1[0,1]$ with $\alpha > 0$ on $(0,1)$ and $\alpha(0) = 0 = \alpha(1)$. If $(A_p, D(A_p))$ is defined by*

$$\begin{aligned} D(A_p) &:= \{u \in W^{1,p}_{loc} \cap L^p(0,1) |\, \alpha u' \in W^{1,p}_o(0,1)\} \\ A_p u &:= (\alpha u')' \end{aligned}$$

then $(A_p, D(A_p))$ generates a C_o - analytic semigroup on $L^p(0,1)$, for $1 < p < \infty$.

Proof. As a consequence of Corollary 2, the assertion is already true for $p = 2$. Now, let us examine the case $p > 2$. If $f \in L^p(0,1)$ and $\lambda \in \mathbf{C}$, $Re\,\lambda > 0$, there exists a unique $u \in D(A_p)$ such that

$$\lambda u - (\alpha u')' = f. \tag{8}$$

Let us multiply (8) by $\overline{u}|u|^{p-2}$ and integrate from 0 to 1. Thus we obtain

$$\lambda \|u\|_p^p - \int_{(0,1)} (\alpha u')'(x)|u|^{p-2}(x)\,dx = \int_{(0,1)} f(x)\overline{u}(x)|u|^{p-2}(x)\,dx. \tag{9}$$

Defining

$$\begin{aligned} \beta &:= \int_{(0,1)} f(x)\overline{u}(x)|u|^{p-2}(x)\,dx \\ \gamma &:= -\int_{(0,1)} (\alpha u')'(x)\overline{u}(x)|u|^{p-2}(x)\,dx, \end{aligned}$$

we can rewrite the equality (9) as follows

$$\lambda \|u\|_p^p + \gamma = \beta.$$

By Hölder's inequality,

$$|\beta| \le \|f\|_p \||u|^{p-1}\|_{p'} = \|f\|_p \|u\|_p^{p-1}$$

where $\frac{1}{p} + \frac{1}{p'} = 1$. Also, integration by parts yields

$$\begin{aligned}
\gamma &= \int_{(0,1)} (\alpha u')(x)(\overline{u}(u\overline{u})^{\frac{p-2}{2}})'(x)\,dx \\
&= \int_{(0,1)} \alpha(x)u'(x)\overline{u}'(x)|u|^{p-2}(x)\,dx + \\
&+ \frac{p-2}{2}\int_{(0,1)} \alpha(x)u'(x)\overline{u}(x)|u|^{p-4}(x)(u'(x)\overline{u}(x) + u(x)\overline{u}'(x))\,dx \\
&= \int_{(0,1)} \alpha(x)|u'(x)|^2|u|^{p-2}(x)\,dx + (p-2)\int_{(0,1)} \alpha(x)|u|^{p-4}(x)(u'\overline{u})(x)\,Re\,(u'\overline{u})(x)\,dx.
\end{aligned}$$

Taking real and imaginary parts in (9), we deduce that, respectively,

$$\begin{aligned}
(Re\,\lambda)\|u\|_p^p + \int_{(0,1)} \alpha(x)|u'|^2(x)|u|^{p-2}(x)\,dx + \int_{(0,1)} \alpha(x)|u|^{p-4}(x)(Re\,(u'\overline{u})(x))^2\,dx = \\
= \int_{(0,1)} Re\,(f\overline{u}|u|^{p-2})(x)\,dx \\
\le \|f\|_p\|u\|_p^{p-1},
\end{aligned}$$

$$\begin{aligned}
|Im\,\lambda|\|u\|_p^p + (p-2)\int_{(0,1)} \alpha(x)|u|^{p-4}(x)sign(Im\,\lambda)Im\,(u'\overline{u})(x)Re\,(u'\overline{u})(x)\,dx = \\
= sign(Im\,\lambda)\int_{(0,1)} Im\,(f\overline{u}|u|^{p-2})(x)\,dx \\
\le \|f\|_p\|u\|_p^{p-1}.
\end{aligned}$$

Thus, for $0 < c \le 1$ it follows that

$$\begin{aligned}
c|\lambda|\|u\|_p^p &\le (Re\,\lambda + c|Im\,\lambda|)\|u\|_p^p \\
&\le (-1 + c(p-2))\int_{(0,1)} \alpha(x)|u'|^2(x)|u|^{p-2}(x)\,dx + (1+c)\|f\|_p\|u\|_p^{p-1} \\
&\le 2\|f\|_p\|u\|_p^{p-1}
\end{aligned}$$

provided that $c \le \frac{1}{p-2}$. Since $p > 2$,

$$\|u\|_p \le \frac{c_p}{|\lambda|}\|f\|_p$$

where $c_p = 2(p-2)$ (hence $c_p \to \infty$ as $p \to \infty$) and this holds for all $f \in L^2(0,1) \cap L^\infty(0,1)$ and all λ with $Re\,\lambda > 0$.

Now, we assume that $1 < p < 2$ and observe that, by duality we have

$$< Au, v >=< u, Av >$$

where $Au = (\alpha u')'$ with boundary conditions

$$\alpha(x)u'(x) \to 0 \text{ as } x \to 0, 1$$

and u, v are in various spaces.

Thus, formally, $A_p^* = A_{p'}$, where A_p (resp. A_p^*) acts on $L^p(0,1)$ (resp. $L^{p'}(0,1)$) and

$$(p')^{-1} + p^{-1} = 1, \text{ for } 1 < p < \infty.$$

Since

$$\|(\lambda - A^*)^{-1}\| = \|(\overline{\lambda} - A)^{-1}\|,$$

this second case on p can be deduced from the first case and the estimates used in its proof. Moreover the fact that $R(\lambda I - A)$ (for $Re\,\lambda > 0$) is dense in $L^p(0,1)$ follows from the L^2 case.

□

Let $1 < p < \infty$ and define $B_p u := \alpha u''$ for $u \in D(B_p)$, where

$$D(B_p) := \{u \in W^{1,p}(0,1) \cap W^{2,p}_{loc}(0,1) |\, B_p u \in W^{1,p}(0,1) \text{ and } \lim_{x \to 0,1} \alpha(x)u''(x) = 0\}$$

i.e. B_p is equipped with the Wentzell boundary conditions.

Thus, we can prove the following

Theorem 2. *$(B_p, D(B_p))$ (for $1 < p < \infty$) generates a C_o analytic semigroup on $W^{1,p}(0,1)$.*

Proof. Let $1 < p < \infty$ and $F \in W^{1,p}(0,1)$ with $F(0) = F(1) = 0$ and take $f = F' \in L^p(0,1)$. From for every $\lambda \in \mathbf{C}, Re\,\lambda > 0$, there is $v \in D(B_p)$ such that

$$\lambda v - (\alpha v')' = f \tag{10}$$

with

$$\alpha(x)v'(x) \to 0, \text{ as } x \to 0, 1.$$

Let $u(x) := \int_0^x v(s)\,ds$. Then $u' = v$ and integrating (10) from 0 to x we deduce that

$$\lambda u - \alpha u'' = F$$

and $(\alpha v')(x) = (\alpha u'')(x) \to 0$ as $x \to 0, 1$.

Thus $u \in D(B_p)$ with $(\lambda - B_p)u = F$ and

$$\|u'\|_p = \|v\|_p \leq \frac{c_p}{|\lambda|}\|f\|_p = \frac{c_p}{|\lambda|}\|F'\|_p.$$

Since $\|u'\|_p$ is equivalent to the usual norm of u in $W_o^{1,p}(0,1)$, by the Poincare' inequality, it follows that B_p (suitably restricted) generates an analytic semigroup on $W_o^{1,p}(0,1)$. The extension of this result from $W_o^{1,p}(0,1)$ to $W^{1,p}(0,1)$ follows by the same argument used to extend the $p = 2$ case from $H_o^1(0,1)$ to $H^1(0,1)$, so we may safely omit the details (see the proof of Corollary 1).

□

Using the same type of approach, new results can be also obtained in the space $C^1[0,1]$, as the following Theorem shows.

Theorem 3. *Under the same assumptions as in Proposition 1, the operator* $(B, D(B))$ *given by*

$$D(B) := \{w \in C^1[0,1] \cap C^2(0,1) |\, \alpha w'' \in C^1[0,1]\}$$
$$Bu := \alpha u''$$

generates a C_o *differentiable semigroup on* $C^1[0,1]$.

Proof. Let us observe that, in view of [5] Theorem 3.3, the operator $(A, D_\infty(A))$, where

$$D(A_\infty) := \{u \in C[0,1] \cap C^1(0,1) |\, \alpha u' \in C^1[0,1]\}$$

and $Au := (\alpha u')'$, generates a C_o differentiable contraction semigroup on $C[0,1]$.
Now, let $F \in C^1[0,1]$, $\lambda > 0$ and consider the equation

$$\lambda u - (\alpha u')' = F'. \tag{11}$$

Since $F' \in C[0,1]$, it has a unique solution $u \in D(A_\infty)$, with

$$\|u\|_{C[0,1]} \leq \frac{1}{\lambda}\|F'\|_{C[0,1]}.$$

Hence, integrating (11) from 0 to x, we deduce that

$$\lambda \int_{(0,x)} u(y)\, dy - \alpha(x)u'(x) = F(x) - F(0),$$

namely

$$\lambda\, [\int_{(0,x)} u(y)\, dy + \frac{F(0)}{\lambda}] - \alpha(x)\frac{d^2}{dx^2}[\int_{(0,x)} u(y)\, dy + \frac{F(0)}{\lambda}] = F(x).$$

Let

$$w(x) := \int_{(0,x)} u(y)\, dy + \frac{F(0)}{\lambda}.$$

Then $w \in C^1[0,1]$. Moreover,

$$\alpha(x)w''(x) = \alpha(x)u'(x) \to 0 \quad \text{as } x \to 0^+, 1^-$$

and $\alpha w'' \in C^1[0,1]$, hence $w \in D(B)$. We also notice that, if

$$\lambda w - \alpha w'' = 0, \quad w \in D(B),$$

then

$$\lambda w' - \frac{d}{dx}\alpha(w')' = 0, \quad w' \in C[0,1]$$

and this implies that $w' = 0$, hence $w(x) = const.$ Consequently $w'' = 0$ and therefore $w = 0$. Thus the uniqueness holds. Now, we must estimate the norm of w. To this aim, we observe that the norm

$$\|w\|_1 := \max\{|w(0)|, \|w'\|_{C[0,1]}\}$$

is equivalent to the usual norm

$$\|w\|_{C^1} := \max\{\|w\|_{C[0,1]}, \|w'\|_{C[0,1]}\}$$

because obviously

$$\|w\|_1 \leq \|w\|_{C^1}$$

and, on the other hand,

$$w(x) = w(0) + \int_{(0,x)} w'(t)\,dt$$

implies that for every $x \in [0,1]$

$$\begin{aligned} |w(x)| &\leq |w(0)| + \int_{(0,x)} |w'(t)|\,dt \\ &\leq |w(0)| + \|w'\|_{C[0,1]} \\ &\leq 2\|w\|_1. \end{aligned}$$

Hence

$$\|w\|_{C^1} \leq \max\{2\|w\|_1, \|w'\|_{C[0,1]}\} \leq 2\|w\|_1.$$

Let us come back to our estimate

$$\begin{aligned} \|w\|_1 &= \max\{|w(0)|, \|w'\|_{C[0,1]}\} \\ &= \max\{|\frac{F(0)}{\lambda}|, \|u\|_{C[0,1]}\} \\ &\leq \max\{\frac{|F(0)|}{\lambda}, \frac{\|F'\|_{C[0,1]}}{\lambda}\} \\ &= \frac{1}{\lambda}\max\{|F(0)|, \|F'\|_{C[0,1]}\} \\ &= \frac{1}{\lambda}\|F\|_1. \end{aligned}$$

Then, $(B, D(B))$ generates a C_o-contraction semigroup on $C^1[0,1]$.

Moreover, for λ in a suitable region Σ as described in [12], Theorem 4.7 p.54, there exists $c > 0$ such that

$$\|u\|_{C[0,1]} \leq c(1 + |Im\,\lambda|)\|F'\|_{C[0,1]}.$$

On the other hand

$$|w(0)| = \frac{1}{|\lambda|}|F(0)| \le c(1 + |Im\,\lambda|)\|F\|_1.$$

Therefore

$$\max\{|w(0)|, \|w'\|_{C[0,1]}\} \le c(1 + |Im\,\lambda|)\|F\|_1.$$

and the semigroup generated by $(B, D(B))$ is differentiable on $C^1[0,1]$, according to the above mentioned result in [12].

□

Let us observe that, if $u \in D(A_\infty)$ (resp. $u \in D(B)$), then $\lim_{x\to 0^+, x\to 1^-} \alpha(x)u'(x) = 0$ (resp. $\lim_{x\to 0^+, x\to 1^-} \alpha(x)u''(x) = 0$).

In particular, all previous results hold for $\alpha(x) := x(1-x)m(x)$, where $x \in [0,1]$ and $m \in C^1[0,1]$, with $m(x) > 0$ in $[0,1]$.

Final remarks.

The long standing conjecture in this area concerns

$$Au := x(1-x)u''$$

with Wentzell boundary conditions.

By Clément and Timmermans [6], A generates a C_o - contraction semigroup on $C[0,1]$. Is this semigroup analytic ? After this work was done, but while the final revisions were being made, G. Metafune kindly provided us with a preprint [11], which states that on $C[0,1]$, A generates a semigroup analytic in the right half plane. Thus, despite the boundary degeneracy, the operator

$$u \to x(1-x)u''$$

with Wentzell boundary conditions generates an analytic semigroup on many spaces of interest.

References

1. F. Altomare - M. Campiti, *Korovkin-type Approximation Theory and its Applications*, de Gruyter Studies in Mathematics, 17 Walter de Gruyter Co., Berlin, New York, 1994.
2. A. Attalienti - S. Romanelli, *On some classes of analytic semigroups on $C([a,b])$ related to* **R** *or* Γ *- admissible mappings*, Evolution Equations, G. Ferreyra - G.R. Goldstein - F. Neubrander (eds) Lect. Notes in Pure and Applied Math. 168, M.Dekker, New York - Basel - Hong Kong, 1995, pp. 29-34.
3. V. Barbu - A. Favini - S. Romanelli, *Degenerate evolution equations and regularity of their associated semigroups*, Funkc. Eqvc. (to appear).
4. H. Brezis - W. Rosenkrantz - B. Singer, *On a degenerate elliptic-parabolic equation occurring in the theory of probability*, Comm. Pure Appl. Math. **24** (1971), 395 - 416.
5. M. Campiti - G. Metafune - D. Pallara, *Degenerate self - adjoint evolution equations on the unit interval*, Semigroup Forum (to appear).
6. Ph. Clément - C.A. Timmermans, *On C_o-semigroups generated by differential operators satisfying Ventcel's boundary conditions*, Indag. Math. **89** (1986), 379 -387.
7. R. F. Curtain - H. Zwart, *An Introduction to Infinite - dimensional Linear Systems Theory*, Springer, 1995.

8. S. N. Ethier - T. G. Kurtz, *Markov Processes, Characterization and Convergence*, Wiley Series in Probability and Mathematical Statistics, J.Wiley, 1986.
9. W. Feller, *The parabolic differential equations and the associated semi-groups of transformations*, Ann. of Math. **(2) 55** (1952), 468-519.
10. G.Fichera, *On a degenerate evolution problem*, Partial Differential Equations with Real Analysis, H. Begehr - A. Jeffrey (eds), Pitman Research Notes in Mathematics Series 263, Longman Scientific and Technical, 1992, pp. 15-42.
11. G. Metafune, *Analyticity for some degenerate evolution equations on the unit interval*, preprint (1996).
12. A. Pazy, *Semigroups of linear Operators and Applications to Partial Differential Equations*, Springer Verlag, Berlin - Heidelberg -Tokyo, 1986.
13. H. Tanabe, *Equations of Evolution*, Pitman Monographs and Studies in Math., London, San Francisco, Melbourne, 1979.

Degenerate Nonlinear Parabolic Problems: The Influence of Probability Theory

JEROME A. GOLDSTEIN* Department of Mathematics, Louisiana State University, Baton Rouge, LA 70803, USA

CHIN-YUAN LIN Department of Mathematics, University of South Carolina, Columbia, S.C. 29208 and Department of Mathematics, National Central University, Chang-Li 320, Republic of China

KUNYANG WANG Department of Mathematics, Louisiana State University, Baton Rouge, La. 70803, USA

1. INTRODUCTION

Of concern are mixed initial-boundary problems for the nonlinear equation

$$u_t = \varphi(x, u_x)u_{xx} + \psi(x, u, u_x) \tag{1}$$

for $x \in [0,1]$ and $t \geq 0$. Here φ is continuous and positive on $(0,1) \times \mathbb{R}$, but $\varphi(x,\xi)$ may approach 0 as x tends to either 0 or 1. Thus the diffusion coefficient may degenerate on the spatial boundary. Problems like this with nonlinear, degenerate diffusion coefficients arise in a variety of contexts in fluid dynamics and elsewhere. The particular example

$$\frac{\partial v}{\partial t} = \frac{1}{y^2}\frac{\partial}{\partial y}[y^4(u + \frac{\partial u}{\partial y} + u^2)] \tag{2}$$

* Partially supported by an NSF grant

Current affiliation: University of Memphis, Memphis, Tennessee (for $0 < y < \infty$ and $t \geq 0$) arises in physics and reduces to (1) when one sets $u(t,x) = v(t, \tan(\frac{\pi}{2}x))$. The theory of the ***Kompaneets equation*** (2) is far from complete; in particular, well-posedness for the Cauchy problem is not yet established. (But see Goldstein [11] for partial results and related references.)

A systematic study of (1) was begun by Goldstein and Lin in [12] in 1987 and continued in [13]-[15], [17], [18]. Among the related articles we cite the interesting work of Dorroh and G. R. Goldstein [7], [5], [6] who allow $\varphi = \varphi(x, u, u_x)$ to depend on u as well. But because this case does not admit a global quasi-dissipative estimate, only local existence is established in general; as global existence is our main concern, we restrict our attention to (1) here.

Suppose $\varphi(x,\xi) \geq \varphi_0(x)$ where $\varphi_0 \in C(0,1)$ and $1/\varphi_0 \in L^1(0,1)$. Then a variety of (linear and nonlinear) boundary conditions can be associated with (1), and the resulting problem is governed by a (nonlinear) contraction semigroup on $C[0,1]$. (See [12] and also [8] and the remarks in [9] for extensions.) But for a very degenerate case, such as the Kompaneets equation (or its integrated version), where $\varphi(x,\xi)$ behaves like x^2 near $x = 0$, the appropriate boundary condition is the ***Wentzell boundary condition***:

$$\varphi(x, u_x)u_{xx} + \psi(x, u, u_x) \to 0 \text{ as } x \to 0, 1$$

for each $t \geq 0$. (See [13].) In other words the Cauchy problem takes the form

$$du/dt = Au, \quad u(0) = u_0,$$

for $u : \mathbb{R}^+ = [0,\infty) \to X = C[0,1]$. Here $f \in \mathfrak{D}(A)$ iff $f \in X \cap C^2(0,1)$ and $Af \in C_0(0,1)$, that is, $Af \in X$ and $(Af)(x) \to 0$ as $x \to 0,1$ (where $Af = \varphi(x,f')f'' + \psi(x,f,f')$). This boundary condition is formally equivalent to $du/dt = 0$ on the spatial boundary $\{0,1\}$, whence the Wentzell boundary condition can be viewed as an inhomogeneous Dirichlet boundary condition where the boundary value (for all $t \geq 0$) is that of the initial function u_0.

In [13] it was shown that A determined by (1) (with the Wentzell boundary condition) is m-dissipative on $X = C[0,1]$, provided $\psi \equiv 0$ and $\varphi(x,\xi) \geq \varphi_0(x)$, $\varphi_0 \in C[0,1]$, and $\varphi_0 > 0$ on $(0,1)$. Thus $\varphi(x,\xi)$ may approach zero with arbitrary speed as $x \to 0$ or 1. In [15], an extension was made to a special class of nonzero ψ. The motivation for the hypothesis on ψ comes from a beautiful linear result of Clément and Timmermans [3], which can be viewed as the final sharp result in a theory begun by W. Feller in the 1950s. This result is as follows.

Let $\alpha, \beta \in C(0,1)$ with $\alpha > 0$ on $(0,1)$. On $X = C[0,1]$ define $Bu = \alpha u'' + \beta u'$, where $u \in \mathfrak{D}(B)$ if $u \in C^2(0,1) \cap X$ and $Bu \in C_0(0,1)$, so that B has the Wentzell boundary condition. Then B is densely defined and dissipative. The Clément-Timmermans' theorem states that B is m-dissipative (i.e. Ran $(I - B) = X$) iff both (A_0) and (A_1) hold: Let

$$W(x) = \exp\{-\int_{\frac{1}{2}}^{x} \frac{\beta(s)}{\alpha(s)} ds\}.$$

$$
\begin{array}{llll}
(A_0) & W \in L^1(0, \frac{1}{2}) & \text{or} & \int_0^{\frac{1}{2}} W(x) \int_0^x (\alpha(s)W(s))^{-1}\, dsdx = \infty, \\
(A_1) & W \in L^1(\frac{1}{2}, 1) & \text{or} & \int_{\frac{1}{2}}^1 W(x) \int_x^1 (\alpha(s)W(s))^{-1}\, dsdx = \infty.
\end{array}
$$

The idea is best explained in terms of the underlying Markov diffusion process. The drift coefficient β is competing with the diffusion coefficient α. The purpose of the boundary condition is to instruct the Markov particle how to proceed after it reaches a boundary point $j \in \{0, 1\}$. If the particle cannot reach j, then no boundary condition should be assigned at j, since doing so restricts the domain of B too much and prevents $I - B$ from being surjective. Condition (A_j) is what ensures that the Markov particles actually reach the boundary point j.

In the case of (1), φ (or perhaps φ_0) corresponds to α and ψ corresponds to β. The paper [13] dealt with $\psi \equiv 0$, and thus one thinks of the nonlinear diffusion as describing particles that diffuse without drift; they must eventually reach each boundary point and so a boundary condition is necessary at each end of $[0, 1]$. The corresponding nonlinear operator is thus expected to be m-dissipative, and in fact it is. In [15] we imposed sign conditions. Thus $\psi(x, \eta, \xi)$ is to have one sign near $x = 0$ (for all η, ξ) and the other sign for x near 1. The choice of sign forces the particle to drift toward the relevant boundary point. Thus the particles reach the two boundaries and the imposition of boundary conditions is necessary. Thus $\mathrm{Ran}(I - A) = X$ was proved in [15], and this is an intuitively expected result. But the conditions on ψ are very restrictive. One wishes to allow $\psi(x, \eta, \xi)$ to grow linearly with ξ (in analogy with the $\beta(x)\partial u/\partial x$ term in the linear problem) and to replace sign conditions (near each endpoint) by integrability conditions (corresponding to $W \in L^1(0, 1)$, which is the useful sufficient condition in the Clément-Timmermans theorem).

Here we want to discuss a new result of this type. Details will appear in [16], but it is still not strong enough to cover the Kompaneets equation. Thus, in his thesis [20], K. Wang has sought to approach the Kompaneets equation by its most general linear version. Below we indicate his extension of the Clément-Timmermans result to the context of a generalized linear version of equation (2), namely

$$
\frac{\partial v}{\partial t} = \frac{1}{\beta(y)} \frac{\partial}{\partial y} \left[\alpha(y) \frac{\partial u}{\partial y} + k(y)u \right]. \tag{3}
$$

The integrability conditions we impose give rather sharp results. They are sharp in the sense that when our nonlinear equation reduces to a linear one, the sufficient condition becomes necessary as well. This can be achieved by relaxing the hypotheses $\varphi \in C([0, 1] \times \mathbb{R})$, $\psi \in C([0, 1] \times \mathbb{R}^2)$ to allow $\varphi(x, \xi)$, $\psi(x, \eta, \xi)$ to be discontinuous at $x = 0, 1$. This extension is contained in Wang [20].

Now we want to point out the influence of probability theory on this work. Probability theory gives us the Wentzell boundary condition, which seems to be intimately tied to our problems in degenerate nonlinear diffusion. The interpretation of what happens to a Markov diffusing particle upon reaching the boundary leads to the integrability conditions of Feller, Clément and Timmermans. Thus probability theory provides us with the boundary conditions and hints on how to make hypotheses giving useful sufficient conditions in terms of integrability conditions. The final results make no mention of probability theory, either in the statements or the proofs. Still, this research could not have been done without the motivational influence of probability theory.

It is a pleasure to dedicate this paper to Professor M.M. Rao on his 65th birthday. M.M. is the mathematical father of the first named author and the mathematical grandfather of the other two authors.

2. DEGENERATE NONLINEAR DIFFUSION WITH DRIFT: INTEGRABILITY CONDITIONS

We want to present a precise statement of a new existence theorem for (1). Details will appear in [16]. Comparing (1) with the linear equation

$$u_t = \alpha(x)u_{xx} + \beta(x)u_x$$

treated by Clément and Timmermans [3] (and discussed in Section 1), we view φ as corresponding to α and ψ as corresponding to βu_x. Since $\varphi(x,\xi) \geq \varphi_0(x)$, it may seem more appropriate to view φ_0 as the analogue of α, but our result will emphasize φ itself. For simplicity of presentation, we considered the simplified Clément-Timmermans' criterion, namely that

$$x \to W(x) = \exp\{-\int_{1/2}^{x} \beta(s)\alpha(s)^{-1}ds\}$$

is in $L^1(0,1)$. (Cf. (A_0) (A_1) of Section 1.)

The operator A is defined on $X = C[0,1]$ by

$$(Au)(x) = \varphi(x, u'(x))u''(x)) + \psi(x, u(x), u'(x))$$

for $x \in [0,1]$ and $u \in \mathfrak{D}(A) = \{v \in C^2(0,1) \cap X : Av \in C_0(0,1)\}$, i.e. Av should be in X and should vanish at the endpoints $x = 0, 1$. Thus we view A as being equipped with the Wentzell boundary condition.

We now state two minimal sets of hypotheses on the coefficients φ and ψ.

(B1) $\varphi \in C([0,1] \times \mathbb{R})$; $\varphi(x,\xi) \geq \varphi_0(x)$ *and* $\varphi_0(x) > 0$ *for all* $(x,\xi) \in (0,1) \times \mathbb{R}$; *and* $\varphi_0 \in C[0,1]$.

(B2) $\psi \in C([0,1] \times \mathbb{R}^2)$; $\psi(x,\eta,\xi)$ *is non-increasing in* η *for each fixed* (x,ξ); $\psi(x,0,0) = 0$ *for all* $x \in [0,1]$;
and for all $r > 0$ *there is a constant* $K(r)$ *such that*

$$|\psi(x,\eta,\xi)| \leq K(r)(1 + |\xi|)$$

for all $(x,\eta,\xi) \in [0,1] \times [-r,r] \times \mathbb{R}$.

The (non-increasing) monotonicity of $\eta \to \psi(x, \eta, \xi)$ can be weakened to monotonicity of $\eta \to \psi(x, \eta, \xi) - w\eta$ for some real ω. The drift coefficient $\psi(x, \eta, \xi)$ can have arbitrary growth in η but ψ is restricted to have linear growth in ξ. But the arbitrary growth in η is illusory. By the maximum principle, solutions u should satisfy

$$\sup_{x,t} |u(x,t)| \leq \sup_{x} |u(x,0)|,$$

and so the only relevant values of η (for fixed $u(\cdot, 0)$) correspond to a bounded interval.

Let (B1) , (B2) hold. Suppose also that $\varphi \geq \varepsilon$ for some $\varepsilon > 0$, i.e., $\varphi_0(0) > 0$ and $\varphi_0(1) > 0$. Then A is densely defined and m-dissipative, and so A determines a semigroup $T = \{T(t) : t \geq 0\}$ by the Crandall-Liggett-Benilan theorem [4], [2], [1]. This implies that for all $f \in \overline{\mathfrak{D}(A)} = X$,

$$u(t) = T(t)f = \lim_{n \to \infty} (I - \frac{t}{n} A)^{-n} f$$

exists (for all $t \geq 0$) and defines the unique ***mild*** solution of

$$du(t)/dt = A(u(t)), \qquad u(0) = f.$$

This is the unique solution (in a suitable generalized sense) of

$$u_t = \varphi(x, u_x) u_{xx} + \psi(x, u, u_x), \tag{4}$$

$$u(x, 0) = f(x),$$

$$\varphi(x, u,) u_{xx} + \psi(x, u, u_x) \to 0 \text{ as } x \to 0, 1. \tag{5}$$

Moreover, $\|u(t) - v(t)\|_\infty$ is non-increasing in t for all solutions u, v of the above problem (4), (5) corresponding to initial conditions f, g respectively. The Wentzell boundary condition (5) means that $u_t = 0$ on the spatial boundary $\{0, 1\}$; hence $u(j, t) = f(j)$ for all $t \geq 0$ and $j = 0, 1$.

To verify the hypotheses of the Crandall-Liggett theorem, we must check three conditions.

(C1) *For some $\lambda > 0$ and all h some dense set in X, there is a u in $\mathfrak{D}(A)$ satisfying*

$$u - \lambda A u = h.$$

(C2) *If $u_i - \lambda A u_i$ holds for $i = 1, 2$ where $h_i \in X$ and $\lambda > 0$, then*

$$\|u_1 - u_2\|_\infty \leq \|h_1 - h_2\|_\infty.$$

(C3) *The graph of A is closed in $X \times X$, and $\mathfrak{D}(A)$ is dense in X.*

The last condition is easy and we will not discuss it further. (C2) is the *dissipativity* of A. (C1) is the *range condition.* (C1) is the hard part so we begin with (C2).

Let (B1), (B2) hold and let h_i, λ, u_i, be as in (C2). Let $u = u_1 - u_2$ (or $u_2 - u_1$ if necessary). Choose $x_0 \in [0,1]$ such that $u(x_0) = \|u\|_\infty$. If $0 < x_0 < 1$ then $u'(x_0) = 0, u''(x_0) \leq 0$, whence

$$\begin{aligned}\|u_1 - u_2\|_\infty = u(x_0) &\leq u(x_0) - \lambda\varphi(x_0, u'(x_0))u''(x_0)\\ &= (u_1 - \lambda Au_1)(x_0) - (u_2 - \lambda Au_2)(x_0)\\ &-\lambda\{\psi(x_0, u_1(x_0),\ u_1'(x_0)) - \psi(x_0, u_2(x_0), u_2'(x_0))\}\end{aligned}$$

since $u_1'(x_0) = u_2'(x_0),\ \lambda > 0,\ \varphi \geq 0,\ (u_1 - u_2)''(x_0) \leq 0,$

$$\leq (u_1 - \lambda Au_1)(x_0) - (u_2 - \lambda Au_2)(x_0) \tag{6}$$

since $\lambda\psi(x_0, \eta, u_1'(x_0))$ is nonincreasing in η

$$= (h_1 - h_2)(x_0) \leq \|h_1 - h_2\|_\infty.$$

If $x_0 \in \{0,1\}$, then equality holds in (6) since $Au_i(x_0) = 0$, and (C2) is verified in all cases.

For the range condition, let $\lambda > 0$ and $h \in X$. We want to solve

$$u - \lambda Au = h.$$

This is easy to do when $\varphi \geq \varepsilon > 0$ on $[0,1] \times \mathbb{R}$. The boundary condition ($Au(j) = 0$) implies $u(j) = h(j)$ for $j = 0, 1$. Let $\ell(x) = \alpha x + \beta$ be the linear function such that $k(x) = h(x) - \ell(x)$ vanishes at both endpoints $x = 0, 1$. Let $v(x) = u(x) - \ell(x)$. Then $u - \lambda Au = h$ is equivalent to

$$v - Bv = k,$$

where

$$(Bv)(x) = \lambda\{\varphi(x, v'(x) + \alpha)v''(x) + \psi(x, v(x) + \ell(x), v'(x) + \alpha)\}$$

for $v \in \mathfrak{D}(B) = \{w \in C^2(0,1) \cap X : Bw \in X,\ w(0) = w(1) = 0\}$. This can be rewritten as

$$-v'' = \frac{h - v - \lambda\psi(x, v + \ell, v' + \alpha)}{\lambda\varphi(x, v' + \alpha)}$$

with (homogeneous) Dirichlet boundary conditions. Using the Green's function for the Dirichlet Laplacian on $[0,1]$ this becomes the integral equation

$$v(x) = \int_0^1 G(x,y)\varphi(y, v'(y) + \alpha)^{-1}[\lambda^{-1}(h(y) - v(y)) + \psi(y, v(y) + \ell(y), v'(y) + \alpha)]dy$$

which can be solved by a fixed point argument (cf. [12], [15]). For φ not bounded away from zero, let $\varphi_n = \max\{\varphi, 1/n\}$ and let u_n be the unique solution of

$$u_n - \lambda Au_n = h.$$

By the above argument u_n exists. We want to show that as $n \to \infty$, u_n converges to the solution u of $u - \lambda Au = h$.

The proof uses a potential theory type argument (cf. [12], [15]). The following two hypotheses allow this to be done.

$$(B3) \qquad \psi(x,\eta,\xi) = M_0(x,\eta,\xi) + \xi M_1(x,\eta,\xi)$$

where $M_0, M_1 \in C^1([0,1] \times \mathbb{R}^2)$, *and for each* $R > 0$,

$$\sup\{|M_i(x,\eta,\xi)| : x \in [0,1],\ |\eta| \le R,\ |\xi| \le R,\ i = 0,1\} < \infty,$$
$$\sup\{|\frac{\partial M}{\partial x}(x,\eta,\xi)| :\ x \in [0,1],\ |\eta| \le R,\ |\xi| \le R,\ i = 0,1\} < \infty,$$

and $M_0(x,\eta,\xi) = 0$ *for all* $(\eta,\xi) \in \mathbb{R}^2$ *and* $x \in \{0,1\}$.

The integrability hypothesis is as follows.

(B4) *There exists a* $\mu \in L^1(0,1)$ *such that*

$$\sup_{v \in C^1[0,1]} \{\exp[-\int_{1/2}^{x} \varphi(s,v'(s))^{-1} M_1(s,v(s),v'(s))ds]\} \le \mu(x) \ a.e. on \ (0,1).$$

In the above integrability condition, φ corresponds to α and M_1 to β (cf. $(A_0),(A_1)$).

Theorem 1. (Goldstein-Lin [16]). *Let (B1)-(B4) hold. Then A is densely defined and m-dissipative on* $X = C[0,1]$. *In particular, (C1)-(C3) holds and A determines a strongly continuous contraction semigroup (given by the Crandall-Liggett exponential formula) which governs the well posed Cauchy problem* $u_t = Au$, $u(0) = f$ *for* $f \in X$.

For a specific example, let

$$\varphi(x,\xi) = \alpha(x)\varphi_1(x,\xi), \qquad M_0 \equiv 0,$$
$$M_1(x,\eta,\xi) = \beta(x)\psi_1(x,\eta,\xi)$$

where for some $\varepsilon_1 > 0$ and all (x,η,ζ),

$$\varepsilon_1 \le \varphi_1(x,\xi),\ \psi_1(x,\eta,\xi) \le 1/\varepsilon_1.$$

In this case, ignoring regularity questions, (B4) holds if and only if $W \in L^1(0,1)$ where W is defined using α, β in the usual way (cf. $(A_0),(A_1)$).

The hypotheses in Theorem 1, especially (B3), (B4), are much less restrictive than the (sign restriction) hypotheses of [15].

The Generalized Kompaneets Equation

Of concern is

$$u_t = \frac{1}{\beta}[\alpha(u_x + ku + F(u))]_x \tag{7}$$

for $0 < t,\ x < \infty$ with initial conditions $u(x, 0) = f(x)$ and boundary conditions

$$\alpha(u_x + ku + F(u)) \to 0$$

as $x \to 0, \infty$. Here $\beta, \alpha, \alpha', k, k' \in C(0, \infty)$ with $\alpha, \beta > 0$ on $(0, \infty)$; and $F \in C^1(\mathbb{R})$. Define the operator A by

$$Au = \beta^{-1}[\alpha(u' + ku + F(u))]',$$

with domain $\mathfrak{D}(A) = \{v \in X = L^1((0, \infty); \beta(x)dx) : Av \in X,\ \alpha(v' + kv + F(v)) \to 0$ as $x \to 0, \infty\}$. By standard nonlinear semigroup theory, it is not difficult to see that A is dissipative on X, (i.e., (C1) holds) if and only if for all $u_1, u_2 \in \mathfrak{D}(A)$,

$$\langle Au_1 - Au_2,\ \text{sign}_0(u_1 - u_2)\rangle \le 0$$

where $\text{sign}_0 r = r/|r|$ or 0, according as $r \neq 0$ or $r = 0$. Let $u = u_1 - u_2$ and write the open set $\{x \in \mathbb{R} : u(x) \neq 0\}$ as a union of open intervals

$$\bigcup_{n=1}^{\infty} (a_n,\ b_n).$$

Then

$$\langle Au_1 - Au_2,\ \text{sign}_0(u)\rangle = \sum_{n=1}^{\infty} \int_{a_n}^{b_n} (Au_1 - Au_2)\ \text{sign}_0(u)\beta dx,$$

which is non-positive, provided that each term is. Consider

$$J_n = \int_{a_n}^{b_n} (Au_1 - Au_2)\ \text{sign}_0(u)\beta dx,$$

where $0 < a_n < b_n < \infty$ and $u > 0$ on $(a_n b_n)$. Then $u(a_n) = u(b_n) = 0$ and $u'(a_n) \ge 0,\ u'(b_n) \le 0$. Consequently

$$\begin{aligned} J_n &= \int_{a_n}^{b_n} [\alpha(u' + ku + F(u_1) - F(u_2))]' dx \\ &= \alpha(u' + ku + F(u_1) - F(u_2))]_{a_n}^{b_n} \\ &= 0 + \alpha(b_n)u'(b_n) - \alpha(a_n)u'(a_n) \le 0 \end{aligned}$$

by the above and since $u_1 = u_2$ at $a_n,\ b_n$. The other possible cases are similar, except that the boundary conditions must be used if $a_n = 0$ or if $b_n = \infty$.

Condition (C3) can be dispensed with by replacing A by it closure. Condition (C1) is the stumbling block. We want to solve $u - \lambda Au = h$ (given $h \in X$ and $\lambda > 0$).

Let

$$v(x) = \int_0^x u(s)\beta(s)ds, \qquad x \in [0, \infty].$$

Then $v \in C[0, \infty],\ v(0) = 0,\ v(\infty) = \int_0^\infty u\beta ds \in \mathbb{R}$ (since $u \in X$). Now replace $x \in [0, \infty]$ by $y = \frac{2}{\pi}\tan^{-1}(x) \in [0, 1]$. Then A induces an operator B on $C[0, 1]$ of the form

$$Bu = \gamma(y)u + \psi(y, u, u')$$

with Wentzell boundary conditions. For the classical Kompaneets equation (2), $\gamma(y) \cong$ const. y^2 near $y = 0$ and $\gamma(y) \to \infty$ as $y \to 1$. In particular the previous work of Goldstein and Lin on (1) should be extended to cover the case when $\varphi(x,\xi)$, $\psi(x,\eta,\xi)$ are allowed to be discontinuous at $x = 0, 1$. This has been done by Wang [20]. But the resulting theory (and the extension of the theorem of Section 2 to this context) seems insufficient to produce a well-posedness result for (2).

Wang [19], [20] has extended the linear theory of (7) (cf. (4)) to the context of the Clément-Timmermans theorem. The nonlinear extensions are still under investigation, so we shall not report on them here.

We make two hypotheses.

(D1) *Let* $\alpha, \alpha', k, k' \in C(0,\infty)$, *with* $\alpha, \beta > 0$ *on* $(0,\infty)$.

Define

$$Au = \frac{1}{\beta}[\alpha(u' + ku)]'$$

for $u \in \mathfrak{D}(A) = \{v \in X = L^1((0,\infty); \beta(x)dx) : v \in C^2(0,\infty),\ Au \in X,\ \alpha(v' + ku) \to 0 \text{ as } x \to 0, \infty\}$.

(D2) $\int_0^\infty \beta(x) \exp\{-\int_1^x k(s)ds\}dx < \infty$.

Theorem 2. (Wang [19]) *Let (D1), (D2) hold. Then the closure of A is densely defined and m-dissipative (i.e. (C1) holds for A). Thus by the Hille-Yosida theorem (cf. [10]), A generates a strongly continuous contraction (linear) semigroup on X.*

Theorem 3. (Wang [19]) *Let (D1),(D2) hold. Let*

$$Y = L^2((0,\infty)); \beta(x) \exp\{\int_1^x k(s)ds\}dx).$$

Let A_2 be the operator A but with its domain modified in the obvious way so that it acts on Y rather than X. Then A_2 is a non-positive essentially selfadjoint operator.

Condition (D2) is the analogue of the Clément-Timmermans condition $W \in L^2(0,1)$ (cf. $(A_0), (A_1)$). For the semigroup $T = \{T(t) : t \geq 0\}$ generated by $\bar{A}$ (or $(\bar{A}_2)$ on both X and Y, T is a positive semigroup in the sense that if $f \geq 0$ (for f in X or Y), $T(t)f \geq 0$. Because of Theorem 3, T is positive in two senses, namely $T(t)$ is a positive operator (as above) in the lattice sense and $T(t)$ is also a positive selfadjoint operator.

Let $\alpha(x) \equiv x^4$, $\beta(x) \equiv x^2$, $k(x) \equiv k =$ constant . When $k > 0$, then (D2) holds and both Theorem 2 and 3 apply. When $k = 0$, $\bar{A}$ is m-dissipative and $\bar{A}_2$ is non-positive selfadjoint, but (D2) fails. (This corresponds to $(A_0), (A_1)$ holding but $W \notin L^1(0,1)$.) When $k < 0$, the closure of A is not m-dissipative; the conclusions of Theorems 2,3 fail in this case.

REFERENCES

1. V. Barbu, *Nonlinear Semigroups and Differential Equations in Banach Space,* Noordhoff, Leyden, 1976.

2. Ph. Bénilan, *Equations d'Evolution dans un Espace de Banach Quelconque et Applications,* Thesis, Université de Paris XI, Orsay, 1972.

3. Ph. Clément and C. A. Timmermans, On C_0-semigroups generated by differential operators satisfying Ventcel's boundary conditions, *Indag. Math.* 89 (1986), 379-386.

4. M.G. Crandall and T. M. Liggett, Generation of semigroups of nonlinear transformations on General Banach spaces, *Amer. J. Math.* 93 (1971), 265-298.

5. J. R. Dorroh and G. R. Goldstein, Quasilinear diffusions, in *Evolution Equations, Control Theory and Biomathematics* (ed. by Ph. Clément and G. Lumer), Dekker, New York (1994), 155-166.

6. J. R. Dorroh and G. R. Goldstein, A singular quasilinear parabolic problem in n dimensions, in preparation.

7. J. R. Dorroh and G. R. Rieder, A singular quasilinear parabolic problem in one space dimension, *J. Diff. Equations* 91 (1991), 1-23.

8. G. R. Goldstein, Nonlinear singular diffusion with nonlinear, boundary conditions, *Math Meth. Appl. Sci.* 16 (1993), 279-298.

9. G. R. Goldstein, J. A. Goldstein, and S. Oharu, The Favard class for a nonlinear parabolic problem, in *Recent Development in Evolution Equations* (ed. by A. C. McBride and G. F. Roach), Longman, Pitman Notes, Harlow (1995), 134-147.

10. J. A. Goldstein, *Semigroups of Linear Operators and Applications,* Oxford University Press, New York and Oxford, 1985.

11. J. A. Goldstein, The Kompaneets equation, in *Differential Equations in Abstract Spaces* (ed. by G. Dore, A. Favini, E. Obrecht, and A. Venni), Dekker, New York (1993), 115-123.

12. J. A. Goldstein and C.-Y. Lin, Singular nonlinear parabolic boundary value problems in one space dimension, *J. Diff. Equations* 68 (1987), 429-443.

13. J. A. Goldstein and C.-Y. Lin, Highly degenerate parabolic boundary value problems, *Diff. Int. Eqns.* 2 (1989), 215-227.

14. J. A. Goldstein and C.-Y.Lin, An L^p-semigroup approach to degenerate parabolic boundary value problems, *Ann. Mat. Pura. Appl.* 159 (1991), 211-227.

15. J. A. Goldstein and C.-Y. Lin, Parabolic problems with strong degeneracy at the spatial boundary, in *Semigroup Theory and Evolution Equations* (ed. by Ph. Clément, E. Mitidieri, and B. de Pagter), Dekker (1991), 181-191.

16. J.A. Goldstein and C.-Y. Lin, in preparation.

17. C.-Y. Lin, Degenerate nonlinear parabolic boundary value problems, *Nonlinear Anal. TMA* 13 (1989), 1303-1315.

18. G. Lumer, R. Redheffer, and W. Walter, Estimates for solutions of degenerate second order differential equations and inequalities with application to diffusion, *Nonlinear Anal. TMA* 12 (1988), 1105-1121.

19. K. Wang, The linear Kompaneets equation, *J. Math. Anal. Appl.*, to appear.

20. K. Wang, *The Generalized Kompaneets Equation*, Ph.D. Thesis, Louisiana State University, 1995.

An Application of Measure Theory to Perfect Competition

NEIL E. GRETSKY Department of Mathematics, University of California, Riverside, CA 92521 (email: neg@math.ucr.edu)

JOSEPH M. OSTROY Department of Economics, University of California, Los Angeles, CA 90024 (email: ostroy@econ.sscnet.ucla.edu)

WILLIAM R. ZAME Department of Economics, University of California, Los Angeles, CA 90024 (email: zame@econ.sscnet.ucla.edu)

Dedicated to Professor M.M. Rao on the occasion of his 65th birthday.

1 Introduction

This talk was given by the first author as part of the Festschrift held in November, 1994 at the University of California, Riverside in honor of M.M. Rao's 65th birthday. The work presented here is a complement to a much larger project; this paper constitutes an alternative approach to a problem discussed in (Gretsky, Ostroy, and Zame a). Other related material and background can be found in (Gretsky, Ostroy, and Zame 1992; Gretsky, Ostroy, and Zame b). Since one of Professor Rao's influences was to instill in his students a love of vector measures, the purpose of this talk is to show how a measure theoretic approach allows an appropriate description of non-manipulation in very large economies and to give a sample result.

2 The Model

We start by describing a very large[1] assignment economy. There are many "stories" for which the assignment economy is an appropriate model; we choose one in which the commodities being traded are houses.

As is the usual practice, an economy will be specified by listing the endowments and preferences of the agents who participate in the economy. Our story may be summarized as follows. There are two kinds of agents: buyers and sellers. Each buyer wishes to buy one house and is assumed to be initially endowed with enough money to buy any house; his preferences consist of a schedule describing what he is willing to pay for each house. Each seller is initially endowed with one house which she will be willing to sell if the price were right. The equilibria of the resulting exchange economy are the subject of study in the present paper[2].

We give a mathematical model for this story. Let the set of *houses* be a compact metric space H. For convenience, we introduce a fictional house $\bar{0}$ to H to indicate no trade. Denote $H \cup \bar{0}$ as H_0. The set of *buyers* is given by $B = \{b : H \to [0,1] |\ b \text{ is continuous}\}$ and the set of *sellers* is given by $S = H \times [0,1]$. Denote the set of agents as $I = B \cup S$. We have assumed continuity for buyers partially for simplicity; a less restrictive condition on buyers leads to a much more complicated set of results as discussed in (Gretsky, Ostroy, and Zame b).

We interpret the information for buyers as meaning that $b(h)$ is the reservation value that a buyer labeled as being of type b places on house h, i.e. $b(h)$ is the maximum amount of money that a buyer of type b is willing to pay for house h. For sellers the interpretation is that a seller labeled as being of type $s = (h, r)$ owns house h and places a reservation value r on this house, i.e. a seller of type (h, r) will not sell her house for an amount of money less than r. The reservation values are normalized to lie in the interval $[0, 1]$.

We can describe the preferences of an agent of type i by a *utility function* defined on houses and money as:

$$\begin{aligned} u_i(h, m) &= v_i(h) + m \\ v_i(h) &= \begin{cases} b(h) & \text{if } i = b, h \in H \\ -r & \text{if } i = s = (h,r), h \in H \\ 0 & \text{if } h = \bar{0} \end{cases} \end{aligned}$$

Finally, the initial data of the economy consists of a *population measure* $\mu \in M_+(I)$ which is compactly supported. Such a measure specifies what types of, and in what relative quantities, agents are present. The condition of compact support is no restriction on the collection of sellers since the space S is *a priori* compact. However, this condition ensures that the collection of continuous functions for the buyers who are actually present in the economy form an equicontinuous family.

The appropriate solution concept for such an exchange economy is Walrasian equilibrium, i.e. a price system on houses and a collection of trades among individuals such that individuals maximize utility, taking the price system as given, and such that markets clear.

[1]Such economies are called *non-atomic* assignment economies since, in the attempt to describe large economies, it is common to assume the measures involved are non-atomic. The results given here do not actually depend on such an assumption.

[2]The goal of the assignment problem is to pair up buyers and sellers so as to maximize the total profit. The equilibria of the exchange economy provide an alternative, but equivalent, formulation to the optimization problem. See (Gretsky, Ostroy, and Zame 1992) for this and other equivalences.

A price system in our context should be a bounded Borel measurable function on the space of houses. For technical reasons[3] we agree to identify price systems as being identical if they agree almost everywhere in the natural measure μ_H induced on houses by the given population measure μ; viz. the measure μ_H is the first marginal of the population measure μ_S defined on $S = H \times [0,1]$. Thus, a price system is a member of $L^\infty(\mu_H)$. The trading allocation of houses to buyers and sellers is described by a *housing distribution* measure $y \subset M_+(I \times H_0)$.

If the support of the housing measure μ_H is a proper subset of H, there is an ambiguity about prices for houses in H which are not in the support of μ_H. It is convenient to insist that prices for such houses be identically 1 so that no buyer ever strictly prefers to buy such a house. Moreover, for the fictional house $\bar{0}$ we extend p by setting $p(\bar{0}) = 0$. It was shown in (Gretsky, Ostroy, and Zame 1992) in the present case of an equicontinuous family of buyer functions that Walrasian price systems can be chosen to be continuous on the support of μ_H. Given any Walrasian price system p, we will without further mention always reduce to the unique price system p' which agrees with p almost everywhere with respect to μ_H, is continuous on the support of μ_H, and is identically 1 on the complement of the support of μ_H.

In order to define Walrasian equilibrium, we introduce an auxiliary concept. Given a price system p for houses, the *indirect utility function* at these prices is defined to be

$$v_i^*(p) = \begin{cases} \max\{\sup_{h\in H}\{b(h) - p(h)\}, 0\} & \text{if } i = b \\ \max\{p(h) - r, 0\} & \text{if } i = s = (h, r) \end{cases}$$

where the supremum in the first expression is the μ_H-essential supremum performed in the space $L^\infty(\mu_B)$ and the second expression is in $L^\infty(\mu_S)$. This gives the maximum utility attainable by an agent of type i subject to the budget constraint

$$m = \begin{cases} -p(h) & \text{if } i = b \\ p(h) & \text{if } i = s \end{cases}$$

DEFINITION: A *Walrasian equilibrium* for the assignment economy μ is a pair (p, y) where $p \in L^\infty(\mu_H)$ is a price system and $y \in M_+(I \times H_0)$ is a housing distribution such that

(i) $y_1 = \mu$,

(ii) $y(B \times G) = y(S \times G)$ for every Borel set $G \subseteq H$, and

(iii)
$$\begin{aligned} y(I \times H_0) &= y\{(b,h) \in B \times H : v_b(h) - p(h) = v_b^*(p)\} \\ &\quad + y\{(s,h) \in S \times H : v_s(h) + p(h) = v_s^*(p)\} \\ &\quad + y\{(i,\bar{0}) : v_i(\bar{0}) = v_i^*(p)\} \end{aligned}$$

These conditions state, respectively, that y is population consistent with μ; that the houses received by the buyers equal those supplied by the sellers; and that, except possibly for a μ-null set, all buyers and sellers are maximizing utility subject to the budget constraint defined by prices p.

[3]If individual point functions are used for price systems then there are several constructions the results of which cannot be guaranteed to be measurable. See (Gretsky, Ostroy, and Zame a) for more detail.

3 Misrepresentation

A natural question in the study of perfect competition is whether any agent or coalition of agents can favorably manipulate the Walrasian equilibria by misrepresenting their type. We consider a *subpopulation* of a given population to be a measure $\nu \in M_+(I)$ such that $0 \leq \nu \leq \mu$. Note that, in an assignment economy, trading really occurs only between pairs of matched individuals – one buyer and one seller. Consequently, without loss of generality, we may restrict attention to subpopulations consisting solely of buyers or sellers. We first consider subpopulations of buyers only. Thus, we start with a population $\mu = (\mu_B, \mu_S)$ and a subpopulation ν of μ_B for which we will measure the results of misrepresentation. Denote by Δ the diagonal set of $B \times B$.

DEFINITION: An *announcement* of misrepresentations for the subpopulation ν of buyers is a measure α on $B \times B$ such that

(i) $\alpha_1 = \mu_B$,

(ii) $\|\alpha_2\| = \|\mu_B\|$, and

(iii) $(\alpha|_{\Delta^c})_1 \leq \nu$.

The interpretation is that α is a distribution on $B \times B$ which describes announcements of types; here, the pair (b, b') means that an agent of type b announces himself to be of type b'. (Notice that an agent may misrepresent his preferences but not his endowment.) The three conditions are, respectively, that the population implicit in the announcement is consistent with the given population, that population mass is conserved, and that the actual misrepresenters are contained in the subpopulation ν. The misrepresentation leads to the "new" economy given by the population (α_2, μ_S) and to a resulting Walrasian equilibrium (y^α, p^α).

DEFINITION: Given a population measure μ, a subpopulation of buyers ν, an announcement α of misrepresentations for ν, and a choice (y^α, p^α) of a Walrasian equilibrium for the economy given by the population measure (α_2, μ_S), an *outcome* of the announcement is a measure $\eta^\alpha \in M_+(B \times B \times H_0)$ such that

(i) $\eta^\alpha_{1,2} = \alpha$, and

(ii) $\eta^\alpha_{2,3} = y^\alpha$.

The measure η^α is a distributional description of house assignments with respect to agents' true and announced types. The total utility of this outcome to all buyers is the aggregate of the outcome that each agent receives in this "new" economy measured in terms of that agent's *true* preferences. This total utility can be expressed as

$$\int_{B\times B\times H_0} \{b(h) - p^\alpha(h)\} d\eta^\alpha(b, b', h)$$

Unfortunately, as it stands, the distribution η is not very useful for two reasons: (i) η lacks descriptive power in that the structure of the misrepresentation is not clear; (ii) η lacks technical power in that there is not a convenient[4] way to prove any characterization theorems about manipulation in this formulation.

We recall a classical result in the theory of vector measures.

PROPOSITION 1 *Let (Ω, Σ, μ) be a finite complete measure space and let $\mathcal{X}$ be a Banach space. If a countably additive vector measure $F : \Sigma \to \mathcal{X}^*$ satisfies $\|F(E)\| \leq K|\mu(E)|$ for all $E \in \Sigma$ for some constant K, then there exists a weak*-measurable function $f : \Omega \to \mathcal{X}^*$ such that for all $x \in \mathcal{X}$ and for all $E \in \Sigma$*

$$\langle F(E), x\rangle = \int_E \langle f(\omega), x\rangle d\mu(\omega)$$

We call the function f a *Gelfand density* (or a *weak* density*) for F with respect to μ. We write $F(E) = \int f d\mu$ weak* in $\mathcal{X}^*$.

This weak* representation theorem is equivalent to the fact that every bounded linear operator T from $L^1(\Omega, \Sigma, \mu)$ to $\mathcal{X}^*$ has a weak* density. In fact, the vector measure F and the operator T which correspond to each other under the map $F(\cdot) = T(\chi.)$ have the same weak* density. The Gelfand integral and its properties were introduced in (Gelfand 1936; Gelfand 1938). The representation theorem was proved by Gelfand in the special case that $\mathcal{X}$ is a *separable* Banach space[5] . The general case is proved by means of a lifting theorem as in (Dinculeanu and Uhl 1973). A more detailed discussion may be found in (Diestel and Uhl 1977).

In the present model, it has been assumed that the collection H of houses is a compact Hausdorff space. Consequently, the Banach spaces of continuous functions $C(H)$, $C(B)$, and $C(S)$ are separable and we may apply Gelfand's original theorem in these cases for misrepresentation outcomes. We start with housing allocations. For clarity, we will write the argument of a measure-valued density as a subscript; e.g. for $\psi : I \to M_+(H_0)$, we will write $\psi_i(E)$ instead of $\psi(i, E)$ or $(\psi(i))(E)$.

THEOREM 1 *Let $y \in M_+(I \times H_0)$ be a Walrasian allocation for the economy given by the population measure μ. Then there is a weak*-measurable function $\psi : I \to M_+(H_0)$ with $\|\psi_i\| = 1$ a.e.$[\mu]$ such that for any $A \subseteq I$*

$$y(A, \cdot) = \int_A \psi d\mu \quad \text{weak* in } M_+(H)$$

In fact,

$$dy(i, h) = \psi_i(dh)d\mu(i)$$

Proof: Define $F : \mathcal{B}_I \to M_+(H_0)$ by $(F(E))(G) = y(E, G)$ for each Borel set $E \subseteq I$ and each Borel set $G \subseteq H_0$. Since H_0 is compact, $M(H_0)$ is the dual of the space of continuous functions $C(H_0)$. It follows easily from $y \in M_+(I \times H_0)$ that F is a countably additive vector measure. Moreover, since y is a Walrasian allocation for μ, we have that $\|F(E)\| = (F(E))(H_0) = y(E, H_0) = y_1(E) = \mu(E)$. Consequently, by Gelfand's theorem there exists a weak* density $\psi : I \to M_+(H_0)$ such that for all continuous functions x on H_0 and for all Borel $E \subseteq I$

$$\langle F(E), x\rangle = \int_E \langle \psi_i, x\rangle d\mu(i)$$

[4]However, see (Gretsky, Ostroy, and Zame a) for an alternative approach which was developed after this talk was given.

[5]The speaker is grateful to J.J. Uhl, Jr. for pointing out this historical fact.

which can be written as

$$\int_{H_0} y(E, dh)x(h) = \int_E \int_{H_0} x(h)d\psi_i(h)d\mu(i)$$

or, equivalently,

$$dy(i,h) = \psi_i(dh)d\mu(i)$$

Note that for any A a Borel subset of I we have $\mu(A) = y(A, H_0) = \int \|\psi_i\| d\mu(i)$ so that $\|\psi_i\| = 1$ a.e.$[\mu]$. □

With this theorem there is a complete description in terms of weak* densities for the outcome of a misrepresentation.

THEOREM 2 *Let (y^α, p^α) be any Walrasian equilibrium and η^α be any outcome for a population measure μ, a buyer subpopulation ν, and an announcement α of misrepresentations for ν. Then there exist a weak*-measurable function $\psi^\alpha : B \to M_+(H_0)$ with $\|\psi^\alpha_{b'}\| = 1$ a.e.$[\mu_B]$ and a weak*-measurable function $a : B \to M_+(B)$ with $\|a_b\| = 1$ a.e.$[\mu_B]$ such that*

$$y^\alpha(E, \cdot) = \int_E \psi^\alpha_{b'} d\mu_B(b') \quad \textit{weak* in } M(H_0)$$

$$\alpha(E, \cdot) = \int_E a_b d\mu_B(b) \quad \textit{weak* in } M(B)$$

and

$$d\eta(b, b', h) = \psi^\alpha_{b'}(dh)a_b(db')d\mu_B(b)$$

Proof: Given the announcement α of misrepresentations, we apply Gelfand's theorem to the vector measure $(Z(E)) = \alpha(E, F)$ and theorem 1 to the Walrasian allocation y^α. □

The total utility to all buyers in the outcome of the announced economy

$$\int_{B\times H\times H_0} \{b(h) - p^\alpha(h)\} d\eta(b, b', h)$$

may be rewritten by Theorem 2 as

$$\int_B \int_B \int_{H_0} \{b(h) - p^\alpha(h)\}\psi^\alpha_{b'}(dh)a_b(db')d\mu_B(b)$$

Thus, the total utility to the misrepresenting subpopulation ν is

$$\int_B \int_B \int_{H_0} \{b(h) - p^\alpha(h)\}\psi^\alpha_{b'}(dh)a_b(db')d\nu(b)$$

With no manipulation allowed, α is μ_B normalized on the diagonal Δ; i.e. a_b is the point mass $\delta_b \in M(B)$. In this case, the utility to the subpopulation ν is

$$\int_B \int_B \int_{H_0} \{b(h) - p(h)\}\psi_{b'}(dh)\delta_b(db')d\nu(b)$$

which simplifies to

$$\int_B \int_{H_0} \{b(h) - p(h)\}\psi_b(dh)d\nu(b)$$

Consequently, the increase in utility to the misrepresenting subpopulation ν of buyers resulting from the announcement α in the economy μ is

$$U_\mu^\alpha(\nu) = \int_B \int_B \int_{H_0} \{b(h) - p^\alpha(h)\}\psi_{b'}^\alpha(dh)a_b(db')d\nu(b) - \int_B \int_{H_0} \{b(h) - p(h)\}\psi_b(dh)d\nu(b)$$

Note that the increase in utility may be negative. Moreover, we have in the notation suppressed the dependence of $U_\mu^\alpha(\nu)$ on the choices of y^α and η; these need not be unique for a given μ, ν, and α.

A similar expression can be developed for subpopulations of sellers. Now the announcement α is defined on $S \times S$ and this leads to a distributional description η on $S \times S \times H_0$ of houses as a result of true and announced types. The utility to the misrepresenting subpopulation ν is

$$\int_{S\times S\times H_0} \{p^\alpha(h) - \sigma(s)\}d\eta(s, s', h)$$

which can be rewritten in a similar manner to that of Theorem 2 as

$$\int_S \int_S \int_{H_0} \{p^\alpha(h) - \sigma(s)\}d\phi_{s'}^\alpha(h)da_s(s')d\nu(s)$$

where $y^\alpha(\cdot, F) = \int_F \phi_{s'}^\alpha d\mu_S(s')$ weak* in $M(H_0)$ with $\|\phi_{s'}^\alpha\| = 1$ a.e.$[\mu_S]$ and $\alpha(E, \cdot) = \int_E a_s d\mu_S(s)$ weak* in $M(S)$ with $\|a_s\| = 1$ a.e.$[\mu_S]$.

Since a seller type can misrepresent only her preferences, viz. the reservation value for her own house, the measure a_s which describes how the type s misrepresents is a measure on $S = H \times [0, 1]$ which is supported on the set $\{\pi s\} \times [0, 1]$ and consequently can be described as a measure ρ_s on $[0, 1]$. Moreover, if a seller of type s were to announce herself as being type s', then it would have to follow that $s' = (h, r')$ where $h = \pi s$ and $r' \in [0, 1]$ since the house endowment cannot be misrepresented. Consequently, $\phi_{s'}^\alpha = u^\alpha(r')\delta_h + \{1 - u^\alpha(r')\}\delta_{\bar{0}}$ is a measure on H_0 in which the latter term represents no-trade activity and thus will not contribute to the utility integral. So we may rewrite the utility integral as

$$\int_S \int_{[0,1]} \{p^\alpha(\pi s) - \sigma(s)\}u^\alpha(r)d\rho_s(r)d\nu(s).$$

We can compute the increase in utility to the misrepresenting subpopulation ν of sellers resulting from the distribution α of announcements in the economy μ to be

$$U_\mu^\alpha(\nu) = \int_S \int_{[0,1]} \{p^\alpha(\pi s) - \sigma(s)\}u^\alpha(r)d\rho_s(r)d\nu(s) - \int_S \{p(\pi s) - \sigma(s)\}u(s)d\nu(s)$$

As we did in the case for subpopulations of buyers, we again suppress in the notation the dependence of $U_\mu^\alpha(\nu)$ on the choices of y^α and η.

4 Manipulation

DEFINITION: The economy μ is *non-manipulable* if there is a Walrasian price $p \in P(\mu)$ with the property that given $\epsilon > 0$ there is $\delta > 0$ such that for any subpopulation ν of buyers or sellers satisfying $\|\nu\| < \delta$, any distribution α of announcements of misrepresentations for ν, any Walrasian equilibrium (y^α, p^α) for the misrepresented economy, and any outcome η of the announcement it follows that

$$U_\mu^\alpha(\nu) < \epsilon\|\nu\|$$

An economy being non-manipulable is a manifestation of it being perfectly competitive in the sense that asymptotically small coalitions[6] cannot favorably manipulate the economy by misrepresentation. This turns out to be equivalent to a number of other conditions including stability of the Walrasian price correspondence and differentiability of the function totaling the gains from all trades. We give a sample result.

The Walrasian price correspondence is a correspondence (a possibly multiply-valued function) $P : M_+(I) \to L_+^\infty(\mu_H)$ which takes each economy described by a population measure to the collection of Walrasian prices for that economy. We will consider the stability question for the subset of population measures which are absolutely continuous with respect to a fixed measure μ. If ν is such a measure then it has Radon-Nikodym derivative $d\nu/d\mu \in L_+^1(\mu)$; moreover, ν is a subpopulation if and only if $d\nu/d\mu \leq 1$ a.e.$[\mu]$. Our restriction gives the correspondence $P_\mu : L_+^1(\mu) \to L_+^\infty(\mu_H)$ where $P_\mu(h) = P(\nu)$ for $h = d\nu/d\mu$.

THEOREM 3 *The economy μ is non-manipulable if and only if the (restricted) price correspondence $P_\mu : L_+^1(\mu) \to L_+^\infty(\mu_H)$ is norm-norm continuous at* $\mathbf{1}$.

Proof: For the sake of argument, we consider a subpopulation of buyers. Assume that P_μ is norm-norm continuous at $\mathbf{1}$. We need to compare the expression

$$\int_B \int_B \int_{H_0} \{b(h) - p^\alpha(h)\}\psi_{b'}^\alpha(dh)a_b(db')d\nu(b)$$

to the expression

$$\int_B \int_{H_0} \{b(h) - p(h)\}\psi_b(dh)d\nu(b).$$

The latter may be written as

$$\int_B \int_B \int_{H_0} \{b(h) - p(h)\}\psi_{b'}(dh)\delta_b(b')d\nu(b).$$

Note that the indirect utility v_b^* is continuous with respect to the sup norm topology on prices. Thus, although the Walrasian allocations for nearby prices need not be close in variation to the given allocation for μ, the utility of such allocations is close to the utility of the given allocation, i.e.

$$\| \int_H \{b(h) - p^\alpha(h)\}\psi_{b'}^\alpha(dh) - \int_H \{b(h) - p(h)\}\psi_{b'}(dh)\|$$

is continuous at 0 with respect to α. Consequently,

$$\lim_{\|\nu\|\to 0} \frac{U_\mu^\alpha(\nu)}{\|\nu\|} = 0$$

as desired.

Conversely, if P_μ is not norm-to-norm continuous at $\mathbf{1}$, then there exists some $\epsilon > 0$ such that for every $\delta > 0$ there is a subpopulation ν and an announcement α consistent with ν

[6]Small coalitions serve as proxies for the individual agents which would be the objects of concern in finite economies.

such that $||\nu|| < \delta$ and $||p^\alpha - p|| \geq \epsilon$ for some $p^\alpha \in P_\mu(\mathbf{1} - d\nu/d\mu)$. Hence for b in the support of ν it follows that

$$\begin{aligned} |v_b^*(p^\alpha) - v_b^*(p)| &= |\sup_h\{b(h) - p^\alpha(h)\} - \sup_h\{b(h) - p(h)\}| \\ &\geq ||p^\alpha - p|| \end{aligned}$$

Consequently, $U_\mu^\alpha(\nu)/||\nu||$ need not converge to 0. □

References

Diestel, J. and J. J. Uhl, Jr. (1977). *Vector Measures.* Number 15 in Mathematical Surveys. Providence: American Mathematical Society.

Dinculeanu, N. and J. J. Uhl, Jr. (1973). A unifying Radon-Nikodym theorem for vector measures. *Journal of Multivariate Analysis 3*, 184–203.

Gelfand, I. M. (1936). Sur un lemme de la theorie des espaces lineaires. *Comm. Inst. Sci. Math. Mec. Univ. de Kharkoff et Soc. Math. Kharkoff (4) 13*, 35–40.

Gelfand, I. M. (1938). Abstrakte funktionen und lineaire operatoren. *Matematicheskii Sbornik (New Series) 4 46*, 235–286.

Gretsky, N. E., J. M. Ostroy, and W. R. Zame. Perfect competition in the nonatomic assignment model: The continuous case. Forthcoming.

Gretsky, N. E., J. M. Ostroy, and W. R. Zame. Perfect competition in the nonatomic assignment model: The discontinuous case. Forthcoming.

Gretsky, N. E., J. M. Ostroy, and W. R. Zame (1992). The nonatomic assignment model. *Economic Theory 2*, 103–127.

Dilations of Hilbert–Schmidt Class Operator-Valued Measures and Applications

YÛICHIRÔ KAKIHARA Department of Mathematics, University of California, Riverside, Riverside, CA 92521-0135, U. S. A.

Dedicated to Professor M. M. Rao on the occasion of his 65th birthday

ABSTRACT The space of Hilbert-Schmidt class operators has a gramian structure, i.e., a trace class operator valued inner product. A gramian orthogonally scattered dilation of a Hilbert-Schmidt class operator valued measure is considered. Several new characterizations of it are given. An application to Hilbert space valued second order stochastic processes is made, where some equivalence conditions are given for a process to have an operator stationary dilation.

1. INTRODUCTION

The orthogonally scattered dilation of Hilbert space valued measures has been fully studied and, in Section 2, we shall state basic results on it. The purpose of this paper is to consider gramian orthogonally scattered dilation of Hilbert-Schmidt class operator valued measures and its application to Hilbert space valued second order stochastic processes.

Let H, K be a pair of complex Hilbert spaces. $B(H)$ denotes the algebra of all bounded linear operators on H and $T(H)$ the Banach space of all trace class operators on H. $S(K,H)$ denotes the Hilbert space of all Hilbert-Schmidt class operators from K into H.

$S(K,H)$ has some nice properties and among them is a gramian structure. That is,

$S(K,H)$ is a left $B(H)$-module with the operator multiplication from left and if we define $[x,y] = xy^* \in T(H)$ for $x, y \in S(K,H)$, then $[\cdot,\cdot]$ satisfies that

(1) $[x,x] \geq 0$, and $[x,x] = 0$ if and only if $x = 0$;
(2) $[x+y,z] = [x,z] + [y,z]$;
(3) $[ax,y] = a[x,y]$;
(4) $[x,y]^* = [y,x]$,

where $x, y, z \in S(K,H)$ and $a \in B(H)$. The $T(H)$-valued inner product $[\cdot,\cdot]$ is called the *gramian* in $S(K,H)$. We say that $S(K,H)$ is a *normal Hilbert $B(H)$-module.*

In Section 3, we characterize those $S(K,H)$-valued measures which have gramian orthogonally scattered dilations. In Section 4, a new necessary and suffifient condition for an H-valued second order stochastic process on a locally compact abelian group to have an operator stationary dilation is given together with known conditions. All the contents of this paper will be included in detail in the monograph [6].

2. ORTHOGONALLY SCATTERED DILATION

Let $(\Theta, \mathfrak{A})$ be a measurable space. $ca(\mathfrak{A}, K)$ denotes the set of all K-valued bounded c.a. (countably additive) measures on $(\Theta, \mathfrak{A})$. The *semivariation* $\|\xi\|(A)$ of $\xi \in ca(\mathfrak{A}, K)$ at $A \in \mathfrak{A}$ is defined by

$$\|\xi\|(A) = \sup\left\{\left\|\sum_{\Delta\in\pi}\alpha_\Delta\xi(\Delta)\right\|_K : |\alpha_\Delta| \leq 1,\, \Delta \in \pi \in \Pi(A)\right\}, \tag{2.1}$$

where $\Pi(A)$ denotes the set of all finite $\mathfrak{A}$-measurable partitions of A and $\|\cdot\|_K$ the norm in K. ξ is said to be *orthogonally scattered* (*o.s.*) if $\big(\xi(A), \xi(B)\big)_K = 0$ for every disjoint pair $A, B \in \mathfrak{A}$, where $(\cdot,\cdot)_K$ is the inner product in K. $caos(\mathfrak{A}, K)$ denotes the set of all o.s. measures in $ca(\mathfrak{A}, K)$.

DEFINITION 2.1 (1) $\xi \in ca(\mathfrak{A}, K)$ is said to have an *orthogonally scattered dilation* (*o.s.d.*) if there exist a Hilbert space $\mathfrak{K}$ containing K as a closed subspace and an $\eta \in caos(\mathfrak{A}, \mathfrak{K})$ such that $\xi = J\eta$, where $J : \mathfrak{K} \to K$ is the orthogonal projection. The triple $\{\eta, \mathfrak{K}, J\}$ is also called an *o.s.d.* of ξ.

(2) ξ is said to have a *spectral dilation* if there exist a Hilbert space $\mathfrak{K}$, a (weakly c.a.) spectral measure $E(\cdot)$ in $\mathfrak{K}$, an operator $S \in B(\mathfrak{K}, K)$ and a vector $\psi \in \mathfrak{K}$ such that $\xi(\cdot) = SE(\cdot)\psi$, where $B(\mathfrak{K}, K)$ is the Banach space of all bounded linear operators from $\mathfrak{K}$ into K.

Let $\mathfrak{X}$ and $\mathfrak{Y}$ be two normed linear spaces. A linear operator $T : \mathfrak{X} \to \mathfrak{Y}$ is said to be *absolutely 2-summing* if $\pi_2(T) < \infty$, where $\pi_2(T) = \inf\{C > 0 : (2.2) \text{ holds}\}$:

$$\left(\sum_{j=1}^n \|Tx_j\|_{\mathfrak{Y}}^2\right)^{\frac{1}{2}} \leq C \sup\left\{\left(\sum_{j=1}^n |x^*(x_j)|^2\right)^{\frac{1}{2}} : x^* \in \mathfrak{X}^*,\, \|x^*\|_{\mathfrak{X}^*} \leq 1\right\} \tag{2.2}$$

for any $n \geq 1$ and $x_1, \ldots, x_n \in \mathfrak{X}$,

where $\|\cdot\|_{\mathfrak{Y}}$ and $\|\cdot\|_{\mathfrak{X}^*}$ are norms in $\mathfrak{Y}$ and $\mathfrak{X}^*$, respectively.

Let $L^0(\Theta)$ be the set of all complex valued $\mathfrak{A}$-simple functions on Θ. For $f \in L^0(\Theta)$ consider the *sup norm* $\|f\|_\infty$ defined by $\|f\|_\infty = \sup_{t\in\Theta} |f(t)|$. Then $(L^0(\Theta), \|\cdot\|_\infty)$ becomes a normed linear space. Let $\xi \in ca(\mathfrak{A}, K)$ and define the integral of $f = \sum_{j=1}^{n} \alpha_j 1_{A_j} \in L^0(\Theta)$ w.r.t. (with respect to) ξ over $A \in \mathfrak{A}$ in an obvious manner by

$$\int_A f\, d\xi = \sum_{j=1}^{n} \alpha_j \xi(A_j \cap A),$$

where 1_A is the indicator function of A. Then the following theorem is known (cf. Niemi [10, 11], Rao [13] and Rosenberg [14]):

THEOREM 2.2 *Let $\xi \in ca(\mathfrak{A}, K)$. Then the following conditions are equivalent:*

(1) *ξ has an o.s.d.*

(2) *ξ has a spectral dilation.*

(3) *There exists a constant $C > 0$ such that for any $n \geq 1$ and $f_1, \dots, f_n \in L^0(\Theta)$*

$$\sum_{j=1}^{n} \left\| \int_\Theta f_j\, d\xi \right\|_K^2 \leq C \left\| \sum_{j=1}^{n} |f_j|^2 \right\|_\infty. \tag{2.3}$$

(4) *The operator $S_\xi : (L^0(\Theta), \|\cdot\|_\infty) \to K$ is absolutely 2-summing, where S_ξ is defined by $S_\xi f = \int_\Theta f\, d\xi$ for $f \in L^0(\Theta)$.*

(5) *There exists a positive finite measure $\nu \in ca(\mathfrak{A}, \mathbb{R}^+)$ such that*

$$\left\| \int_\Theta f\, d\xi \right\|_K^2 \leq \int_\Theta |f|^2\, d\nu = \|f\|_{2,\nu}^2, \qquad f \in L^0(\Theta),$$

where $\mathbb{R}^+ = [0, \infty)$. In this case, ν is called a 2-majorant of ξ.

It follows from the Grothendieck's inequality (cf. Grothendieck [3] and Lindenstrauss and Pełczyński [7]) that the inequality (2.3) holds with $C = K_G \|\xi\|(\Theta)^2$ where K_G is the Grothendieck constant and $\|\xi\|(\Theta)$ is the total semivariation of ξ (cf. Rosenberg [14]). Therefore we have:

COROLLARY 2.3 *Every $\xi \in ca(\mathfrak{A}, K)$ has an o.s.d.*

3. GRAMIAN ORTHOGONALLY SCATTERED DILATION

We consider $S(K, H)$-valued c.a. measures on $(\Theta, \mathfrak{A})$ and assume that H is separable, so that H has a countable CONS (complete orthonormal system). As in the Introduction, the $T(H)$-valued gramian $[\cdot, \cdot]$ in $S(K, H)$ is considered.

DEFINITION 3.1 An $S(K, H)$-valued measure $\xi \in ca(\mathfrak{A}, S(K, H))$ is said to be *gramian orthogonally scattered (g.o.s.)* if $[\xi(A), \xi(B)] = 0$ for every disjoint $A, B \in$

$\mathfrak{A}$. $cagos(\mathfrak{A}, S(K,H))$ denotes the set of all g.o.s. measures in $ca(\mathfrak{A}, S(K,H))$. $\xi \in ca(\mathfrak{A}, S(K,H))$ is said to have a *gramian orthogonally scattered dilation* (*g.o.s.d.*) if there exist a Hilbert space $\mathfrak{K}$ containing K as a closed subspace and an $\eta \in cagos(\mathfrak{A}, S(\mathfrak{K}, H))$ such that $\xi = P\eta$, where $P : S(\mathfrak{K}, H) \to S(K,H)$ is the gramian orthogonal projection, i.e., P is an orthogonal projection which satisfies that

$$[P^2x, y] = [Px, y] = [x, Py], \qquad x, y \in S(\mathfrak{K}, H).$$

The triple $\{\eta, \mathfrak{K}, P\}$ is also called a *g.o.s.d.* of ξ.

As we mentioned in Section 2, every Hilbert space valued bounded c.a. measure has an o.s.d. But not every Hilbert-Schmidt class operator valued measure has a g.o.s.d. Thus we shall give some necessary and sufficient conditions for the g.o.s.d.

Let $\xi \in ca(\mathfrak{A}, S(K,H))$ and $L^0(\Theta; B(H))$ be the set of all $B(H)$-valued $\mathfrak{A}$-simple functions on Θ. The integral of $\Phi = \sum_{j=1}^{n} a_j 1_{A_j} \in L^0(\Theta; B(H))$ w.r.t. ξ over $A \in \mathfrak{A}$ is defined by

$$\int_A \Phi\, d\xi = \sum_{j=1}^{n} a_j \xi(A_j \cap A),$$

which is in $S(K,H)$. Let $F \in ca(\mathfrak{A}, T(H))$. For $\Phi = \sum_{j=1}^{m} a_j 1_{A_j}, \Psi = \sum_{k=1}^{n} b_k 1_{B_k} \in L^0(\Theta; B(H))$ the integral of (Φ, Ψ) w.r.t. F over $A \in \mathfrak{A}$ is defined by

$$\int_A (\Phi, \Psi)\, dF = \int_A \Phi\, dF\, \Psi^* = \sum_{j=1}^{m}\sum_{k=1}^{n} a_j F(A_j \cap B_k \cap A) b_k^*.$$

Let us put $T^+(H) = \{a \in T(H) : a \geq 0\}$. With these preparations we prove the following proposition which was mentioned without proof in Kakihara [4, 3.9. Proposition].

PROPOSITION 3.2 *Let $\xi \in ca(\mathfrak{A}, S(K,H))$. Then, ξ has a g.o.s.d. if and only if it has a $T^+(H)$-valued 2-majorant $F \in ca(\mathfrak{A}, T^+(H))$. That is,*

$$\left[\int_\Theta \Phi\, d\xi, \int_\Theta \Phi\, d\xi\right] \leq \int_\Theta \Phi\, dF\, \Phi^*, \qquad \Phi \in L^0(\Theta; B(H)). \tag{3.1}$$

Proof: Suppose that ξ has a g.o.s.d. $\{\eta, \mathfrak{K}, P\}$. Put $F(\cdot) = [\eta(\cdot), \eta(\cdot)] = \eta(\cdot)\eta(\cdot)^*$, then $F \in ca(\mathfrak{A}, T^+(H))$. For $\Phi = \sum_{j=1}^{n} a_j 1_{A_j} \in L^0(\Theta; B(H))$ we have that

$$P\int_\Theta \Phi\, d\eta = \int_\Theta \Phi\, d(P\eta) = \int_\Theta \Phi\, d\xi$$

since $\xi = P\eta$ and P commutes with the module action of $B(H)$ and hence

$$\begin{aligned}\left[\int_\Theta \Phi\, d\xi, \int_\Theta \Phi\, d\xi\right] &= \left[P\int_\Theta \Phi\, d\eta, P\int_\Theta \Phi\, d\eta\right] \\ &\leq \left[\int_\Theta \Phi\, d\eta, \int_\Theta \Phi\, d\eta\right] = \int_\Theta \Phi\, dF\, \Phi^*.\end{aligned}$$

Conversely, suppose that (3.1) holds and define $M : \mathfrak{A} \times \mathfrak{A} \to T(H)$ by

$$M(A,B) = F(A \cap B) - M_\xi(A,B), \qquad A, B \in \mathfrak{A},$$

where $M_\xi(A,B) = \big[\xi(A), \xi(B)\big]$. Then we see that M is a positive definite kernel on $\mathfrak{A} \times \mathfrak{A}$ in the sense that

$$\sum_{j,k} a_j M(A_j, A_k) a_k^* \geq 0$$

for any $n \geq 1$, $a_1, \dots, a_n \in B(H)$ and $A_1, \dots, A_n \in \mathfrak{A}$. Thus there exist a reproducing kernel normal Hilbert $B(H)$-module Y of M containing $S(K,H)$ as a closed submodule and an $\eta \in cagos(\mathfrak{A}, Y)$ such that $\xi = P\eta$, where $P : Y \to S(K,H)$ is the gramian orthogonal projection (cf. [4]). By the structure theorem (cf. Ozawa [12]) there exists a Hilbert space $\mathfrak{K}$ such that $Y \cong S(\mathfrak{K}, H)$, i.e., Y and $S(K,H)$ are isomorphic as normal Hilbert $B(H)$-modules, and K can be regarded as a closed subspace of $\mathfrak{K}$. Therefore, $\{\eta, \mathfrak{K}, P\}$ is a g.o.s.d. of ξ.

Let $\xi \in ca\big(\mathfrak{A}, S(K,H)\big)$. The *operator semivariation* $\|\xi\|_o(A)$ of ξ at $A \in \mathfrak{A}$ is defined by

$$\|\xi\|_o(A) = \sup\left\{ \left\| \sum_{\Delta \in \pi} a_\Delta \xi(\Delta) \right\|_\sigma : a_\Delta \in B(H),\ \|a_\Delta\| \leq 1,\ \Delta \in \pi \in \Pi(A) \right\},$$

where $\|\cdot\|_\sigma$ is the Hilbert-Schmidt norm. Now ξ^* defined by $\xi^*(\cdot) = \xi(\cdot)^*$ is in $ca\big(\mathfrak{A}, S(H,K)\big)$. The *strong semivariation* $\|\xi^*\|_s(A)$ of ξ^* at $A \in \mathfrak{A}$ is defined by

$$\|\xi^*\|_s(A) = \sup\left\{ \left\| \sum_{\Delta \in \pi} \xi^*(\Delta)\phi_\Delta \right\|_K : \phi_\Delta \in H,\ \|\phi\|_H \leq 1 \right\}.$$

Note that $\xi^*\phi$ defined by $(\xi^*\phi)(\cdot) = \xi^*(\cdot)\phi$ is in $ca(\mathfrak{A}, K)$ for each $\phi \in H$. Then we have the following proposition (cf. Kakihara [5, 5.2 Theorem, 5.3 Corollary and 5.7 Theorem]). The part (2) $\Rightarrow$ (1) was proved in Truong-van [17].

PROPOSITION 3.3 *For $\xi \in ca\big(\mathfrak{A}, S(K,H)\big)$ the following statements are equivalent:*

(1) *ξ has a g.o.s.d.*

(2) $\|\xi\|_o(\Theta) < \infty$.

(3) $\|\xi^*\|_s(\Theta) < \infty$.

(4) *For some CONS $\{\phi_k\}_{k=1}^\infty$ in H there exists a family $\{\eta_k, \mathfrak{K}_k, J_k\}_{k=1}^\infty$ of o.s.d.'s of $\{\xi^*\phi_k\}_{k=1}^\infty \subseteq ca(\mathfrak{A}, K)$ such that $\sum\limits_{k=1}^{\infty} \|\eta_k(\Theta)\|_{\mathfrak{K}_k}^2 < \infty$, where $\|\cdot\|_{\mathfrak{K}_k}$ is the norm in $\mathfrak{K}_k$ for $k \geq 1$.*

$L^0(\Theta\,; K)$ denotes the set of all K-valued $\mathfrak{A}$-simple functions on Θ and $fa(\mathfrak{A}, K)$ the Banach space of all K-valued finitely additive (f.a.) measures on $(\Theta, \mathfrak{A})$ with the total semivariation norm $\|\cdot\|(\Theta)$, where the semivariation $\|\zeta\|(\cdot)$ of $\zeta \in fa(\mathfrak{A}, K)$ is defined as in (2.1). For $\zeta \in fa(\mathfrak{A}, K)$ and $\varphi = \sum\limits_{j=1}^{n} \psi_j 1_{A_j} \in L^0(\Theta\,; K)$ we define the integral of φ w.r.t. ζ by

$$\int_\Theta \langle \varphi, d\zeta \rangle = \sum_{j=1}^{n} \big(\psi_j, \zeta(A_j)\big)_K.$$

Moreover, the norm $\|\varphi\|_*$ is defined by

$$\|\varphi\|_* = \sup\left\{\left|\int_\Theta \langle\varphi, d\zeta\rangle\right| : \zeta \in fa(\mathfrak{A}, K),\ \|\zeta\|(\Theta) \le 1\right\}.$$

It is known (see e.g. Makagon and Salehi [8]) that the dual space of $\big(L^0(\Theta\,;K), \|\cdot\|_*\big)$ can be identified with the Banach space $fa(\mathfrak{A}, K)$, where the isomorphism $U : fa(\mathfrak{A}, K) \to L^0(\Theta\,;K)^*$ is given by

$$(U\zeta)(\varphi) = \int_\Theta \langle\varphi, d\zeta\rangle, \qquad \varphi \in L^0(\Theta\,;K).$$

If $\xi \in ca\big(\mathfrak{A}, S(K,H)\big)$, then the integral of φ w.r.t. ξ is defined by

$$\int_\Theta d\xi\,\varphi = \sum_{j=1}^n \xi(A_j)\psi_j.$$

Now we get other characterizations of g.o.s.d. as follows:

PROPOSITION 3.4 *For $\xi \in ca\big(\mathfrak{A}, S(K,H)\big)$ the following statements are equivalent:*

(1) *ξ has a g.o.s.d.*

(2) *For every CONS $\{\phi_k\}_{k=1}^\infty$ in H there exists a family $\{\eta_k, \mathfrak{K}_k, J_k\}_{k=1}^\infty$ of o.s.d.'s of $\{\xi^*\phi_k\}_{k=1}^\infty \subseteq ca(\mathfrak{A}, K)$ such that $\sum\limits_{k=1}^\infty \|\eta_k(\Theta)\|^2_{\mathfrak{K}_k} < \infty$, where $\|\cdot\|_{\mathfrak{K}_k}$ is the norm in $\mathfrak{K}_k$ for $k \ge 1$.*

(3) *$\sum\limits_{k=1}^\infty \|\xi^*\phi_k\|(\Theta)^2 < \infty$ for every ON sequence $\{\phi_k\}_{k=1}^\infty$ in H.*

(4) *The operator $S_\xi : \big(L^0(\Theta\,;K), \|\cdot\|_*\big) \to H$ defined by $S_\xi\varphi = \int_\Theta d\xi\,\varphi$ for $\varphi \in L^0(\Theta\,;K)$ is absolutely 2-summing.*

(5) *The adjoint operator $S_\xi^* : H \to \big(fa(\mathfrak{A}, K), \|\cdot\|(\Theta)\big)$ of S_ξ defined in (4) is absolutely 2-summing.*

Proof: (1) $\Rightarrow$ (2) can be shown in a same manner as in [5, 5.7 Theorem].

(2) $\Rightarrow$ (3): Let $\{\phi_k\}_{k=1}^\infty$ be any ON sequence in H and $\{\psi_k\}_{k=1}^\infty$ be a CONS in H containing $\{\phi_k\}_{k=1}^\infty$. Let $\{\eta_k, \mathfrak{K}_k, J_k\}_{k=1}^\infty$ be a family of o.s.d.'s of $\{\xi^*\psi_k\}_{k=1}^\infty$ such that $\sum\limits_{k=1}^\infty \|\eta_k(\Theta)\|^2_{\mathfrak{K}_k} < \infty$, which exists by assumption. Then, for each $k \ge 1$, we have

$$\|\xi^*\psi_k\|(\Theta) = \|J_k\eta_k\|(\Theta) \le \|J_k\|\|\eta_k\|(\Theta) \le \|\eta_k(\Theta)\|_{\mathfrak{K}_k},$$

since η_k is o.s., and hence we obtain

$$\sum_{k=1}^\infty \|\xi^*\phi_k\|(\Theta)^2 \le \sum_{k=1}^\infty \|\xi^*\psi_k\|(\Theta)^2 \le \sum_{k=1}^\infty \|\eta_k(\Theta)\|^2_{\mathfrak{K}_k} < \infty.$$

(3) ⇒ (5): For any ON sequence $\mathbf{e} = \{\phi_k\}_{k=1}^{\infty}$ in H we have

$$\begin{aligned}
\|S_\xi^*\|_{\mathbf{e}} &\equiv \sum_{k=1}^{\infty} \|S_\xi^*\phi_k\|(\Theta)^2 \\
&= \sum_{k=1}^{\infty} \sup\left\{|\langle\varphi, S_\xi^*\phi_k\rangle|^2 : \varphi \in L^0(\Theta\,;K),\ \|\varphi\|_* \le 1\right\} \\
&= \sum_{k=1}^{\infty} \sup\left\{|(S_\xi\varphi, \phi_k)_H|^2 : \varphi \in L^0(\Theta\,;K),\ \|\varphi\|_* \le 1\right\} \qquad (3.2)\\
&= \sum_{k=1}^{\infty} \sup\left\{\left|\left(\int_\Theta d\xi\,\varphi, \phi_k\right)_H\right|^2 : \varphi \in L^0(\Theta\,;K),\ \|\varphi\|_* \le 1\right\} \\
&= \sum_{k=1}^{\infty} \sup\left\{\left|\int_\Theta \langle\varphi, d\xi^*\phi_k\rangle\right|^2 : \varphi \in L^0(\Theta\,;K),\ \|\varphi\|_* \le 1\right\} \\
&\le \sum_{k=1}^{\infty} \sup\left\{\|\varphi\|_*^2\|\xi^*\phi_k\|(\Theta)^2 : \varphi \in L^0(\Theta\,;K),\ \|\varphi\|_* \le 1\right\} \\
&= \sum_{k=1}^{\infty} \|\xi^*\phi_k\|(\Theta)^2 < \infty,
\end{aligned}$$

by assumption. By Słowikowski [16, Theorem 1] we see that S_ξ^* is absolutely 2-summing.

(5) ⇒ (4): Let $\mathbf{e} = \{\phi_k\}_{k=1}^{\infty}$ be any ON sequence in H and define $\sigma_{\mathbf{e}}(S_\xi)$ by

$$\sigma_{\mathbf{e}}(S_\xi) = \left(\sum_{k=1}^{\infty} \sup_{\|\varphi\|_* \le 1} |(S_\xi\varphi, \phi_k)_H|^2\right)^{\frac{1}{2}}.$$

Then, $\sigma_{\mathbf{e}}(S_\xi)$ is equal to the right hand side of (3.2), so that $\sigma_{\mathbf{e}}(S_\xi) = \|S_\xi^*\|_{\mathbf{e}}$. Since $\sigma_{\mathbf{e}}(S_\xi) < \infty$ for every ON sequence $\mathbf{e}$ in H by assumption, it follows from Słowkowski [16, Proposition 1] that there exist a Hilbert space $\mathfrak{K}$, $T_1 \in B(L^0(\Theta\,;K), \mathfrak{K})$ with $\|T_1\| = 1$ and a Hilbert-Schmidt class operator $T_2 \in S(\mathfrak{K}, H)$ such that $S_\xi = T_2T_1$. Since every Hilbert-Schmidt class operator is absolutely 2-summing, S_ξ is also absolutely 2-summing.

(4) ⇒ (1): Since S_ξ is absolutely 2-summing, it follows from Makagon and Salehi [8, 5.4 Theorem] that ξ has a weakly c.a. spectral dilation. Then, ξ^* also has a weakly c.a. spectral dilation, so that the proof of 5.2 Theorem in [5] can be applied to conclude the proof.

Combining Propositions 3.2, 3.3 and 3.4, we get the following characterization of g.o.s.d. for a measure in $ca(\mathfrak{A}, S(K,H))$.

THEOREM 3.5 *Let $\xi \in ca(\mathfrak{A}, S(K,H))$. Then the following conditions are equivalent:*

(1) *ξ has a g.o.s.d.*

(2) *ξ has a $T^+(H)$-valued 2-majorant $F \in ca(\mathfrak{A}, T^+(H))$, i.e.,*

$$\left[\int_\Theta \Phi\,d\xi, \int_\Theta \Phi\,d\xi\right] \le \int_\Theta \Phi\,dF\,\Phi^*, \qquad \Phi \in L^0(\Theta\,;B(H)).$$

(3) $\|\xi\|_o(\Theta) < \infty$.

(4) $\|\xi^*\|_s(\Theta) < \infty$.

(5) *For every CONS* $\{\phi_k\}_{k=1}^\infty$ *in* H *there exists a family* $\{\eta_k, \mathfrak{K}_k, J_k\}_{k=1}^\infty$ *of o.s.d.'s of* $\{\xi^*\phi_k\}_{k=1}^\infty \subseteq ca(\mathfrak{A}, K)$ *such that* $\sum_{k=1}^\infty \|\eta_k(\Theta)\|_{\mathfrak{K}_k}^2 < \infty$, *where* $\|\cdot\|_{\mathfrak{K}_k}$ *is the norm in* $\mathfrak{K}_k$ *for* $k \geq 1$.

(6) *For some CONS* $\{\phi_k\}_{k=1}^\infty$ *in* H *the conclusion of* (5) *holds.*

(7) $\sum_{k=1}^\infty \|\xi^*\phi_k\|(\Theta)^2 < \infty$ *for every ON sequence* $\{\phi_k\}_{k=1}^\infty$ *in* H.

(8) *The operator* $S_\xi : L^0(\Theta\,; K) \to H$ *defined by* $S_\xi\varphi = \int_\Theta d\xi\,\varphi$ *for* $\varphi \in L^0(\Theta\,; K)$ *is absolutely 2-summing.*

(9) *The adjoint operator* $S_\xi^* : H \to fa(\mathfrak{A}, K)$ *of* S_ξ *defined in* (8) *is absolutely 2-summing.*

4. APPLICATION

In this section, we consider Hilbert space valued second order stochastic processes. So let $(\Omega, \mathfrak{F}, \mu)$ be a probability measure space and $L_0^2(\Omega\,; H)$ be the Hilbert space of all H-valued strong random variables $x(\cdot)$ on Ω such that $\int_\Omega x(\omega)\,\mu(d\omega) = 0$ and $\int_\Omega \|x(\omega)\|_H^2\,\mu(d\omega) < \infty$. We put $X = L_0^2(\Omega\,; H)$. The inner product and the norm in X are given respectively by

$$(x, y)_X = \int_\Omega \big(x(\omega), y(\omega)\big)_H\,\mu(d\omega), \quad \|x\|_X = (x, x)_X^{\frac{1}{2}} \qquad x, y \in X.$$

X is a left $B(H)$-module with the natural action of $B(H)$ and has a $T(H)$-valued inner product, which we called a *gramian,* defined by

$$[x, y] = \int_\Omega x(\omega) \otimes \overline{y(\omega)}\,\mu(d\omega), \qquad x, y \in X,$$

where $\otimes$ is in the sense of Schatten [15]. In fact, X and $S\big(L_0^2(\Omega), H\big)$ are isomorphic as normal Hilbert $B(H)$-modules, where $L_0^2(\Omega) = \big\{f \in L^2(\Omega) : \int_\Omega f\,d\mu = 0\big\}$ and the correspondence between these spaces is given by $X \ni x \leftrightarrow T_x^* \in S\big(L_0^2(\Omega), H\big)$, T_x being defined by

$$T_x\phi = (\phi, x)_H, \qquad \phi \in H. \tag{4.1}$$

Let G be a locally compact abelian group and $\widehat{G}$ be its dual. $\mathfrak{B}_{\widehat{G}}$ denotes the Borel σ-algebra of $\widehat{G}$. $\xi \in ca\big(\mathfrak{B}_{\widehat{G}}, L_0^2(\Omega)\big)$ is said to be *regular* if for any $\varepsilon > 0$ and $A \in \mathfrak{A}$ there exist a compact set C and an open set O such that $C \subseteq A \subseteq O$ and $\|\xi\|(O\backslash C) < \varepsilon$. Regularity of X-valued measures is defined similarly. $rca\big(\mathfrak{B}_{\widehat{G}}, L_0^2(\Omega)\big)$, $rcaos\big(\mathfrak{B}_{\widehat{G}}, L_0^2(\Omega)\big)$, $rca(\mathfrak{B}_{\widehat{G}}, X)$ and $rcagos(\mathfrak{B}_{\widehat{G}}, X)$ denote the sets of all regular measures in $ca\big(\mathfrak{B}_{\widehat{G}}, L_0^2(\Omega)\big)$, $caos\big(\mathfrak{B}_{\widehat{G}}, L_0^2(\Omega)\big)$, $ca(\mathfrak{B}_{\widehat{G}}, X)$ and $cagos(\mathfrak{B}_{\widehat{G}}, X)$, respectively. The *strong semivariation* $\|\xi\|_s(\cdot)$ of $\xi \in ca(\mathfrak{B}_{\widehat{G}}, X)$ is defined to be that of $T_{\xi(\cdot)} \in ca\big(\mathfrak{B}_{\widehat{G}}, S(H, L_0^2(\Omega))\big)$ (cf. (4.1)).

Let $\{x(t)\}$ be an $L_0^2(\Omega)$-valued process on G. That is, $x(\cdot) : G \to L_0^2(\Omega)$ is a mapping. Then, $\{x(t)\}$ is said to be *weakly stationary* if its (scalar) *covariance function* $\gamma(s, t) = \big(x(s), x(t)\big)_2$ $(s, t \in G)$ depends only on the difference st^{-1} and if, putting $\tilde{\gamma}(st^{-1}) = \gamma(s, t)$,

$\tilde{\gamma}$ is continuous on G, where $(\cdot,\cdot)_2$ is the inner product in $L_0^2(\Omega)$. In this case, there exists a unique $\xi \in rcaos(\mathfrak{B}_{\widehat{G}}, L_0^2(\Omega))$ such that

$$x(t) = \int_{\widehat{G}} \langle t, \chi\rangle\, \xi(d\chi), \qquad t \in G, \tag{4.2}$$

where $\langle\cdot,\cdot\rangle$ is the duality pair of G and $\widehat{G}$. ξ is called the *representing measure* of the process $\{x(t)\}$. $\{x(t)\}$ is said to be *weakly harmonizable* if it has an integral representation (4.2) for some $\xi \in rca(\mathfrak{A}, L_0^2(\Omega))$.

DEFINITION 4.1 Let $\{x(t)\}$ be an X-valued process on G, i.e., $x(\cdot) : G \to X$ is a mapping.

(1) $\{x(t)\}$ is said to be *scalarly weakly harmonizable* if $\{(x(t), \phi)_H\}$ is an $L_0^2(\Omega)$-valued weakly harmonizable process for each $\phi \in H$.

(2) $\{x(t)\}$ is said to be *weakly harmonizable* if it has an integral representation (4.2) for some $\xi \in rca(\mathfrak{B}_{\widehat{G}}, X)$.

(3) $\{x(t)\}$ is said to be *weakly operator harmonizable* if it has an integral representation (4.2) for some $\xi \in rca(\mathfrak{B}_{\widehat{G}}, X)$ of bounded operator semivariation, i.e., $\|\xi\|_o(\widehat{G}) < \infty$.

(4) $\{x(t)\}$ is said to be *operator stationary* if its *operator covariance function* $\Gamma(s,t) = [x(s), x(t)]$ $(s,t \in G)$ depends only on the difference st^{-1} and, putting $\tilde{\Gamma}(st^{-1}) = \Gamma(s,t)$, $\tilde{\Gamma}$ is a $T(H)$-valued weakly continuous function on G, i.e., $\mathrm{Tr}(a\tilde{\Gamma}(\cdot))$ is continuous for every $a \in B(H)$. Here $\mathrm{Tr}(\cdot)$ is the trace in $T(H)$. In this case, $\{x(t)\}$ has an integral representation (4.2) for a unique $\xi \in rcagos(\mathfrak{B}_{\widehat{G}}, X)$.

We are interested in stationary dilations of $L_0^2(\Omega)$-valued or X-valued processes on G. An $L_0^2(\Omega)$-valued process $\{x(t)\}$ is said to have a *weakly stationary dilation* if there exist a Hilbert space $\mathfrak{K}$ containing $L_0^2(\Omega)$ as a closed subspace and a $\mathfrak{K}$-valued weakly stationary process $\{y(t)\}$ on G such that $x(t) = Jy(t)$ for $t \in G$, where $J : \mathfrak{K} \to L_0^2(\Omega)$ is the orthogonal projection. The triple $\{\{y(t)\}, \mathfrak{K}, J\}$ is also called a *weakly stationary dilation* of $\{x(t)\}$. Note that $\mathfrak{K}$ can be taken as an L^2-space $L_0^2(\tilde{\Omega})$ for some probability measure space $(\tilde{\Omega}, \tilde{\mathfrak{F}}, \tilde{\mu})$. Moreover, if $\eta \in rcaos(\mathfrak{B}_{\widehat{G}}, \mathfrak{K})$ is the representing measure of $\{y(t)\}$, then $\{x(t)\}$ is weakly harmonizable with the representing measure $\xi \in rca(\mathfrak{B}_{\widehat{G}}, L_0^2(\Omega))$ for which η is an o.s.d. of ξ.

Since every $\xi \in ca(\mathfrak{B}_{\widehat{G}}, L_0^2(\Omega))$ has an o.s.d. by Corollary 2.3, every $L_0^2(\Omega)$-valued weakly harmonizable process has a weakly stationary dilation, which was proved by Niemi [9] (see also Chang and Rao [1] and Rao [13]). If $\{x(t)\}$ is an X-valued scalarly weakly harmonizable process, then there exists a family $\{\{y_\phi(t)\}, L_0^2(\Omega_\phi), J_\phi\}_{\phi\in H}$ of weakly stationary dilations of $L_0^2(\Omega)$-valued processes $\{(x(t),\phi)_H\}_{\phi\in H}$. That is, for each $\phi \in H$, $(\Omega_\phi, \mathfrak{F}_\phi, \mu_\phi)$ is a probability measure space, $L_0^2(\Omega_\phi)$ may be assumed to contain $L_0^2(\Omega)$ as a closed subspace and $(x(t), \phi)_H = J_\phi y_\phi(t)$ for $t \in G$, where $J_\phi : L_0^2(\Omega_\phi) \to L_0^2(\Omega)$ is the orthogonal projection. In this case, $\{\{y_\phi(t)\}, L_0^2(\Omega_\phi), J_\phi\}_{\phi\in H}$ is called a *scalarly weakly stationary dilation* of $\{x(t)\}$.

DEFINITION 4.2 An X-valued process $\{x(t)\}$ on G is said to have an *operator stationary dilation* if there exist a normal Hilbert $B(H)$-module $Y = L_0^2(\tilde{\Omega}; H)$ containing X as a closed submodule and a Y-valued operator stationary process $\{y(t)\}$ such that $x(t) = Py(t)$ for $t \in G$, where $P : Y \to X$ is the gramian orthogonal projection and

$(\tilde{\Omega}, \tilde{\mathfrak{F}}, \tilde{\mu})$ is a probability measure space. The triple $\{\{y(t)\}, Y, P\}$ is also called an *operator stationary dilation* of $\{x(t)\}$.

DEFINITION 4.3 An $L_0^2(\Omega)$-valued process $\{x(t)\}$ on G is said to be *V-bounded* if

a) $x(\cdot) : G \to L_0^2(\Omega)$ is norm continuous,

b) $\{x(t)\}$ is bounded, i.e., $\sup\{\|x(t)\|_2 : t \in G\} < \infty$,

c) There exists a constant $C > 0$ such that

$$\left\| \int_G \varphi(t) x(t)\, dt \right\|_2 \le C \|\hat{\varphi}\|_\infty, \qquad \varphi \in L^1(G), \tag{4.3}$$

where $L^1(G)$ is the L^1-group algebra of G with the Haar measure dt, $\|\hat{\varphi}\|_\infty$ is the sup norm of the Fourier transform of φ, and the integral is in the sense of Bochner.

An X-valued process $\{x(t)\}$ on G is said to be *scalarly V-bounded* if, for each $\phi \in H$, the $L_0^2(\Omega)$-valued process $\{(x(t), \phi)_H\}$ is V-bounded in the above sense.

An X-valued process $\{x(t)\}$ is said to be *operator V-bounded* if

d) $x(\cdot) : G \to X$ is norm continuous,

e) $\{x(t)\}$ is bounded,

f) There exists a constant $C > 0$ such that

$$\left\| \int_G \Phi(t) x(t)\, dt \right\|_X \le C \|\hat{\Phi}\|_\infty, \qquad \Phi \in L^1(G\,; B(H)),$$

where $L^1(G\,;B(H))$ is the Banach space of all $B(H)$-valued functions on G which are Bochner integrable w.r.t. dt and, for $\Phi \in L^1(G\,;B(H))$, $\hat{\Phi} \in C_0(\widehat{G}\,;B(H))$ (the Banach space of all $B(H)$-valued norm continuous functions on G vanishing at infinity with the sup norm $\|\cdot\|_\infty$) is its Fourier transform, i.e.,

$$\hat{\Phi}(\chi) = \int_G \langle t, \chi\rangle\, \Phi(t)\, dt, \qquad \chi \in \widehat{G}.$$

If $\{x(t)\}$ is an X-valued weakly harmonizable process with the representing measure $\xi \in rca(\mathfrak{B}_{\widehat{G}}, X)$, then it has an operator stationary dilation if and only if ξ satisfies one of the conditions of (1) – (9) in Theorem 3.5. More generally, we have the following proposition.

PROPOSITION 4.4. *Let $\{x(t)\}$ be an X-valued pocess on G. Then the following statements are equivalent:*

(1) *$\{x(t)\}$ has an operator stationary dilation.*

(2) *$\{x(t)\}$ is weakly operator harmonizable.*

(3) *$\{x(t)\}$ is operator V-bounded.*

(4) *For some CONS $\{\phi_k\}_{k=1}^\infty$ in H there exists a scalarly weakly stationary dilation $\{\{y_k(t)\}, L_0^2(\Omega_k), J_k\}_{k=1}^\infty$ of $\{(x(t), \phi_k)_H\}_{k=1}^\infty$ such that $\sum_{k=1}^\infty \|y_k(t)\|_{2,k}^2 < \infty$ for $t \in G$, where $\|\cdot\|_{2,k}$ is the norm in $L_0^2(\Omega_k)$ for $k \ge 1$.*

(5) *For every CONS $\{\phi_k\}_{k=1}^\infty$ in H the conclusion of (4) holds.*

(6) $\{x(t)\}$ *is scalarly V-bounded and* C_{ϕ_k}*'s are chosen so that* $\sum_{k=1}^{\infty} C_{\phi_k}^2 < \infty$ *for every ON sequence* $\{\phi_k\}_{k=1}^{\infty}$ *in H, where, for each* $k \geq 1$, $C_{\phi_k} > 0$ *is the constant in (4.3) for the* $L_0^2(\Omega)$*-valued process* $\{(x(t), \phi_k)_H\}$.

Proof: The equivalence (1) $\Leftrightarrow$ (2) was proved in [5, 5.5 Theorem], (2) $\Leftrightarrow$ (3) in [4, 3.7. Theorem] and (1) $\Leftrightarrow$ (4) in [5, 5.7 Theorem]. (1) $\Leftrightarrow$ (5) is proved similarly.

(1) $\Rightarrow$ (6): Suppose that $\{x(t)\}$ has an operator stationary dilation $\{\{y(t)\}, Y, P\}$. We may assume that $Y = L_0^2(\tilde{\Omega}; H)$ for some probability measure space $(\tilde{\Omega}, \tilde{\mathfrak{F}}, \tilde{\mu})$. Let $\eta \in rcagos(\mathfrak{B}_{\widehat{G}}, Y)$ be the representing measure of $\{y(t)\}$ and put $\xi = P\eta \in rca(\mathfrak{B}_{\widehat{G}}, X)$, which is of bounded operator semivariation and the representing measure of $\{x(t)\}$. For $\phi \in H$ and $t \in G$ we put $x_\phi(t) = (x(t), \phi)_H$. Then, $\{x_\phi(t)\}$ is an $L_0^2(\Omega)$-valued weakly harmonizable process with the representing measure $\xi_\phi(\cdot) = (\xi(\cdot), \phi)_H \in rca(\mathfrak{B}_{\widehat{G}}, L_0^2(\Omega))$. It is well-known that an $L_0^2(\Omega)$-valued process is weakly harmonizable if and only if it is V-bounded (see e.g. [13]). Hence, $\{x(t)\}$ is scalarly V-bounded.

Let $\{\phi_k\}_{k=1}^{\infty}$ be any ON sequence in H and $\{\psi_k\}_{k=1}^{\infty}$ be a CONS in H containing $\{\phi_k\}_{k=1}^{\infty}$. Then we get for each $k \geq 1$

$$\begin{aligned}
\sup&\left\{\left\| \int_G \varphi(t) x_{\psi_k}(t)\, dt \right\|_2 : \varphi \in L^1(G),\ \|\hat{\varphi}\|_\infty \leq 1\right\} \\
&= \sup\left\{\left\| \int_{\widehat{G}} \hat{\varphi}(\chi)\, \xi_{\psi_k}(d\chi) \right\|_2 : \varphi \in L^1(G),\ \|\hat{\varphi}\|_\infty \leq 1\right\} \\
&= \sup\left\{\left\| \int_{\widehat{G}} f(\chi)\, \xi_{\psi_k}(d\chi) \right\|_2 : f \in C_0(\widehat{G}),\ \|f\|_\infty \leq 1\right\} \\
&= \|\xi_{\psi_k}\|(\widehat{G}) = \|J\eta_{\psi_k}\|(\widehat{G}) \qquad\qquad (4.4)\\
&\leq \|\eta_{\psi_k}\|(\widehat{G}) = \|\eta_{\psi_k}(\widehat{G})\|_2,
\end{aligned}$$

where $C_0(\hat{G})$ is the Banach space of all complex valued continuous functions on $\widehat{G}$ vanishing at infinity, $J : L_0^2(\tilde{\Omega}) \to L_0^2(\Omega)$ is the orthogonal projection and $\eta_\phi(\cdot) = (\eta(\cdot), \phi)_H \in rcaos(\mathfrak{B}_{\widehat{G}}, L_0^2(\tilde{\Omega}))$ for $\phi \in H$. Putting $C_{\phi_k} = \|\eta_{\phi_k}(\widehat{G})\|_2$ for $k \geq 1$, we see that

$$\sum_{k=1}^{\infty} C_{\phi_k}^2 = \sum_{k=1}^{\infty} \|\eta_{\phi_k}(\widehat{G})\|_2^2 \leq \sum_{k=1}^{\infty} \|\eta_{\psi_k}(\widehat{G})\|_2^2 = \|\eta(\widehat{G})\|_Y^2 < \infty,$$

since $\eta(\cdot) = \sum_{k=1}^{\infty} (\eta(\cdot), \psi_k)_H \psi_k$, where $\|\cdot\|_Y$ is the norm in Y.

(6) $\Rightarrow$ (1): Let $\{\psi_k\}_{k=1}^{\infty}$ be a CONS in H. Since $\{x(t)\}$ is scalarly V-bounded, it is scalarly weakly harmonizable. For each $k \geq 1$ let $\xi^{\psi_k} \in rca(\mathfrak{B}_{\widehat{G}}, L_0^2(\Omega))$ be the representing measure of the $L_0^2(\Omega)$-valued process $\{(x(t), \psi_k)_H\}$. Then, we have by (4.4) that $\|\xi^{\psi_k}\|(\widehat{G}) \leq C_{\psi_k}$. Define $\xi(\cdot) = \sum_{k=1}^{\infty} \xi^{\psi_k}(\cdot)\psi_k$. Then ξ is well-defined and in $rca(\mathfrak{B}_{\widehat{G}}, X)$

since (cf. Diestel and Uhl [2, p. 4])

$$\begin{aligned}\|\xi\|(\widehat{G})^2 &\le 16 \sup_{A\in\mathfrak{B}_{\widehat{G}}} \|\xi(A)\|_X^2 = 16 \sup_{A\in\mathfrak{B}_{\widehat{G}}} \sum_{k=1}^{\infty} \|\xi^{\psi_k}(A)\|_2^2 \\ &\le 16 \sum_{k=1}^{\infty} \|\xi^{\psi_k}\|(\widehat{G})^2 \le 16 \sum_{k=1}^{\infty} C_{\psi_k}^2 < \infty\end{aligned}$$

by assumption. Now we see that $\{x(t)\}$ is weakly harmonizable with the representing measure ξ since

$$\begin{aligned}x(t) &= \sum_{k=1}^{\infty} \big(x(t), \psi_k\big)_H \psi_k = \sum_{k=1}^{\infty} \int_{\widehat{G}} \langle t, \chi\rangle\, \xi^{\psi_k}(d\chi) \\ &= \int_{\widehat{G}} \langle t, \chi\rangle \sum_{k=1}^{\infty} \xi^{\psi_k}(d\chi)\psi_k = \int_{\widehat{G}} \langle t, \chi\rangle\, \xi(d\chi), \qquad t \in G.\end{aligned}$$

Let $\{\phi_k\}_{k=1}^{\infty}$ be any ON sequence in H. Then, we have $\sum\limits_{k=1}^{\infty} \|\xi_{\phi_k}\|(\widehat{G})^2 \le \sum\limits_{k=1}^{\infty} C_{\phi_k}^2 < \infty$, where $\xi_\phi(\cdot) = \big(\xi(\cdot), \phi\big)_H \in rca\big(\mathfrak{B}_{\widehat{G}}, L_0^2(\Omega)\big)$ for each $\phi \in H$. Consequently, by Proposition 3.4, ξ has a g.o.s.d., i.e., $\{x(t)\}$ has an operator stationary dilation.

References

[1] D. K. Chang and M. M. Rao, Bimeasures and nonstationary processes, *in: Real and Stochastic Analysis,* Ed. by M. M. Rao, John Wiley & Sons, New York, pp. 7–118, 1986.

[2] J. Diestel and J. J. Uhl, Jr., *Vector Measures,* Amer. Math. Soc., Providence, R. I., 1977.

[3] A. Grothendiek, Résumé de la théorie métrique des produits tensoriels topologiques, *Bol. Soc. Mat. Sao Paulo* **8** (1956), 1–79.

[4] Y. Kakihara, A note on harmonizable and V-bounded processes, *J. Multivar. Anal.* **16** (1985), 140–156.

[5] Y. Kakihara, A classification of vector harmonizable processes, *Stochastic Anal. Appl.* **10** (1992), 277–311.

[6] Y. Kakihara, *Multidimensional Second Order Stochastic Processes,* World Scientific, in preparation.

[7] J. Lindenstrauss and A. Pełczyński, Absolutely summing operators in $\mathfrak{L}_p$-spaces and their applications, *Studia Math.* **29** (1968), 275–326.

[8] A. Makagon and H. Salehi, Spectral dilation of operator-valued measures and its application to infinite dimensional harmonizable processes, *Studia Math.* **85** (1987), 257–297.

[9] H. Niemi, On stationary dilations and the linear prediction of certain stochastic processes, *Soc. Sci. Fenn. Comment. Phys.-Math.* **45** (1975), 111-130.

[10] H. Niemi, On orthogonally scattered dilations of bounded vector measures, *Ann. Acad. Sci. Fenn. Ser. A I Math.* **3** (1977), 43–52.

[11] H. Niemi, Orthogonally scattered dilations of finitely additive vector measures with values in a Hilbert space, *in: Prediction Theory and Harmonic Analysis, The Pesi*

Masani Volume, Ed. by V. Mandrekar and H. Salehi, North-Holland, New York, pp. 233–251, 1983.

[12] M. Ozawa, Hilbert $B(H)$-modules and stationary processes, *Kodai Math. J.* **3** (1980), 26–39.

[13] M. M. Rao, Harmonizable processes: structure theory, *L'Enseign. Math.* **28** (1982), 295–351.

[14] M. Rosenberg, Quasi-isometric dilations of operator-valued measures and Grothendieck's inequality, *Pacific J. Math.* **103** (1981), 135–161.

[15] R. Schatten, *Norm Ideals of Completely Continuous Operators,* Springer, New York, 1960.

[16] W. Słowikowski, Absolutely 2-summing mappings from and to Hilbert spaces and a Sudakov Theorem, *Bull. Acad. Polon. Sci. Sér. Sci. Math. Astronom. Phys.* **17** (1969), 381–386.

[17] B. Truong-van, Une généralisation du théorème de Kolmogorov-Aronszajn processus V-bornés q-dimensionnels: domaine spectral § dilatations stationnaires, *Ann. Inst. Henri Poincaré Sec.* **B 17** (1981), 31–49.

Transient Solution of the M/M/1 Queueing System via Randomization

ALAN KRINIK — Department of Mathematics, California State Polytechnic University, Pomona, California 91768

DANIEL MARCUS — Department of Mathematics, California State Polytechnic University, Pomona, California 91768

DAN KALMAN — Department of Mathematics and Statistics, American University, Washington D. C. 20016

TERRY CHENG — Department of Mathematics, Irvine Valley Community College, Irvine, California 92720

1 INTRODUCTION

A. K. Erlang's [3] single server queueing system M/M/1 is the prototype of modern queueing models. Students of queueing theory generally begin their studies of queueing systems by learning about the M/M/1 system. Mathematical and statistical research on M/M/1 is quite extensive (see, for example, [7,8,10,14,15]) and remains a rich source of active research interest today. The total number of customers in the M/M/1 queueing system is a birth-death process having constant birth (arrival) rates λ and constant death (service) rates μ. The state space is the natural numbers which represents the possible number of customers in the system at any point in time. Transitions within the M/M/1 queueing sytem are pictured according to the following state rate diagram.

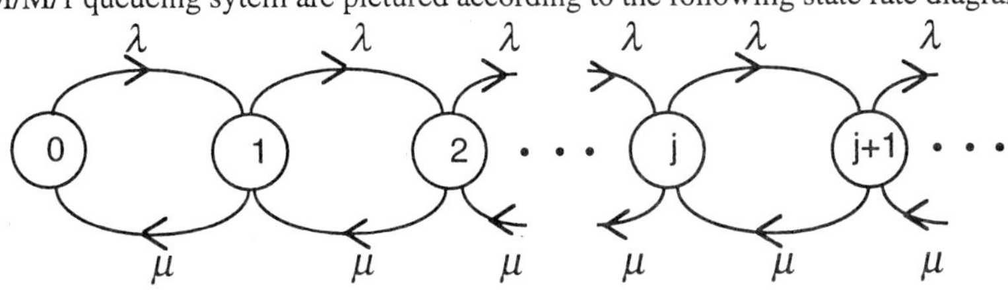

Diagram 1

Alternatively, M/M/1 may be characterized as having independent and identically distributed interarrival times that are exponentially distributed with parameter λ and independent and identically distributed service times that are exponentially distributed with parameter μ .

In this article we consider the fundamental problem of determining the transient probability functions, $p_{i,j}(t)$, which represent the probability of going from state i to state j in time t. These functions are known [7,10] to satisfy the following infinite system of differential equations.

$$p'_{i,0}(t) = -\lambda p_{i,0}(t) + \mu p_{i,1}(t)$$

$$p'_{i,j}(t) = \lambda p_{i,j-1}(t) - (\lambda + \mu) p_{i,j}(t) + \mu p_{i,j+1}(t)$$

for i=0,1,2,3,...and j=1,2,3,... where λ and μ are positive constants. This problem for the M/M/1 queueing system was first solved in the early 1950's [1,4,6,13] - decades after Erlang's pioneering work. Even after the solution was known, researchers have continued to develop new solution methods (cf. [8, 12, 15]) to achieve better understanding of this process. Interests in the transient probabilities of M/M/1 is strong for a number of reasons: the central (and influential) role played by M/M/1 in queueing theory, increasing realization of the importance of transient information, the inherent complicated (and interesting) nature of existing solution methods which challenge us to develop, when possible, simpler solution methods and alternative solution forms that are numerically stable and inviting to practitioners.

Interestingly, also starting in the early 1950's, a general method to numerically determine transient probability functions in Markov processes known as randomization (or uniformization) was introduced; see [9] for history. Since its inception, randomization has been gaining recognition as an effective, numerically stable method to calculate transient probabilities. The importance of randomization is its use in numerically solving for transient probabilities in complicated queueing systems where analytic solutions are unknown or intractable. Essentially, randomization shifts the problem of determining transition probability functions in time t to finding n-step transition probabilities of a suitable, embedded Markov chain (see [9,10,11,14]). Our purpose in this article is to report a solution of the transition probability functions of M/M/1 using the randomization method. The key idea is to count certain sets of sample paths of the embedded Markov chain associated with M/M/1 by making a one to one correspondence with another class of sample paths which can be counted using the reflection principle. Our solution has a different form than the usual solution of M/M/1 given in terms of Bessel functions (cf. [7,10]). In place of using Bessel function theory, Laplace transforms, generating functions and complex analysis, we rely upon having the randomization form of the solution and counting sample paths via coding and the reflection principle.

2 SOLUTIONS OF M/M/1 VIA RANDOMIZATION

The governing system of differential equations for $p_{i,j}(t)$ may be written in infinite matrix notation as

$$P'(t) = P(t)\,Q$$

where

$$P(t) = \left[p_{i,j}(t)\right] \qquad i, j = 0, 1, 2, \ldots \qquad \text{and}$$

$$Q = \begin{bmatrix} -\lambda & \lambda & 0 & \cdots & \cdots \\ \mu & -(\lambda+\mu) & \lambda & 0 & \cdots \\ 0 & \mu & -(\lambda+\mu) & \lambda & \cdots \\ \vdots & \vdots & \vdots & \vdots & \\ \vdots & \vdots & \vdots & \vdots & \end{bmatrix}$$

A solution of this system by randomization (cf. [10,14]) is given as

$$P(t) = e^{-(\lambda+\mu)t} \sum_{N=0}^{\infty} \frac{(\lambda+\mu)^N t^N}{N!} P^N$$

where

$$P = \frac{1}{\lambda+\mu} \begin{bmatrix} \mu & \lambda & 0 & 0 & \cdots \\ \mu & 0 & \lambda & 0 & \cdots \\ 0 & \mu & 0 & \lambda & \cdots \\ \vdots & \vdots & \vdots & \vdots & \vdots \end{bmatrix}$$

is a stochastic matrix corresponding to the following embedded Markov chain.

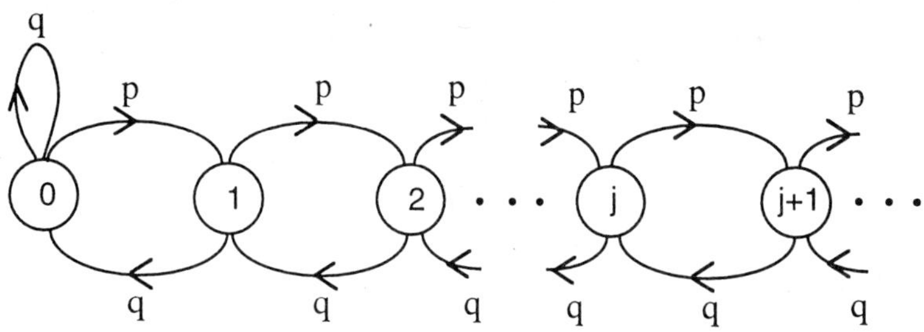

Diagram 2

$$\text{with} \quad p = \frac{\lambda}{\lambda + \mu} \quad \text{and} \quad q = \frac{\mu}{\lambda + \mu} \, .$$

In particular,

$$p_{i,j}(t) = e^{-(\lambda+\mu)t} \sum_{N=0}^{\infty} \frac{(\lambda+\mu)^N t^N}{N!} p_{i,j}^{(N)}$$

for i, j = 0, 1, 2,... where $p_{i,j}^{(N)}$ is the probability of going from state i to state j in N steps for the embedded Markov chain pictured in Diagram 2. Therefore, we have an explicit solution of $p_{i,j}(t)$ for the M/M/1 queueing system once we have an expression for $p_{i,j}^{(N)}$. This is accomplished in the following theorem.

THEOREM 1 Suppose i and j are any two states: i, j = 0, 1, 2,... Assume N = 1, 2, 3,... is chosen and let $c = i + j$ and $d = j - i$ and $p = \frac{\lambda}{\lambda + \mu}$ and $q = \frac{\mu}{\lambda + \mu}$. Then $p_{i,j}^{(N)}$, the N-step transition probability from state i to state j for the embedded Markov chain in Diagram 2 is given by

$$p_{i,j}^{(N)} = \left\{ \binom{N}{\frac{N+d}{2}} - \binom{N}{\frac{N+c+2}{2}} \right\} q^{\frac{N-d}{2}} p^{\frac{N+d}{2}} +$$

$$\sum_{m=1}^{\frac{N-d}{2}-i} \left\{ \left\{ \binom{N}{\frac{N+c+2m}{2}} - \binom{N}{\frac{N+c+2m+2}{2}} \right\} q^{\frac{N-d}{2}+m} p^{\frac{N+d}{2}-m} \right\}$$

when N + d is even and

$$p_{i,j}^{(N)} = \sum_{m=0}^{\frac{N-d-1}{2}-i} \left\{ \left\{ \binom{N}{\frac{N+c+2m+1}{2}} - \binom{N}{\frac{N+c+2m+3}{2}} \right\} q^{\frac{N-d+1}{2}+m} p^{\frac{N+d-1}{2}-m} \right\}$$

when N + d is odd.

Note, we adopt the convention that $\binom{N}{M} = 0$ if $M > N$ or $M < 0$.

PROOF. The number of sample paths in Diagram 2 from state i to state j in N steps having either 2m or 2m+1 loops at state zero is provided in the following lemma. This counting result allows one to determine the contribution to $p_{i,j}^{(N)}$ from sample paths having a prescribed number of loops at state zero.

For any finite sequence of numbers $S = s_1 s_2 s_3 \ldots s_N$ we adopt the notation that S_k means $\sum_{n=1}^{k} s_n$. The following definitions are motivated by classifying sample paths from i to j in Diagram 2 according to whether we loop at zero exactly m times, merely touch state zero without looping or never even visit state zero during our journey from i to j.

DEFINITION 2 Suppose i, j are two arbitrary states from 0,1,2,3, ... and assume N=1,2,3,... is given. A sample path of length N from state i to state j in Diagram 2 may be represented by a finite sequence of N elements $A = a_1 a_2 a_3 \ldots a_N$ such that $A_0 = 0$ by convention, $A_N = j - i$ and

$$a_n = \begin{cases} 0 \text{ or } 1 & \text{if } A_{n-1} = -i \\ -1 \text{ or } 1 & \text{if } A_{n-1} > -i \end{cases} \qquad \text{for } n=1,2,\ldots N.$$

The set of all sample paths, A, of length N from i to j having exactly m of its elements equal to zero (that is, $a_{n_k} = 0$ for k=1,2,...m for some subsequence of a_n) is denoted by $S_{i,j}^N(m)$ where $m \geq 1$. $S_{i,j}^N(0)$ will denote those sample paths A of length N from i to j for which $a_n \neq 0$ for each n=1,2,...,N and yet $A_n = -i$ for some n=1,2,...,N-1. Finally, let $R_{i,j}^N$ represent all the remaining sample paths A of length N from i to j; that is, those sample paths A that go from i to j in N steps such that $A_n \neq -i$ for each n=1,2,...,N-1.

The next proposition establishes that $S_{i,j}^N(m)$ and $S_{i,j+m}^N(0)$ have the same number of sample paths for m=0,1,..., N-i-j. The following graphs illustrate how this one-to-one correspondence may be visualized when N=9, i=1, j=2 and m=0,1 or 2. Start with a sample path in $S^9_{1,2}$ (0) such as shown in Figure 1. If we change the right most "zero to one" segment (dashed in Figure 1) into a horizontal segment and lower the graph to the right of this segment one unit, we obtain the sample path in $S^9_{1,1}$ (1) pictured in Figure 2. Repeat the process again by changing the right most "zero to one" segment of Figure 2 (the dashed segment) into another horizontal segment and once again lower the part of the graph to the right of the new horizontal segment one unit to obtain the sample path in $S^9_{1,0}$ (2) of Figure 3. These operations are easy to reverse by successively replacing horizontal segments by "zero to one" segments and raising the graph to the right of each replacement one unit. The reasoning as to why this procedure is a one-to-one correspondence is contained within the proof of the following proposition.

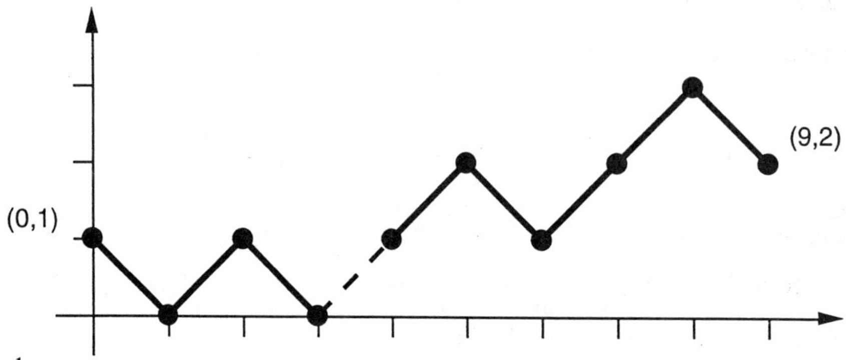

Figure 1

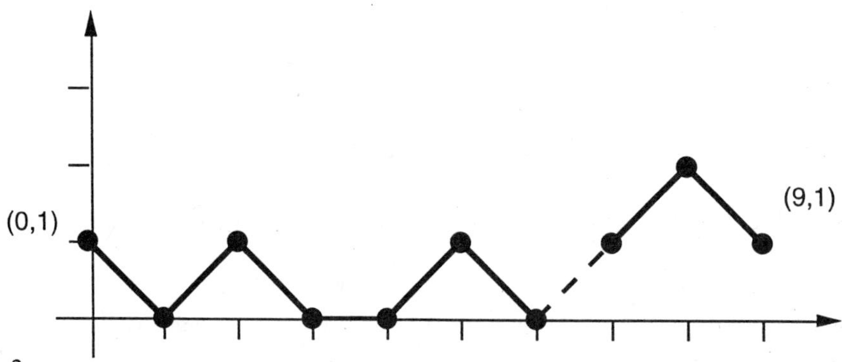

Figure 2

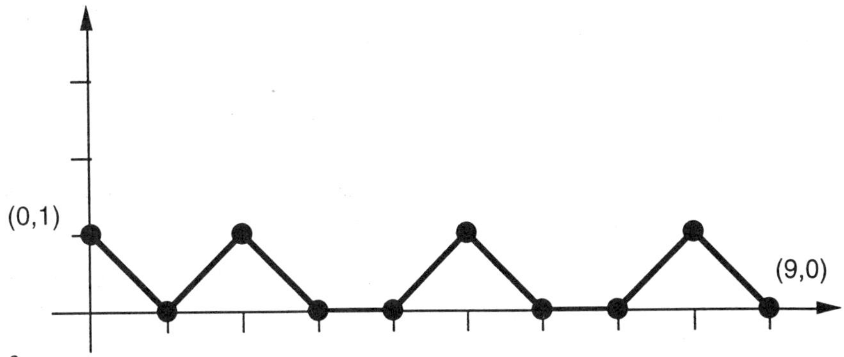

Figure 3

PROPOSITION 3 Suppose i, j are arbitrary states of the embedded Markov chain in Diagram 2. Assume N = 1, 2, 3,..., then

(a) $S^N_{i,0}$ (k) and $S^N_{i,k}$ (0) have the same number of sample paths for k=0,1,..., N-i.

(b) $S^N_{i,j}$ (m) and $S^N_{i,j+m}$ (0) have the same number of sample paths for m=0,1,..., N-i-j.

PROOF.

(a) Define functions f and g as follows.

$f: S^N_{i,0}(k) \rightarrow S^N_{i,k}(0)$ such that for $A \in S^N_{i,0}(k)$, $f(A) = B = b_1\, b_2 \ldots b_N$ where

$b_n = \begin{cases} -1 & \text{if} \quad a_n = -1 \\ 1 & \text{otherwise} \end{cases}$. In words, f changes all zero entries of A into ones. Define

$g: S^N_{i,k}(0) \rightarrow S^N_{i,0}(k)$ so that for $B \in S^N_{i,k}(0)$, $g(B) = A = a_1\, a_2 \ldots a_N$ where

$$a_n = \begin{cases} -1 & \text{if} \quad b_n = -1 \\ 0 & \text{if} \quad b_n = 1 \text{ and } B_{n-1} < B_r \quad \text{for each } r = n, n+1, \ldots, N \\ 1 & \text{otherwise} \end{cases}.$$

The function g changes k one entries to zero and keeps everything else the same.

To see that f is one to one, argue by contradiction. That is, suppose there exists $A \neq A'$ with $f(A) = f(A')$. Let $A = a_1\, a_2 \ldots a_N$ and $A' = a'_1\, a'_2 \ldots a'_N$ and let h be the largest subscript such that $a_h \neq a'_h$. Neither a_h nor a'_h can be -1 because $f(A) = f(A')$ and the definition of f would force both to be -1 which contradicts $a_h \neq a'_h$. So $a_h = 0$ and $a'_h = 1$ or $a_h = 1$ and $a'_h = 0$. Without loss of generality assume $a_h = 0$ and $a'_h = 1$. However, $A_N = A'_N = -i$, implying $A_h = A'_h$ and $A_{h-1} = -i$ by Definition 2 of $S^N_{i,0}$ (k). But this means $A'_{h-1} = -i-1$, an impossibility since $A'_{h-1} \geq -i$ by Definition 2. Therefore, $f(A) \neq f(A')$ and f is one to one. Clearly $f(g(B)) = B$

for all B $\in S^N_{i,k}$ (0) which means f is also onto. This completes the proof that f is a one to one correspondence between $S^N_{i,0}$ (k) and $S^N_{i,k}$ (0).

(b) Define f and g as in part (a). Again f: $S^N_{i,j}$ (m) $\rightarrow S^N_{i,j+m}$ (0) changes zeroes to ones and g: $S^N_{i,j+m}$ (0) $\rightarrow S^N_{i,0}$ (j+m) is as before. The same argument given in part (a) applies to give that f is one to one. To check that f is also onto, define a new function

h: $S^N_{i,0}$ (j+m) $\rightarrow S^N_{i,j}$ (m) as follows. If A $\in S^N_{i,0}$ (j+m) then A has j+m zero entries. That is, there is a subsequence $n_1 < n_2 < \ldots < n_{j+m}$ such that $a_{n_r} = 0$ if and only if r=1,2,...,j+m. Define h(A)=B where B=$b_1 b_2 \ldots b_N$ such that

$$b_n = \begin{cases} 1 & \text{if} \quad n = n_r \quad \text{where} \quad r = 1+m, 2+m, \ldots, j+m \\ a_n & \text{otherwise} \end{cases}$$

In other words, h changes the last j zeroes into ones. Then for any A $\in S^N_{i,j+m}$ (0), f(h(g(A))) = A , so f is also onto. This establishes the one to one correspondence between $S^N_{i,j}$ (m) and $S^N_{i,j+m}$ (0).

REMARK 4 A related result to the preceding proposition appears in [8, pp. 10-17].

LEMMA 5 Suppose i, j are arbitrary states for the embedded Markov chain drawn in Diagram 2 with i, j = 0, 1, 2, 3,... . Assume N=1, 2, 3,... and let c = i + j and d = j - i, then

(a) $S^N_{i,j}(0) \cup R^N_{i,j}$ has $\left\{ \binom{N}{\frac{N+d}{2}} - \binom{N}{\frac{N+c+2}{2}} \right\}$ sample paths, provided N+d is even.

(b) $S^N_{i,j}(2m)$ has $\left\{ \binom{N}{\frac{N+c+2m}{2}} - \binom{N}{\frac{N+c+2m+2}{2}} \right\}$ sample paths, for N+c even

and where m=1,2,..., $\frac{N-c}{2}$ and

$$S^{N}_{i,j}\,(2m+1) \text{ has } \left\{ \binom{N}{\frac{N+c+2m+1}{2}} - \binom{N}{\frac{N+c+2m+3}{2}} \right\} \text{ sample paths,}$$

for $N + c$ odd, where $m=0,1,\ldots,\dfrac{N-c-1}{2}$.

PROOF. Part (a) follows directly from the reflection principle. The number of sample paths within $S^{N}_{i,j}$ (2m), and $S^{N}_{i,j}$ (2m+1) equals the number of sample paths within $S^{N}_{i,j+2m}$ (0), and $S^{N}_{i,j+2m+1}$ (0) respectively by Proposition 3, part (b). The results then follow using the reflection principle; see [5] for additional details.

REMARK 6 The use of lattice path combinatorics to determine the n-step transition probabilities of embedded Markov chains for certain birth-death processes is very promising. The preceding randomization/combinatoric approach also produces analytic solutions of the transition probabilities for the M/M/2 queueing system. These results and related matters will be presented elsewhere.

REFERENCES

1. Bailey, N. T. J. (1954). A continuous time treatment of a simple queue using generating functions, J. Roy. Statist. Soc. Ser. B, 16: 288-291.
2. Bhattacharya, R. N. and Waymire, E. C. (1990). Stochastic Processes with Applications, Wiley, New York.
3. Brockmeyer, E., Halstrom, H. L. and Jensen, A. (1948). The life and works of A.K. Erlang, Transactions of the Danish Academy of Technical Sciences, 2: 1-277.
4. Champernowne, D. G. (1956). An elementary method of solution of the queueing problem with a single server and continuous parameter, J. Roy. Statist. Soc. Ser. B, 18: 125-128.
5. Cheng, T. (1994). Reflecting Paths in the Single Server Queueing System, Masters Thesis, California State Polytechnic University, Pomona.
6. Clark, A. B. (1953). The time dependent waiting line problem, Report M729-IR39, Univ. Michigan.
7. Cohen, J. W. (1982) The Single Server Queue, North-Holland Ser. Appl. Math. Mech. 8, North-Holland, Amsterdam, 2nd edition.
8. Conolly, B. (1975). Lecture Notes on Queueing Systems, Wiley, New York.
9. Grassmann, W.K. (1991). Finding transient solutions in Markovian event systems through randomization, Numerical Solution of Markov Chains (W. J. Stewart, ed.), Marcel Dekker, New York.
10. Gross, D. and Miller, D. R. (1984). The randomization technique as a modeling tool and solution procedure for transient Markov processes, Oper. Res. 32 : 343-361.
11. Gross, D. and Harris, C. M. (1985). Fundamentals of Queueing Theory, Wiley, New York, 2nd edition.
12. Krinik, A. (1992). Taylor series solution of the M/M/1 queueing system, J. Comp. Appl. Math. 44: 371-380.
13. Ledermann, W. and Reuter, G. E. H. (1954). Spectral theory for the differential equations of simple birth and death processes, Phil. Trans. Roy. Soc. London Ser. A 246: 321-369.
14. Medhi, J. (1991). Stochastic Models in Queueing Theory, Academic Press, Boston.
15. Parthasarathy, P. R. (1987). A transient solution to an M/M/1 queue: a simple approach, Adv. Appl. Prob., 19: 997-998.

A Characterization of Hida Measures

HUI-HSIUNG KUO* Department of Mathematics, Louisiana State University, Baton Rouge, LA 70803 (kuo@math.lsu.edu)

1. WHAT IS A HIDA MEASURE?

Let $\mathcal{S}$ be the Schwartz space of real-valued rapidly decreasing functions on $\mathbb{R}$. It is a nuclear space with the topology given by the family of norms

$$|\xi|_p = |A^p \xi|_0, \quad p \geq 1,$$

where A is the operator $A = -d^2/dx^2 + x^2 + 1$ and $|\cdot|_0$ is the $L^2(\mathbb{R})$-norm. Let $\mathcal{S}'$ denote the dual space of $\mathcal{S}$. The σ-fields on $\mathcal{S}'$ generated by the weak topology, the strong topology, and the inductive limit topology are all the same, which is regarded as the Borel field of $\mathcal{S}'$. A Hida measure is a certain Borel measure on $\mathcal{S}'$ to be defined below.

For each $p \geq 1$, let $\mathcal{S}_p$ denote the completion of $\mathcal{S}$ with respect to the norm $|\cdot|_p$. The dual space $\mathcal{S}'_p$ of $\mathcal{S}_p$ can be identified as the completion of $L^2(\mathbb{R})$ with respect to the norm

$$|x|_{-p} = |A^{-p} x|_0$$

so that $\mathcal{S}'$ is the inductive limit of $\{\mathcal{S}'_p; p \geq 1\}$. Let $\mathcal{S}'_c$ and $\mathcal{S}'_{p,c}$ denote the complexification of $\mathcal{S}'$ and $\mathcal{S}'_p$, respectively.

For any $p \geq 1$ and $0 \leq \beta < 1$, let $\mathcal{A}_{p,\beta}$ denote the space of all complex-valued function φ on $\mathcal{S}'_c$ satisfying the following two conditions:

* Research supported by the U.S. Army Research Office grant DAAH04-94-G-0249

(a) φ is an analytic function on $\mathcal{S}'_{p,c}$.
(b) There exists some constant $C \geq 0$ such that

$$|\varphi(x)| \leq C \exp\Big[\frac{1}{2}(1+\beta)|x|_{-p}^{\frac{2}{1+\beta}}\Big], \quad \forall x \in \mathcal{S}'_{p,c}.$$

For $\varphi \in \mathcal{A}_{p,\beta}$, define $\|\varphi\|_{p,\beta}$ by

$$\|\varphi\|_{p,\beta} = \sup_{x \in \mathcal{S}'_{p,c}} |\varphi(x)| \exp\Big[-\frac{1}{2}(1+\beta)|x|_{-p}^{\frac{2}{1+\beta}}\Big].$$

Obviously, $\mathcal{A}_{p,\beta}$ is a Banach space with the norm $\|\cdot\|_{p,\beta}$. Moreover, $\mathcal{A}_{p,\beta} \subset \mathcal{A}_{q,\beta}$ for any $p \geq q \geq 1$ and the inclusion is continuous. For fixed $0 \leq \beta < 1$, define

$$\mathcal{A}_\beta = \bigcap_{p \geq 1} \mathcal{A}_{p,\beta}.$$

We endow $\mathcal{A}_\beta$ with the projective limit topology, i.e., the coarsest topology on $\mathcal{A}_\beta$ such that the inclusion $\mathcal{A}_\beta \subset \mathcal{A}_{p,\beta}$ is continuous for each $p \geq 1$. It is easy to check that $\mathcal{A}_\beta$ is a Fréchet space, i.e., a complete metrizable locally convex space (see Reed and Simon [5]). Let $\mathcal{A}_\beta^*$ and $\mathcal{A}_{p,\beta}^*$ denote the dual spaces of $\mathcal{A}_\beta$ and $\mathcal{A}_{p,\beta}$, respectively. Note that we have

$$\mathcal{A}_\beta^* = \bigcup_{p \geq 1} \mathcal{A}_{p,\beta}^*.$$

We call $\mathcal{A}_\beta$ the space of *test functions* on $\mathcal{S}'$. Its dual $\mathcal{A}_\beta^*$ is the space of *generalized functions* on $\mathcal{S}'$. For the case $\beta = 0$, these spaces have been introduced by Lee in [4].

A function φ in $\mathcal{A}_{p,\beta}$ is defined on $\mathcal{S}'_c$. We will use the same notation φ to denote its restriction to $\mathcal{S}'$.

Definition. A Borel measure ν on $\mathcal{S}'$ is called a *Hida measure* of order β if $\varphi \in L^1(\nu)$ for all $\varphi \in \mathcal{A}_\beta$ and the linear functional $\varphi \to \int_{\mathcal{S}'} \varphi(x)\, d\nu(x)$ is continuous on $\mathcal{A}_\beta$.

If ν is a Hida measure of order β, then it induces a generalized function, denoted by Φ_ν, in $\mathcal{A}_\beta^*$ such that

$$\Phi_\nu(\varphi) = \int_{\mathcal{S}'} \varphi(x)\, d\nu(x), \quad \varphi \in \mathcal{A}_\beta.$$

To give a simple example of Hida measure, note that by the Minlos theorem there exists a unique probability measure μ on $\mathcal{S}'$ such that

$$\int_{\mathcal{S}'} e^{i\langle x,\xi\rangle}\, d\mu(x) = \exp\Big[-\frac{1}{2}|\xi|_0^2\Big], \quad \xi \in \mathcal{S}.$$

This measure μ is referred as the standard Gaussian measure on $\mathcal{S}'$. It is easy to check that μ is a Hida measure of order $\beta = 0$.

A Hida measure of nonzero order has been introduced by Kondratiev and Streit [1] (see also Kuo [3]). Let $L_\lambda(t)$, $0 < \lambda \leq 1$, be the Mittag-Leffler function

$$L_\lambda(t) = \sum_{n=0}^{\infty} \frac{(-t)^n}{\Gamma(1+\lambda n)}, \quad t \geq 0,$$

where Γ is the Gamma function. The function $L_\lambda(|\xi|_0^2)$, $\xi \in \mathcal{S}$, is positive definite. Hence by the Minlos theorem there exists a unique probability measure ν_λ such that

$$\int_{\mathcal{S}'} e^{i\langle x,\xi\rangle}\, d\nu_\lambda(s) = L_\lambda(|\xi|_0^2), \quad \xi \in \mathcal{S}.$$

It can be shown that ν_λ is a Hida measure of order $1-\lambda$. For details, see Kondratiev and Streit [1] or Kuo [3].

2. A CHARACTERIZATION THEOREM

Note that the inclusion $L^2(\mathbb{R}) \subset \mathcal{S}'_p$ is a Hilbert-Schmidt operator for any $p > \frac{1}{2}$. Hence $(L^2(\mathbb{R}), \mathcal{S}'_p)$ is an abstract Wiener space for such p (see Kuo [2]). Thus the standard Gaussian measure μ is supported on $\mathcal{S}'_p$ for any $p > \frac{1}{2}$. Then by the Fernique theorem (see Kuo [2]), there exists some constant $a > 0$ such that

$$\int_{\mathcal{S}'_p} e^{a|x|_{-p}^2}\, d\mu(x) < \infty.$$

This type of integrability can be used to characterize Hida measures on $\mathcal{S}'$.

THEOREM. *A Borel measure ν on $\mathcal{S}'$ is a Hida measure of order β if and only if ν is supported in $\mathcal{S}'_p$ for some $p \geq 1$ and*

$$\int_{\mathcal{S}'_p} \exp\left[\frac{1}{2}(1+\beta)|x|_{-p}^{\frac{2}{1+\beta}}\right] d\nu(x) < \infty.$$

This theorem has been proved for the case $\beta = 0$ by Lee in [4]. However, the proof for the necessity part in [4] cannot be adapted to the case $\beta \neq 0$.

To prove the sufficiency of the above theorem, let $\varphi \in \mathcal{A}_\beta$. Then

$$\begin{aligned}
&\int_{\mathcal{S}'_p} |\varphi(x)|\, d\nu(x) \\
&= \int_{\mathcal{S}'_p} \Big(|\varphi(x)| \exp\Big[-\frac{1}{2}(1+\beta)|x|_{-p}^{\frac{2}{1+\beta}}\Big]\Big) \exp\left[\frac{1}{2}(1+\beta)|x|_{-p}^{\frac{2}{1+\beta}}\right] d\nu(x) \\
&\leq \|\varphi\|_{p,\beta} \int_{\mathcal{S}'_p} \exp\left[\frac{1}{2}(1+\beta)|x|_{-p}^{\frac{2}{1+\beta}}\right] d\nu(x).
\end{aligned}$$

This implies that the linear functional

$$\Phi_\nu(\varphi) \equiv \int_{\mathcal{S}'_p} |\varphi(x)|\, d\nu(x), \quad \varphi \in \mathcal{A}_\beta,$$

is continuous on $\mathcal{A}_\beta$. Hence $\Phi_\nu \in \mathcal{A}_\beta^*$ and so ν is a Hida measure of order β.

For the necessity part, we only sketch the proof. For details, see Kuo [3]. Suppose ν is a Hida measure on $\mathcal{S}'$ of order β. Then it induces a generalized function Φ_ν in $\mathcal{A}_\beta^*$. Since $\mathcal{A}_\beta^* = \cup_{q\geq 1}\mathcal{A}_{q,\beta}^*$, there exists some $q \geq 1$ such that $\Phi_\beta \in \mathcal{A}_{q,\beta}^*$ and

$$\Phi_\nu(\varphi) = \int_{\mathcal{S}'} \varphi(x)\, d\nu(x), \quad \varphi \in \mathcal{A}_{q,\beta}. \tag{1}$$

Define a function ψ on $\mathcal{S}'_{q,c}$ by

$$\psi(x) = \sum_{n=0}^{\infty} \frac{1}{(2^n n!)^{1+\beta}} |x|_{-q}^{2n}.$$

It is easy to check that ψ is an analytic function on $\mathcal{S}'_{q,c}$. By Lemma 6.6 in Kuo [3] we have the following inequality for any $\gamma \geq 1$:

$$\sum_{n=0}^{\infty} \left(\frac{u^n}{n!}\right)^\gamma \leq e^{\gamma u}, \quad \forall u \geq 0.$$

We can use this inequality to obtain that

$$\begin{aligned}\psi(x) &= \sum_{n=0}^{\infty} \left(\frac{1}{n!}\left[\frac{1}{2}|x|_{-q}^{\frac{2}{1+\beta}}\right]^n\right)^{1+\beta} \\ &\leq \exp\left[\frac{1}{2}(1+\beta)|x|_{-q}^{\frac{2}{1+\beta}}\right].\end{aligned}$$

Therefore, $\psi \in \mathcal{A}_{q,\beta}$ and by equation (1) we have

$$\Phi_\nu(\psi) = \int_{\mathcal{S}'} \psi(x)\, d\nu(x). \tag{2}$$

On the other hand, by Lemma 15.16 in Kuo [3] we have the following inequality for any $r > 0$:

$$e^{ru/4} \leq e^{3r/2} \sum_{n=0}^{\infty} \left(\frac{u^n}{2^n n!}\right)^r, \quad \forall u \geq 0.$$

By letting $r = 1+\beta$ and $u = |x|_{-q}^{\frac{2}{1+\beta}}$, we get

$$\exp\left[\frac{1+\beta}{4}|x|_{-q}^{\frac{2}{1+\beta}}\right] \leq e^{3(1+\beta)/2}\,\psi(x), \quad x \in \mathcal{S}'_{q,c}. \tag{3}$$

It follows from equations (2) and (3) that

$$\int_{\mathcal{S}'} \exp\left[\frac{1+\beta}{4}|x|_{-q}^{\frac{2}{1+\beta}}\right] d\nu(x) \leq e^{3(1+\beta)/2}\,\Phi_\nu(\psi). \tag{4}$$

Now note that the inverse operator of $A = -d^2/dx^2 + x^2 + 1$ acting on $L^2(\mathbb{R})$ is continuous and its operator norm is given by $\|A^{-1}\| = 2^{-1}$. Hence for any $p \geq q \geq 1$ we have

$$\begin{aligned}|x|_{-p} &= |A^{-p}x|_0 \\ &= |A^{-(p-q)}A^{-q}x|_0 \\ &\leq 2^{-(p-q)}|A^{-q}x|_0 \\ &= 2^{-(p-q)}|x|_{-q}.\end{aligned}$$

We can choose large p such that $p > q + \frac{1+\beta}{2}$. Then

$$\int_{S'} \exp\left[\frac{1}{2}(1+\beta)|x|_{-p}^{\frac{2}{1+\beta}}\right] d\nu(x) \le \int_{S'} \exp\left[\frac{1+\beta}{4}|x|_{-q}^{\frac{2}{1+\beta}}\right] d\nu(x).$$

Thus by equation (4) we have

$$\int_{S'} \exp\left[\frac{1}{2}(1+\beta)|x|_{-p}^{\frac{2}{1+\beta}}\right] d\nu(x) \le e^{3(1+\beta)/2}\, \Phi_\nu(\psi) \ < \infty.$$

But $|x|_{-p} = \infty$ for any $x \in \mathcal{S}' \setminus \mathcal{S}_p'$. Thus the last inequality implies that the measure ν is supported on $\mathcal{S}_p'$ and

$$\int_{\mathcal{S}_p'} \exp\left[\frac{1}{2}(1+\beta)|x|_{-p}^{\frac{2}{1+\beta}}\right] d\nu(x) \ < \infty.$$

This completes the proof of the theorem.

Example. The probability measure ν_λ in Section 1 is a Hida measure of order $1-\lambda$. Hence by the above theorem, it is supported in $\mathcal{S}_p'$ for some $p \ge 1$ and we have

$$\int_{\mathcal{S}_p'} \exp\left[\frac{1}{2}(2-\lambda)|x|_{-p}^{\frac{2}{2-\lambda}}\right] d\nu_\lambda(x) \ < \infty.$$

REFERENCES

[1] Kondratiev, Yu. G. and Streit, L.: Spaces of white noise distributions: Constructions, Descriptions, Applications. I; *Reports on Math. Phys.* **33** (1993) 341–366

[2] Kuo, H.-H.: *Gaussian Measures in Banach Spaces.* Lecture Notes in Math. **463**, Springer-Verlag, 1975

[3] Kuo, H.-H.: *White Noise Distribution Theory.* CRC Press, 1996

[4] Lee, Y.-J.: Analytic version of test functionals, Fourier transform and a characterization of measures in white noise calculus; *J. Funct. Anal.* **100** (1991) 359–380

[5] Reed M. and Simon, B.: *Methods of Modern Mathematical Physics I: Functional Analysis.* Academic Press, 1972

New Results in the Simplex Method in Linear Programming

ROGER N. PEDERSEN Department of Mathematics, Carnegie Mellon University, Pittsburgh, PA 15213

"Notation is important. It can even solve problems. But, at some point, you must do some work yourself." K. O. Friedrichs.

1. INTRODUCTION AND STATEMENT OF THE PROBLEM.

Without using any symbols at all, we can give a precise statement of the problem by saying that it is to find the maximum, if it exists, of a linear function of a finite number of real variables on a convex plane polyhedron of the same variables. The simplex method of solving the problem is then to find a vertex of the polyhedron and then to proceed along edges from one vertex to the next, in a manner that the linear function increases, until the maximum is reached. All the data needed to state and solve the problem can be stored in an $(m+1) \times (n+1)$ matrix $\mathcal{A}$.

The analytical statement of the problem then is to find the maximum of the objective function

$$\sum_{j=1}^{n} \mathcal{A}_{m+1,j} X_j + \mathcal{A}_{m+1,n+1} \tag{1.1}$$

subject to the constraints

$$\sum_{j=1}^{n} \mathcal{A}_{ij} X_j + \mathcal{A}_{i,n+1} \geq 0, \quad i = 1, ..., m. \tag{1.2}$$

By defining A to be the matrix comprising the first m rows and first n columns of $\mathcal{A}$ and b to be the transpose of $\mathcal{A}_{1,n+1,..,}\mathcal{A}_{m,n+1}$, the constraint (1.2) takes the simpler form

$$Ax + b \geq 0, \tag{1.3}$$

meaning, of course, that each component of the column vector is non-negative. The vector x is superfluous for the purpose of applying the simplex algorithm. But, working only with the matrix $\mathcal{A}$, can lead to misconceptions as we shall see in the next section. But first, let us find another notation for the constraint set by using $\mathcal{A}_i$ to denote the rows of A. Then (1.3) can be replaced by

$$L_i(x) = (\mathcal{A}_i, x) + b_i \geq 0, \quad i = 1, ..., m \tag{1.4}$$

where b_i is the ith coordinate of b and $(,)$ represents the canonical inner product.

2. BUT, THOSE SLACK VARIABLES ARE UNNECESSARY.

Let us re-write (1.3) as

$$(A\,C)\left(C^{-1}X\right) + b \geq 0 \tag{2.1}$$

where C is any non-singular $n \times n$ matrix, noting that this does not require an equality. Now, assuming A has rank n, we may apply elementary column operations to reduced echelon form. If C is the product of the corresponding elementary column matrices and $y = C^{-1}X$, the first n coordinate of (2.1) are

$$y_i + b_i \geq 0. \tag{2.2}$$

Then, by making the translation $z_i = y_i + b_i$, we may assume the constraint set to be in, what is commonly called, canonical form. Furthermore, if for one $j, 1 \leq j \leq n$, we put $x_j = z_j - \beta_j$ in (1.1),(1.2) we see that this corresponds to multiplying the jth column of the full matrix $\mathcal{A}$ by β_i and subtracting it from the $(n+1)^{\text{th}}$ column; that is, it is an elementary column operation. I prefer doing elementary row operations on the transpose. Thus the simplex method reduces to transposing the matrix $\mathcal{A}$ and applying elementary row operations until the first n column are in reduced echelon form, with the restriction that the pivots are to be picked from the first n rows of $\mathcal{A}^T$. The only question that remains is when to start using the simplex pivoting strategy. After the system is in canonical form, we must use the simplex strategy; before that we may use instead the standard Gaussian Elimination Strategy. Note that the simplex strategy requires picking the maximum positive element of the current column and hence is a partial pivoting strategy. We shall have more to say about this in Section 5.

3. EMPTY SETS, REDUNDANT CONSTRAINTS AND LOWER DIMENSIONAL SETS.

Let us now suppose that the normals of the first n constraints form a linearly independent set. Then, for any $k > n$,

$$\mathcal{A}_k = \sum_{i=1}^{n} \alpha_{ki} \mathcal{A}_i \tag{3.1}$$

and hence

$$L_k(x) = \sum_{i=1}^{n} \alpha_{ki} L_i(x) + \triangle_k \tag{3.2}$$

with

$$\triangle_k = \mathcal{A}_{k,n+1} - \sum_{i=1}^{n} \alpha_{ki}\mathcal{A}_{i,n+1}. \tag{3.3}$$

It follows from (1.4) and (3.2) that if $(\alpha_{k1}, ., ..., \alpha_{k,n}, \triangle_k)$ are all non-negative, the kth constraint is redundant and that if they are all negative the set is empty. If for some $i \leq n$, $\alpha_{ki} > 0$, $\alpha_{kj} \leq 0$ for $j \neq i$ and $\triangle_k < 0$, then the ith constraint is redundant.

In all other cases where none of the numbers $(\alpha_{k1}, ., ..., \alpha_{k,n}, \triangle_k)$ is zero it is easily shown that the set formed for the first n and the kth, is non-empty. The other important special case occurs when $\triangle_k = 0$ and $\alpha_{ki} \leq 0$ for $i = 1, ..., n$. Then the entire constraint set is contained in the set where $L_k(x) = 0$. Hence, we may use this constraint to eliminate a variable and obtain a lower dimensional set. This means that, by reducing the number of dimensions, we may assume that this case does not occur.

We note from (3.2) and (3.3) that, when the constraint set is in canonical form, $\mathcal{A}_{i,n+1} = 0$, $i = 1, ..., n$, so the α_{ki}'s and $\triangle_k$ are just the coefficients of the constraint equation. From this point on we shall assume that the set is in canonical form. The origin will be called the basic vertex, the first n constraints the basic constraints and the rest of the constraints the non-basic constraints.

4. THE SIMPLEX ALGORITHM WITH A NON-DEGENERATE BASIC VERTEX.

A vertex which is the intersection of more than n-planes is called a degenerate vertex. This means that, when the basic vertex is non-degenerate, all of the non-basic constraints have non-zero constants. The simplex strategy then is to increase by one the number of positive constants among these until they are all positive and then to increase the constant in the objective function.

Let us assume that the constraints are ordered so that

$$\begin{aligned} &\mathcal{A}_{i,n+1} > 0, \quad i < p \qquad \text{and if } p < m \\ &\mathcal{A}_{i,n+1} < 0, \quad p \leq i \leq m. \end{aligned} \tag{4.1}$$

Our first objective is to increase p by one when it is less than m. The first step is to choose k to maximize

$$\{\mathcal{A}_{p,j} : \mathcal{A}_{p,j} > 0, \ 1 \leq j \leq n\}. \tag{4.2}$$

When $p \leq m$, the results of Section 3 insure that we may assume the above set to be non-empty; when $p = m + 1$, it is only empty when we have found the maximum.

Next, we choose ℓ to maximize the negative numbers

$$\left\{ \frac{\mathcal{A}_{\nu,n+1}}{\mathcal{A}_{\nu,k}} : \nu \leq p-1, \ \mathcal{A}_{\nu,k} < 0 \right\}. \tag{4.3}$$

Suppose that the above set is empty. If $p = m + 1$ and $\mathcal{A}_{m+1,k} > 0$ there is no maximum while if $\mathcal{A}_{m+1,k} < 0$ we may set $x_k = 0$ and continue in one less dimension. If $p \leq m$ we simply set $\ell = p$, observing that the simplex method requires only one step.

Next, we interchange the ℓth non-basic constraint with the kth basic constraint and put the constraint set back into canonical form. This requires applying Gaussian elimination to the kth column of $\mathcal{A}^T$. The new elements of the matrix then are

$$\mathcal{A}'_{\ell,k} = \frac{1}{\mathcal{A}_{\ell,k}} \tag{4.4}$$

$$\mathcal{A}'_{\ell,j} = -\frac{\mathcal{A}_{\ell j}}{\mathcal{A}_{\ell,k}}, \quad j \neq k \tag{4.5}$$

and when $i \neq \ell$,

$$\mathcal{A}'_{i,k} = \frac{\mathcal{A}_{i,k}}{\mathcal{A}_{\ell,k}}, \tag{4.6}$$

$$\mathcal{A}'_{ij} = \mathcal{A}_{ij} - \frac{\mathcal{A}_{ik}\mathcal{A}_{\ell j}}{\mathcal{A}_{\ell,k}}, \quad j \neq k. \tag{4.7}$$

In particular,

$$\mathcal{A}'_{\ell,n+1} = -\frac{\mathcal{A}_{\ell,n+1}}{\mathcal{A}_{\ell,k}} > 0 \tag{4.8}$$

since, whether $\ell = p$ or $\ell < p, \mathcal{A}_{\ell,n+1}$ and $\mathcal{A}_{\ell k}$ have opposite signs. If $i \neq \ell$ and $\ell < p-1$, we see from (4.7) that $\mathcal{A}'_{i,n+1}$ is the sum of two positive numbers when $\mathcal{A}_{ik} > 0$ and that when $\mathcal{A}_{ik} < 0$ it is positive as a consequence of the choice (4.3) of ℓ. Hence, in any case, the first $p-1$ constants remain positive and if $\ell = p, \mathcal{A}_{p,n+1}$ is also positive and we have increased p by one. But we also see from (4.7) that if $\ell < p$,

$$\mathcal{A}'_{p,n+1} > \mathcal{A}_{p,n+1} \tag{4.9}$$

Since the constraint set has only a finite number of vertices, we shall, in a finite number of steps either find the set to be empty, prove that $\mathcal{A}'_{p,n+1} > 0$ or arrive at a degenerate vertex.

5. THE CASE OF A DEGENERATE VERTEX.

The case of a degenerate vertex occurs when there are zero constants $\mathcal{A}_{i,n+1} = 0$. Suppose that we apply the previous strategy to the basic constraints and the non-basic constraints with non-zero constants. Then we see from (4.7) that when $\mathcal{A}_{i,n+1} = 0$,

$$\mathcal{A}'_{i,n+1} = -\left(\frac{\mathcal{A}_{\ell j}}{\mathcal{A}_{\ell k}}\right)\mathcal{A}_{ik} \tag{5.1}$$

and since $\mathcal{A}_{\ell j} < 0$, $\mathcal{A}_{\ell k} > 0$, we have $\mathcal{A}'_{i,n+1} > 0$ whenever $\mathcal{A}_{i,k} > 0$. There is no reason that this should be the case, but, by applying the simplex strategy to the first n columns of $\mathcal{A}$, with the kth playing the roll of the constants, we can use the simplex strategy to achieve this. Because the algorithm is slightly more complicated when the degeneracy is of higher order, it is convenient to introduce constants α_k, β_k satisfying, after reordering the constraints and variables

$$\begin{aligned} \mathcal{A}_{i,k} &= 0, \quad n+1 \leq i < \alpha_k \\ &> 0, \quad \alpha_k \leq i < \beta_k \\ &< 0, \quad \beta_k \leq i < \alpha_{k+1} \end{aligned} \tag{5.2}$$

with $\alpha_{n+2} = m$. The cases $\alpha_k = n+1$, $\beta_k = \alpha_k$ and $\beta_k = \alpha_{k+1}$ are used to indicate that the corresponding set is empty.

Now we apply the following algorithm to the constraint set in canonical form.

[1] $k = n+1$

[2] Reorder the constraints so that (5.2) is satisfied.

Now, we are ready to pick the current constraint indexed by p. The choice agrees with (4.1) when $k = n + 1$.

[3] If $k = n + 1$ or $\beta_k < \alpha_{k+1}$, set $p = \beta_k$ and proceed to [5].

Now, when we arrive at line [4], we have $k < n + 1$ and $\beta_k = \alpha_{k+1}$. This means that the elements of the pivot column below the zeros in the kth row of $\mathcal{A}^T$ are all zero so we can take advantage of the remark preceding (5.1) noting that, because $k < n$ the current pivot row has already been chosen in the line [5].

[4] Replace k by $k + 1$ and proceed to [7].

Now we are ready to pick the current pivot row of $\mathcal{A}^T$.

[5] Reorder the variables so that $\mathcal{A}_{p,k-1}$ maximizes the positive coefficients $\mathcal{A}_{p,i}$, $i \leq i \leq k - 1$ when it is non-empty. If it is empty proceed to [10].

If $\alpha_k = n + 1$, we are ready to begin the updating subroutine. Otherwise, we decrease k by 1 and return to [2].

[6] If $\alpha_k > n + 1$, decrease k by 1 and return to [2].

When we arrive at line [7] we know that the $k - 1^{\text{th}}$ row of $\mathcal{A}^T$ is the current pivot row and, before updating, we must find the current pivot column.

[7] If the set $\{i < p : \mathcal{A}_{i,k-1} < 0\}$ is non-empty, choose ℓ to maximize the ratios $\mathcal{A}_{i,k}|\mathcal{A}_{i,k-1}$. Otherwise set $\ell = p$.

Now, we are ready to interchange the constraints indexed by $k - 1$ and ℓ and then put the matrix back into canonical form.

[8] Return the matrix to canonical form by applying Gaussian elimination to reduced echelon form to $\mathcal{A}^T$ using the element indexed by $\ell, k - 1$ as pivot.

We note that, since the elements $\mathcal{A}_{\ell,j}$, $j > k - 1$, are all zero the elementary row operation correspond to adding zero to the rows of A^T indexed by $j > k$. Hence the α_j's and β_j's, $j > k$ so they are unchanged. We now redefine the α_j's and β_j's for $j \leq k$ returning to [2].

[9] Return to [2].

The program will terminate at [10].

[10] The maximum is $\mathcal{A}_{m+1,n+1}$.

We have tacitly assumed the maximum to exist, leaving to the reader the task of adding the lines, explained in Section 3, regarding empty sets, redundant constraints, lower dimensional problems and problems with no maximum.

6. SMALL PIVOTS AND DEGENERATE VERTICES.

In running the above algorithm, it is crucial that one distinguish between non-zero numbers and zeros represented by round-off errors. The author has studied this problem extensively on the Radio Shack Color Computer and on the Tandy 1000. Computing, respectively, to 9 and 16 places, base 10, the Random Number generator was used to supply the data and, computing to p places base 10, the test for determining whether or not a number is zero was by comparison with $10^{\gamma-p}$, $2 \leq \gamma \leq p/2$. In order to increase the probability that the set is not empty, the probability that the origin satisfying a constraint is set at π, $0 \leq \pi \leq 1$. With no other restriction, a degenerate vertex has never been found. By building in the condition of degeneracy, e.g. by applying a similarity transformation to a known degenerate situation and adding more constraints, the program seems to work as well as in the non-degenerate case. The problem, in each case, is checked by re-running the program on the constraints forming the final basic vertex and by evaluating the objective function at the intersection of their planes.

We have also never found an ill-conditioned matrix with the random number generator. By putting in the Hilbert matrix [2], prob. 169, p. 337, we find the obvious difficulty. However, by computing to a sufficient number of places, we have always been able to overcome the difficulty.

7. FURTHER METHODS OF SPEEDING UP THE PROGRAM.

The Simplest Method of Speeding Up the Program is to remove the redundant constraint using the test of Section 3, noting that the test requires only sign-tests of quantities that are computed anyway. Its disadvantage is that a constraint that shows up as redundant in one coordinate system does not necessarily in another. The number of degenerate constraints can be increased by adding the condition that the objective function be greater than its value at the current basic vertex.

Another method of possibly speeding up the program is to use the fact that once a vertex has been found we know that the constraint set is non-empty. Then we can eliminate a variable using any of the constraints. If the constraint used was redundant, the new set will be empty. Otherwise, we obtain the maximum on an $(n-1)$-dimensional face. The weakness of this method is that we lose time when we use a redundant constraint to eliminate a variable.

8. THE STATEMENT OF THE CONDITION THAT THE SET BE EMPTY OR CONTAIN A REDUNDANT CONSTRAINT.

In this section we iterate the formulas (4.4) - (4.8) for the constraint set

$$\sum_{j=1}^{n} \mathcal{A}_{ij}x_j + \mathcal{A}_{i,n+1} \geq 0,\ i = 1, ..., m \tag{8.1}$$

in canonical form. That is,

$$\mathcal{A}_{ij} = \delta_{ij},\ i = 1, ..., n+1,\ \ j = 1, ..., n. \tag{8.2}$$

Specifically, we generalize the condition that the set is empty when $\mathcal{A}_{kj} < 0$ for all $j = 1, ..., n+1$ and contains a redundant constraint when the set $\{\mathcal{A}_{k1}, ..., \mathcal{A}_{kn}, \mathcal{A}_{k,n+1}\}$ consists only or non-negative elements or $\mathcal{A}_{k,n+1} < 0$ and $\mathcal{A}_{kj} > 0$ for exactly one $j \leq n$.

In this section we shall use the above stated condition to obtain a result for appropriate union by obtaining explicit formulas for the coefficients in the constraints when the constraints

$$k_1, ..., k_r,\ \ k_i \geq n+1,\ r \leq n \tag{8.3}$$

have been interchanged with the constraints

$$\ell_1, ..., \ell_r,\ \ell_i \leq n, \tag{8.4}$$

in the order $k_i, \ell_i,\ i = 1, ..., r$ and the constraint set is returned to canonical form at each step. In order to state the formulas, we denote by

$$f_\sigma\left(i_1, ..., i_\sigma : j_1, ..., j_\sigma\right) \tag{8.5}$$

the minor determinant of $\mathcal{A}_{ij},\ i = m+1, ..., m,\ j = 1, ..., n$ indexed by the rows $i_1, ..., i_n$ and the columns $j_1, ..., j_r$. Then, with $\mathcal{A}^0_{ij}$ representing the original matrix and $\mathcal{A}^r_{ij}$ the matrix after the constraints indexed by $k_1, ..., k_r$ have replaced those indexed by $\ell, ..., \ell_r$,

$$\begin{aligned} \mathcal{K}_r &= \{k_1, ..., k_r\}, \mathcal{K}'_r = [1, m] - \mathcal{K}_r,\ \mathcal{L}_r = \{\ell_1, ..., \ell_r\},\ \mathcal{L}'_r = [1, n+1] - \mathcal{L}_r, \\ D_r &= f_r\left(k_1, ..., k_i : \ell_1, ..., \ell_r\right) \end{aligned} \tag{8.6}$$

we have the formulas for $i \in \mathcal{K}'_r$,

$$\mathcal{A}^r_{ij} = f_{r+1}\left(k_1, ..., k_r, i : \ell_1, ..., \ell_r, j\right)/D_r,\quad j \in \mathcal{L}'_r, \tag{8.7}$$

$$\mathcal{A}^r_{i,\ell_j} = (-1)^{r-j} f_r\left(k_1, ..., k_{j-1}, k_{j+1}, ..., k_r, i : \ell_1, ..., \ell_r\right)/D_r,\quad \ell_j \in \mathcal{L}_r, \tag{8.8}$$

and for $k_i \in \mathcal{K}_r$,

$$\mathcal{A}^r_{ki,j} = (-1)^{r+1-i} f_r\left(k_1, ..., k_r : \ell_1, ..., \ell_{i-1}, \ell_{i+1}, ..., \ell_r, j\right)/D_r,\quad j \in \mathcal{L}'_r, \tag{8.9}$$

$$\mathcal{A}^r_{k_i,\ell_j} = (-1)^{i+j} f_{r-1}\left(k_1, ..., k_{j-1}, k_{j+1}, ..., k_r : \ell_1, ...\ell_{i-1}, \ell_{i+1}, ..., \ell_r\right)/D_r,\quad \ell_j \in \mathcal{L}_r. \tag{8.10}$$

Before stating the condition for redundant constraints or empty sets, we shall prove the following theorem.

Theorem 8.1. *The formulas (8.7) - (8.10) are invariant under permutation of $k_1, ..., k_r$ or $\ell_1, ..., \ell_r$ in the sense the sign of either (8.7), (8.8) or (8.9), (8.10) for fixed i and $j = 1, ..., n+1$ are invariant.*
This makes it possible to state the condition for empty sets or redundant constraints using only the pair (8.7), (8.8) in the order $r = 1, 2, ..., n$.

Proof. First let us note that we may assume that the k's and l's are in increasing order. This follows from the fact that when $k_1, ..., k_n$ are permutations of the same set, then $k_1, ..., k_{j-1}, k_{j+1}, ...k,, j = 1, ..., n$ are merely written down in a different order.

To prove this by induction, let $\sigma = (k_1, ..., k_r)$ and $\sigma_j = (k_1, ..., k_{j-1}, k_{j+1}, ..., k_r)$ and suppose that the largest element y of σ is indexed by ℓ. Then after interchanging the y with the last elements of σ and $\sigma_j, j \neq \ell$, the sign of the ratio σ_j/σ is retained when $j < \ell$, changes when $j > \ell$ and is multiplied by $(-1)^{r-\ell}$ when $j = \ell$. Hence by moving the ℓth ratio to the end of the list and decreasing the order of those indexed by $k, \ell+1 \le k \le r$, we obtain a valid induction proof. Similarly for the l's. ■

Theorem 8.2. *In applying the empty set or redundant constraint test, it is sufficient to scan (8.7), (8.8) for all permutation $(k_1, ..., k_r)$ and $(\ell_1, ..., \ell_r)$ in increasing order of r.*

Proof. In proceeding from r to $r+1$ we interchange the constraints indexed by k_{r+1} and ℓ_{r+1}. A simple computation shows that in an $(n+1)$ constraint set in canonical form, an interchange of the $(n+1)^{\text{st}}$ constraint with a basic constraint can't change the sign test indicating an empty set or redundant constraint[1]. But, by Theorem 8.1, we may assume that any k_i and ℓ_i were interchanged.

[1]See Section 13, #2.

9. THE RECURSION FORMULA.

Assuming that we have computed the matrix $\mathcal{A}_{ij}^{r}$, the matrix $\mathcal{A}_{ij}^{r+1}$ is obtained by interchanging the constraints indexed by k_{r+1}, ℓ_{r+1} and updating the matrix as in [1]. The result is

$$\mathcal{A}_{k_{r+1},\ell_{r+1}}^{r+1} = 1/\mathcal{A}_{k_{r+1},\ell_{r+1}}^{r} \tag{9.1}$$

$$\mathcal{A}_{k_{r+1},j}^{r+1} = -\mathcal{A}_{k_{r+1},j}^{r}/\mathcal{A}_{k_{r+1},\ell_{r+1}}^{r}, \quad j \neq \ell_{r+1} \tag{9.2}$$

and for $i \neq k_{r+1}$,

$$\mathcal{A}_{i,\ell_{r+1}}^{r+1} = \mathcal{A}_{i,\ell_{r+1}}^{r} / \mathcal{A}_{k_{r+1},\ell_{r+1}}^{r}, \tag{9.3}$$

$$\mathcal{A}_{ij}^{r+1} = \mathcal{A}_{ij}^{r} - \frac{\mathcal{A}_{i,\ell_{r+1}}^{r}\mathcal{A}_{k_{r+1},j}^{r}}{\mathcal{A}_{k_{r+1},\ell_{r+1}}^{r}}, \quad j \neq \ell_{r+1}. \tag{9.4}$$

Note, in particular, that (9.4) is the ratio of a 2×2 minor and a 1×1 minor and when $r = 0$, it agrees with (8.10). Also, when $r = 0$, (9.2), (9.3) agree with (8.8) (8.9). In order to make (9.1) agree with (8.6) we make the convention $f_0 = 1$. Before proving the general result, we shall develop some lemmas on determinants.

10. SOME LEMMAS ON DETERMINANTS.

Let us use the usual convention that $\mathcal{B}_{ij}$ is the co-factor of b_{ij}. Then our first and main lemma is:

Lemma 10.1. *Let $\mathcal{B} = (b_{ij})$ be a $k\times k$ matrix and let $\mathcal{C}$ be the $(k-1)\times(k-1)$ matrix*

$$\mathcal{C} = \left(b_{ij} - \frac{b_{i,k}b_{kj}}{b_{kk}}\right), \quad 1 \leq i, \ j \leq k-1. \tag{10.1}$$

Then

$$\det \mathcal{C} = \det \mathcal{B}/b_{kk}. \tag{10.2}$$

Proof. Define:

$$\varphi(\epsilon) = \det\left(b_{ij} - \epsilon\, b_{i,k}b_{kj}\right), \quad 1 \leq i, \ j \leq k-1. \tag{10.3}$$

Now we use the fact that the derivative of a determinant is the sum of the determinants obtained by differentiating one row of the matrix. When we differentiate the ith row of $\mathcal{C}$, the new i^{th} row is

$$-b_{ik}\left(b_{k1}, b_{k2}, ..., b_{k,k-1}\right). \tag{10.4}$$

■

If we interchange this row with each of those indexed by $i+1, ..., k-1$, we have the matrix obtained by deleting the ith row from the first k columns of $\mathcal{B}$. Hence, when we take the determinant, we obtain

$$b_{ik}\mathcal{B}_{ik}. \tag{10.5}$$

It follows that

$$\varphi'(0) = \sum_{i=1}^{k-1} b_{ik}\mathcal{B}_{ik}. \tag{10.6}$$

When we differentiate twice we obtain a sum of determinants of matrices having two rows equal. Hence $\varphi''(\epsilon) \equiv 0$ so $\varphi^{(j)}(\epsilon) = 0$ for $j \geq 2$.

Since $\varphi(0) = \mathcal{B}_{kk}$ we then have

$$\varphi(\epsilon) = \mathcal{B}_{kk} + \epsilon \sum_{i=1}^{k-1} b_{ik}\mathcal{B}_{ik}. \tag{10.7}$$

Putting $\epsilon = 1/b_{kk}$ gives (10.2).

The recursion formula (9.4) with $i > k_{r+1},\ j > \ell_{r+1}$ can be rewritten

$$\mathcal{A}_{ij}^{r+1} = \frac{\det\left(\mathcal{A}_{\mu\nu}^{r}\right)_2}{\mathcal{A}_{k_{r+1},\ell_{r+1}}^{r}} \tag{10.8}$$

where the numerator is the determinant of the 2×2 matrix indexed by $\mu = k_{r+1}, i$ and $\nu = \ell_{r+1,j}$ and is, in fact, just the Lemma 10.1 with $k = 2$ after a change of indices. More generally, we can use Lemma 10.1 to prove inductively that, for $1 \leq p \leq r+1$,

$$\mathcal{A}_{ij}^{r+1} = \frac{\det\left(\mathcal{A}_{\mu\nu}^{r+1-\rho}\right)_{\rho+1}}{\mathcal{A}_{k_{r+2-\rho},\ell_{r+2-\rho}}^{r+1-\rho} \cdots \mathcal{A}_{k_{r+1},\ell_{r+1}}^{r}} \tag{10.9}$$

where the numerator is the determinant of the $(\rho+1) \times (\rho+1)$matrix indexed by $\mu = k_{r+2-\rho}, ..., k_{r+1}, i$ and $\nu = \ell_{r+1-\rho}, ..., \ell_{r+1}, j$. In particular, when $\rho = r+1$, (10.9) reduces, in view of (8.5), to

$$\mathcal{A}_{ij}^{r+1} = \frac{f_{r+2}\left(k_1, ..., k_{r+1},\ i : \ell_1, ..., \ell_{r+1}, j\right)}{\mathcal{A}_{k_{r+1},\ell_{r+1}}^{r}\mathcal{A}_{k_r,\ell_r}^{r-1} \cdots \mathcal{A}_{k_1,\ell_1}^{0}} \tag{10.10}$$

for any $i \in \mathcal{K}_{r+1},\ j \in \mathcal{L}_{r+1}'$. In particular,

$$\mathcal{A}_{k_{r+2},\ell_{r+2}}^{r+1} = \frac{D_{r+2}}{\mathcal{A}_{k_{r+1},\ell_{r+1}}^{r} \cdots \mathcal{A}_{k_1,\ell_1}^{0}} \tag{10.11}$$

or

$$D_{r+2} = \mathcal{A}_{k_1,\ell_1}^{0}\mathcal{A}_{k_2,\ell_2}^{1} \cdots \mathcal{A}_{k_{r+2},\ell_{r+2}}^{r+1}. \tag{10.12}$$

Since this is true for each r, we have proved (8.7) with $i\ \mathcal{K}_{r+1}',\ j \in \mathcal{L}_{r+1}'$ as a consequence of (10.10) and (10.11) with r replaced by $r-1$.

By eliminating $\mathcal{A}_{ij}^{r+1}$ between (10.9), (10.10), setting $i = k_{r+2},\ j = \ell_{r+2}$, and using 8.6 for $\mathcal{A}_{\mu\nu}^{r+1-\rho}$, we obtain the interesting identity

$$\begin{aligned} & f_{r+2}\left(k_1, ..., k_{r+2} : \ell_1, ..., \ell_{r+2}\right)\left(D_{r+1-\rho}\right)^{\rho} \\ = \ & \det\left(f_{r+2-\rho}\left(k_1, ..., k_{r+1-\rho}, \mu : \ell_1, ..., \ell_{r+1-\rho}, \nu\right)\right)_{\rho+1} \end{aligned} \tag{10.13}$$

with μ and ν ranging over the indices $k_{r+2-\rho}, ..., k_{r+2}$ and $\ell_{r+2-\rho}, ..., \ell_{r+2}$. The main use we make of this identity is:

Theorem 10.2. *Consider the identity (10.13) with $\rho = 1$. If three of the four minors comprising the determinant on the right have sign opposite the fourth then $D_r \neq 0$ and the sign of f_{r+2} is determined by the identity.*

We shall also need the following identity

$$f_{r+1}(k_1,...,k_{r+1}:\ell_1,...,\ell_{i-1},\ell_{i+1},...,\ell_{r+1,j})\,f_r(k_1,...,k_r:\ell_i,...\ell_r)$$

$$-\quad f_{r+1}(k_1,...,k_{r+1}:\ell_1,...,\ell_r,j)\,f_r(k_1,...,k_r:\ell_1,...,\ell_{i-1},\ell_{i+1},...,\ell_{r+1}) \tag{10.14}$$

$$+\quad f_{r+1}\left(k_1,...,k_{r+1}:\ell_1,...,\dot{\ell}_{r+1}\right)f_r(k_1,...,k_r:\ell_1,...,\ell_{i-1},\ell_{i+1},...,\ell_r,j)=0.$$

If we suppress the dependence on $k_1,...,k_{r-1}$ and $\ell_1,...,\ell_{r-1}$, the left side of (10.14) is the 3×3 determinant of the matrix with rows indexed by (k_1,k_1,k_2) and column indexed by ℓ_r,ℓ_{r+1},j. Since the first two rows are equal the determinant is zero.

11. COMPLETION OF THE PROOFS OF THE IDENTITIES.

We now have the main tools sufficient for the proofs of (8.7),...,(8.10) by induction. Note that we have proved (8.7) for all r and $i\notin\mathcal{K}_r,\ j\notin\mathcal{L}_r$, the proofs of the cases (8.8), (8.9), (8.10). Hence, we may use $i=k_{r+1},\ j=\ell_{r+1}$ in (8.7) to express (9.1) as

$$\mathcal{A}^{r+1}_{k_{r+1},\ell_{r+1}}=\frac{f_r(k_1,...,k_r:\ell_1,...,\ell_r)}{D_{r+1}}. \tag{11.1}$$

This is the promoted version of (8.10) with $i=r{+}1,\ j=r{+}1$. By putting $i=k_{r+1},\ j\notin\mathcal{L}_{r+1}$ into (8.7) and substituting (11.1) into (8.10), we obtain

$$\mathcal{A}^{r+1}_{k_{r+1},j}=-\frac{f_{r+1}(k_1,...,k_{r+1}:\ell_1,...,\ell_r,j)}{D_r} \tag{11.2}$$

which is the formula (8.9) corresponding to the pair $k_{r+1,j}$ with $j\notin\mathcal{L}_r$. It follows from (9.4) that for $k_i\in\mathcal{K}_r,\ j\notin\mathcal{L}_{r+1}$.

$$\mathcal{A}^{r+1}_{k_i,j}=\mathcal{A}^r_{k_i,j}-\frac{\mathcal{A}^r_{k_i,\ell_{r+1}}\mathcal{A}^r_{k_{r+1},j}}{\mathcal{A}^r_{k_{r+1},\ell_{r+1}}}. \tag{11.3}$$

After substituting (8.7) and (8.9), and setting the result equal to (8.9) with r replaced by $r+1$, we obtain the identity (10.14). This completes the proof of the remaining cases in (8.9). The proof of the promoted version of (8.8) is isomorphic.

There remains the case indexed by $k_i\in\mathcal{K}_r$ and $\ell_j\in\mathcal{L}_r$. We obtain from (9.4), (8.8), (8.9), (8.10) and (10.11) with r replaced by $r-1$ into (11.4), we obtain

$$\mathcal{A}^{r+1}_{k_i,k_j}=\tfrac{(-1)^{i+j}}{D_r}\{f_{r-1}(k_1,...,k_{j-1},k_{i+1},...,k_r:\ell_1,...,\ell_{i-1},\ell_{i+1},...,\ell_r)$$

$$+\quad\frac{f_r(k_1,...,k_r:\ell_1,...,\ell_{i-1},\ell_{i+1},...,\ell_{r-1})\,f_r(k_1,k_{i-1},k_{i+1},...,k_{i-1}:\ell_1,...,\ell_{r+1})}{f_{r+1}(k_1,...,k_{r+1}:\ell_1,...,\ell_{r+1})}\}. \tag{11.4}$$

We now apply (10.13) in the form

$$f_{r+1}(k_1,...,k_{r+1}:\ell_1,...,\ell_{r+1})\,f_{r-1}(k_1,...,k_{j-1},k_{j+1},...,k_r:\ell_1,...,\ell_{i-1},\ell_{i+1},...\ell_r)$$

$$=\quad f_r(k_1,...,k_{j-1},k_{j+1},...,k_{r+1}:\ell_1,...,\ell_{i-1},\ell_{i+1},\ell_{r+1})\,f_r(k_1,...,k_r:\ell_1,...,\ell_r) \tag{11.5}$$

$$-f_r(k_1,...,k_{j-1},k_{j+1},...,k_{r+1}:\ell_1,...,\ell_r)\,f_r(k_1,...,k_r:\ell_1,...,\ell_{i-1},\ell_{i+1},...,\ell_r).$$

After substituting (11.5) into (11.4), we have the promoted version of (8.10) for $k_i\in\mathcal{K}_i,\ \ell_j\in\mathcal{L}_j$. Since the case of $k_{r+1}\subset\mathcal{K}_{r+1},\ell_{r+1}\in\mathcal{L}_{r+1}$ has already been disposed of, the proof is complete.

12. DUPLICATIONS OF CONSTRAINTS.

The formulas (8.7) - (8.10) are derived under the assumption that the sets $(k_1, ..., k_r)$ and $(\ell_1, ..., \ell_r)$ are distinct. In particular the ℓ's are a subset of $(1, ..., n)$ so we must have $r \leq n$. On the other hand, it follows from the recursion formulas (9.1) - (9.4) that we may, at any time, start over with a new matrix and continue until there is a duplication in either the k's or the ℓ's. In this section we resolve the question of such a duplication in the second step.

The new matrix coefficients, after interchanging the ith non-basic constraint with the ℓth basic constraint and then returning to reduced echelon form by the use of elementary column operations, are

$$\mathcal{A}'_{i\ell} = 1/\mathcal{A}_{i\ell} \tag{12.1}$$

$$\mathcal{A}'_{ij} = -\mathcal{A}_{ij}/\mathcal{A}_{i\ell}, \quad j \neq \ell \tag{12.2}$$

and for $k \neq i$,

$$\mathcal{A}'_{k\ell} = \mathcal{A}_{k\ell}/\mathcal{A}_{i\ell}, \tag{12.3}$$

$$\mathcal{A}'_{kj} = \mathcal{A}_{kj} - \mathcal{A}_{k\ell}\ \mathcal{A}_{ij}/\mathcal{A}_{i\ell}, \quad j \neq \ell. \tag{12.4}$$

Now let us interchange the new k^{th} constraint with the ℓth basic constraint. By analogy with (12.1), (12.2) the coefficients for the new k^{th} constraint are

$$\mathcal{A}''_{k\ell} = 1/\mathcal{A}'_{k\ell} \tag{12.5}$$

and

$$\mathcal{A}''_{kj} - \mathcal{A}'_{kj}/\mathcal{A}'_{k\ell}, \quad j \neq \ell. \tag{12.6}$$

After substituting from (12.1) - (12.4) there becomes

$$\mathcal{A}''_{k\ell} = \mathcal{A}_{i\ell}/\mathcal{A}_{k\ell} \tag{12.7}$$

and

$$\mathcal{A}''_{kj} = \mathcal{A}_{ij} - \mathcal{A}_{i\ell}\ \mathcal{A}_{kj}/\mathcal{A}_{k\ell}, \quad j \neq \ell. \tag{12.8}$$

These are just the parameters obtained after interchanging the ith constraint with the ℓth and returning to reduced echelon form. But they are in the position of the kth. The new coefficients for the ith constraint are

$$\mathcal{A}''_{i\ell} = \mathcal{A}'_{i\ell}/\mathcal{A}'_{k\ell}, \tag{12.9}$$

$$\mathcal{A}''_{ij} = \mathcal{A}'_{ij} - \mathcal{A}'_{i\ell}\ \mathcal{A}'_{kj}/\mathcal{A}'_{k\ell}, \quad j \neq \ell. \tag{12.10}$$

Again, after substituting from (12.1) - (12.4) and taking into account cancellations, these become

$$\mathcal{A}''_{i\ell} = 1/\mathcal{A}_{k\ell}, \tag{12.11}$$

$$\mathcal{A}''_{ij} = -\mathcal{A}_{kj}\ /\ \mathcal{A}'_{k\ell}, \quad j \neq \ell. \tag{12.12}$$

They are the coefficients for the kth constraint after interchanging the kth constraint with the ℓth, and they are in the position of the ith.

For $r \neq k$ or i, $r > n$, the new coefficients for the rth constraint are

$$\mathcal{A}''_{r\ell} = \mathcal{A}'_{k\ell}\ /\ \mathcal{A}'_{k\ell}, \tag{12.13}$$

$$\mathcal{A}''_{rj} = \mathcal{A}'_{rj} - \mathcal{A}'_{r\ell}\,\mathcal{A}'_{kj} \,/\, \mathcal{A}'_{k\ell}, \;\; j \neq \ell \tag{12.14}$$

After substituting from (12.1) - (12.4), these become

$$\mathcal{A}''_{r\ell} = \mathcal{A}_{r\ell} \,/\, \mathcal{A}_{k\ell}, \tag{12.15}$$

$$\mathcal{A}''_{rj} = \mathcal{A}_{rj} - \mathcal{A}_{r\ell}\,\mathcal{A}_{rj} \,/\, \mathcal{A}_{k\ell}, \;\; j \neq \ell \tag{12.16}$$

which are just the coefficient obtained after interchanging the rth constraint with the ℓth in the original matrix. This together with the remarks following (12.8) and (12.12) yields a proof of the following theorem.

Theorem 12.1. *Interchanging the ith non-basic constraint with the ℓth, updating and then interchanging the kth with the ℓth and updating is equivalent to merely interchanging the kth with the ℓth in the original matrix, updating and then interchanging the ith and k^{th}.*

Now let us determine the effect of interchanging one non-basic constraint with two different basic constraints. If after obtaining the formulas (12.1) - (12.4), we interchange the ith constraint with the qth basic constraint, $q \neq i$, the new parameter for the ith constraint are

$$\mathcal{A}''_{iq} = 1/\mathcal{A}'_{iq}, \tag{12.17}$$

$$\mathcal{A}''_{i\ell} = -\mathcal{A}'_{i\ell} \,/\, \mathcal{A}'_{iq}, \tag{12.18}$$

and

$$\mathcal{A}'_{qj} = -\mathcal{A}'_{ij} \,/\, \mathcal{A}'_{iq}, \; j \neq q, \ell. \tag{12.19}$$

The formulas (12.17) - (12.19), after substituting from (12.1) - (12.4) are just the formulas obtained after interchanging the ith with the qth in the original matrix. For $k \neq q$,

$$\mathcal{A}''_{kq} = \mathcal{A}'_{kq} \,/\, \mathcal{A}'_{iq} \tag{12.20}$$

$$\mathcal{A}'_{k\ell} = \mathcal{A}'_{k\ell} - \mathcal{A}'_{kq}\,\mathcal{A}'_{i\ell} \,/\, \mathcal{A}'_{iq} \tag{12.21}$$

and

$$\mathcal{A}'_{kj} = \mathcal{A}'_{kj} - \mathcal{A}'_{kq}\,\mathcal{A}'_{ij} \,/\, \mathcal{A}'_{iq}, \;\; j \neq q, \ell. \tag{12.22}$$

Again, after substituting from (12.1) - (12.4), these are just the formula for the kth constraint after interchanging the kth with the qth in the original matrix except that the qth and ℓ^{th} variables have been interchanged.

Theorem 12.2. *If we interchange the ith non-basic constraint with the ℓth basic constraint, update and then interchange the new i^{th} constraint with the qth, $q \neq i$, and update, this is equivalent to merely interchanging the ith with the qth updating and permuting the qth and ℓ^{th} variables.*

13. THE CASE OF $(n+2)$ CONSTRAINTS.

Let us assume the constraint set to be in canonical form. If α is any subset of the non-basic indices, we shall denote by S_α the corresponding set of non-basic constraints and by $\bar{\mathcal{S}}_\alpha$ the set S_α together with the basic constraints. For a single index i we define

$$S_i^+ = \{j \leq n : \mathcal{A}_{ij} > 0\} \tag{13.1}$$

and

$$S_i^- = \{j \leq n : \mathcal{A}_{ij} < 0\}\,. \tag{13.2}$$

The cardinality of set S shall be denoted by $|S|$. We shall assume that our constraint set contains no degenerate vertices and that minor determinants used in counting are always non-zero.

Our $(n+1)$ constraint set S_i is empty when $a_{i,n+1} < 0$ and $\left|S_i^-\right| = n$, hence $\left|S_i^+\right| = 0$. It contains a redundant constraint when $a_{i,n+1} > 0$ and $\left|S_i^+\right| = n$ or $a_{i,n+1} < 0$ and $\left|S_i^+\right| = 1$. Now suppose that $\left|S_i^+\right| = \sigma, 0 \leq \sigma \leq n$. If $\sigma < n$ there exists an $\ell \leq n$ such that $\mathcal{A}_{i\ell} < 0$. If we interchange the ith and ℓ^{th} constraints and put the set back into canonical form we obtain the constraint set $\hat{\mathcal{S}}_i$ with $\left|\hat{\mathcal{S}}_i\right| = \sigma$. If $\sigma > 0$ there exists an index $\ell \leq n$ with $a_{i,\ell} > 0$. After interchanging the ith and ℓ^{th} constraint and putting the set back into canonical form the set $\hat{\mathcal{S}}_i$ has $\left|\hat{\mathcal{S}}_i\right| = n+1-\sigma$. It follows that interchanging two constraints in an $(n+1)$ - constraint set cannot change its status relative to being empty, or having a redundant constraint. Hence, if neither $\mathcal{S}_i$ nor $\mathcal{S}_j$ has this property, we can find an empty set or redundant constraint in an $(n+1)$ constraint set only by interchanging S_i with a basic constraint and examining $\bar{\mathcal{S}}_j, j \neq i$ or conversely. In particular, after making this interchange, the new constant term is

$$\begin{aligned} \mathcal{A}'_{j,n+1} &= \mathcal{A}_{j,n+1} - \frac{\mathcal{A}_{j\ell}}{\mathcal{A}_{i\ell}}\mathcal{A}_{i,n+1} \\ &= -\frac{\mathcal{A}_{j\ell}}{\mathcal{A}_{i\ell}}\left(\mathcal{A}_{i,n+1} - \frac{\mathcal{A}_{i\ell}}{\mathcal{A}_{j\ell}}\mathcal{A}_{j,n+1}\right) \end{aligned} \tag{13.3}$$

Hence, that constant term in the ith constraint, after interchanging the jth and the ℓth, has the same or opposite sign as the jth constant, after interchanging the ith and the ℓth, according to whether $\mathcal{A}_{j\ell}$ and $\mathcal{A}_{i\ell}$ have the opposite or the same signs.

Let us now study the jth constraint after interchanging the ith and the ℓth with $\mathcal{A}_{i\ell} < 0$ and $\mathcal{A}_{j,\ell} > 0$. This requires analyzing the signs of

$$\mathcal{A}'_{j,\ell} = \frac{\mathcal{A}_{j\ell}}{\mathcal{A}_{i\ell}} \tag{13.4}$$

and

$$\mathcal{A}'_{j,k} = \mathcal{A}_{j,k} - \frac{\mathcal{A}_{j\ell}}{\mathcal{A}_{i\ell}}\mathcal{A}_{ik}. \tag{13.5}$$

Since $\mathcal{A}_{j\ell}$ and $\mathcal{A}_{i\ell}$ have opposite signs, it follows from (13.4) that

$$\mathcal{A}'_{j\ell} < 0 \tag{13.6}$$

and from (13.5) that

$$\mathcal{A}'_{jk} > 0, \quad k \in S_i^+ \cap S_j^+ \tag{13.7}$$

and

$$\mathcal{A}'_{jk} < 0, \quad k \in S_i^- \cap S_j^-. \tag{13.8}$$

For $k \in S_i^- \cap S_j^+ - \{\ell\}$, we may make the signs of

$$\mathcal{A}'_{jk} = \left(\frac{\mathcal{A}_{jk}}{\mathcal{A}_{ik}} - \frac{\mathcal{A}_{j\ell}}{\mathcal{A}_{i\ell}}\right) \cdot \mathcal{A}_{ik} \tag{13.9}$$

all negative or all positive, without violating (13.6), by choosing ℓ to minimize or maximize the ratios

$$\frac{\mathcal{A}_{jk}}{\mathcal{A}_{ik}}, \quad k \in S_i^- \cap S_j^+. \tag{13.10}$$

Since, by (13.6), $\mathcal{A}'_{j\ell} < 0$, we can't achieve an empty set or redundant constraint unless $\mathcal{A}'_{j,n+1} < 0$. This rules out the possibility $\mathcal{A}_{i,n+1} > 0$ and $\mathcal{A}_{j,n+1} > 0$. When $\mathcal{A}_{i,n+1} < 0$, $\mathcal{A}_{j,n+1} > 0$, this cannot be the case unless it is true for either $\mathcal{S}_i$ or $\mathcal{S}_j$.

There remains the cases where $\mathcal{A}_{i,n+1}$ and $\mathcal{A}_{j,n+1}$ have opposite signs. By applying the results (13.6) - (13.10), we see that $\mathcal{S}_{ij}$ is empty or contains a redundant constraint if ℓ minimizes the ratios (13.10),

$$\left|S_i^+ \cap S_k^+\right| + \left|\left\{k \in S_i^+ \cap S_j^- : \mathcal{A}_{jk} > 0\right\}\right| \leq 1 \tag{13.11}$$

and either

$$\mathcal{A}_{j,n+1} < 0, \quad \mathcal{A}_{i,n+1} > 0, \frac{\mathcal{A}_{j,n+1}}{\mathcal{A}_{i,n+1}} < \frac{\mathcal{A}_{j\ell}}{\mathcal{A}_{i\ell}} \tag{13.12}$$

or

$$\mathcal{A}_{j,n+1} > 0, \quad \mathcal{A}_{i,n+1} < 0, \frac{\mathcal{A}_{j,n+1}}{\mathcal{A}_{i,n+1}} > \frac{\mathcal{A}_{j\ell}}{\mathcal{A}_{i\ell}}. \tag{13.13}$$

14. THE CASE OF THREE VARIABLES.

We shall focus our attention on the case of proving that the constraint set is non-empty by finding a vertex that satisfies all constraints. Thus, suppose that

$$\mathcal{A}_{i,n+1} > 0, \quad n+1 \leq i \leq p-1, \quad \mathcal{A}_{p,n+1} < 0. \tag{14.1}$$

Definition 14.1. *For each $k \in S_p^+$, we define*

$$\mathcal{T}_k^+ = \left\{i : \mathcal{A}_{i,k} < 0, \quad \mathcal{A}_{p,n+1} - \frac{\mathcal{A}_{pk}}{\mathcal{A}_{ik}}\mathcal{A}_{i,n+1} > 0\right\} \tag{14.2}$$

and

$$\mathcal{T}_k^- = \left\{i : \mathcal{A}_{ik} < 0, \quad \mathcal{A}_{p,n+1} - \frac{\mathcal{A}_{pk}}{\mathcal{A}_{ik}}\mathcal{A}_{i,n+1} < 0\right\}. \tag{14.3}$$

The simplex strategy makes $\mathcal{A}_{p,n+1}$ increase until it is either positive or the set has been demonstrated to be empty. That this strategy requires more than one step requires that $\mathcal{T}_k^-$ be non-empty for each $k \in S_p^+$. Otherwise, if $S_k^- = \emptyset$, we may achieve our objective by interchanging the kth basic constraint with the pth and putting the matrix back into reduced echelon form.

From now on we shall assume the number of dimensions to be three. By making one Simplex step and permuting the variables, we assume that

$$\mathcal{A}_{p1} > 0, \ \ \mathcal{A}_{p2} > 0, \ \ \mathcal{A}_{p3} < 0, \ \ \mathcal{A}_{p4} < 0. \tag{14.4}$$

If there is a constraint indexed by $i < p$ for which $1 \in S_p^-$ and $2 \in S_p^-$, it follows from (13.11) - (13.13) that there is a redundant constraint. Let us then assume that there exist two non-basic constraints indexed by i and j for which

$$i \in \mathcal{T}_1^- \cap \mathcal{T}_2^+, \ \ j \in \mathcal{T}_1^+ \cap \mathcal{T}_2^-. \tag{14.5}$$

The constraints indexed by i, j, p have the following sign configuration

$$\begin{array}{ccccc} & 1 & 2 & 3 & 4 \\ i & \ominus & - & + & + \\ j & - & \ominus & + & + \\ k & + & + & - & - \end{array} . \tag{14.6}$$

The circled and uncircled minuses referring to $\mathcal{T}_k^-$ and $\mathcal{T}_k^+$ respectively. We then have

$$\mathcal{A}_{p4} - \frac{\mathcal{A}_{p1}}{\mathcal{A}_{i1}}\mathcal{A}_{i4} < 0, \tag{14.7}$$

$$\mathcal{A}_{p4} - \frac{\mathcal{A}_{p2}}{\mathcal{A}_{i2}}\mathcal{A}_{i4} > 0, \tag{14.8}$$

$$\mathcal{A}_{p4} - \frac{\mathcal{A}_{p2}}{\mathcal{A}_{j2}}\mathcal{A}_{j4} < 0 \tag{14.9}$$

and

$$\mathcal{A}_{p4} - \frac{\mathcal{A}_{p1}}{\mathcal{A}_{j1}}\mathcal{A}_{j4} > 0. \tag{14.10}$$

These are equivalent to

$$\begin{aligned} f_2(i, p: 1, 4) &> 0, \ \ f_2(i, p: 2, 4) < 0, \ \ f_2(j, p: 2, 4) > 0, \\ f_2(j, p: 1, 4) &< 0. \end{aligned} \tag{14.11}$$

It follows from (14.7), (14.8); (14.9), (14.10) that

$$f_2(i, p: 1, 2) < 0, \ \ f_2(j, p: 1, 2) > 0. \tag{14.12}$$

By writing (14.12) in the form

$$\mathcal{A}_{p1} - \frac{\mathcal{A}_{p2}}{\mathcal{A}_{i2}}\mathcal{A}_{i1} < 0, \ \ \mathcal{A}_{p1} - \frac{\mathcal{A}_{p2}}{\mathcal{A}_{j2}}\mathcal{A}_{j1} > 0, \tag{14.13}$$

we have

$$f_2(i, j: 1, 2) > 0. \tag{14.14}$$

Similarly, it is a consequence of (14.7), (14.10) and (14.8), (14.9) that

$$f_2(i, j: 1, 4) < 0, \ \ f_2(i, j: 2, 4) > 0. \tag{14.15}$$

After interchanging the ith and first constraints and updating, we have the coefficient matrix

$$\frac{1}{f_1(i:1)}\begin{pmatrix} 1 & -f_1(i,2) & -f_1(i:3) & -f_1(i:4) \\ f_1(j,1) & f_2(i,j:1,2) & f_2(i,j:1,3) & f_2(i,j:1,4) \\ f_1(p:1) & f_2(i,p:1,2) & f_2(i,p:1,3) & f_2(i,p:1,4) \end{pmatrix} \tag{14.16}$$

It follows from the imposed signs (14.6) - (14.15) that the matrix (14.16) has the sign configuration

$$\begin{array}{ccccc} & 1 & 2 & 3 & 4 \\ i & - & - & + & + \\ j & + & - & \pm & + \\ p & - & + & \pm & - \end{array} \tag{14.17}$$

If the coefficient indexed by $p, 3$ were negative the third basic constraint would be redundant. Therefore, we impose the sign

$$f(i,p:1,3) < 0 \tag{14.18}$$

leaving the configuration

$$\begin{pmatrix} - & - & + & + \\ + & - & \pm & + \\ - & + & + & - \end{pmatrix} \tag{14.19}$$

with only the 2, 3 element having an arbitrary sign. In any case the interchange of the jth and second constraints is admissible. After this interchange, we have the matrix with $D_2 = f_2(i,j:1,2) > 0$

$$\frac{1}{D_2}\begin{pmatrix} f_1(j:2) & -f_1(i:2) & f_2(i,j:2,3) & f_2(i,j:2,4) \\ -f_1(j:1) & f_1(i:1) & -f_2(i,j:1,3) & -f_2(i,j,1,4) \\ -f_2(j,p:1,2) & f_2(i,p:1,2) & f_3(i,j,p:1,2,3) & f_3(i,j,p:1,2,4) \end{pmatrix} \tag{14.20}$$

Now let the coefficients of the pth constraint be denoted by $\mathcal{A}''_{pj}$. It follows from (14.12) that

$$\mathcal{A}''_{p1} < 0 \text{and } \mathcal{A}''_{p2} < 0. \tag{14.21}$$

Hence, if $\mathcal{A}''_{p4} < 0$ the set is either empty or there is a redundant constraint. If $\mathcal{A}''_{p4} > 0$, this configuration does not contribute to the promoted version of $\mathcal{T}_k^+$. Of course, this statement does not apply if the interchange is made with respect to some other constraint. Let us now examine the other admissible exchanges within the present matrix. From (14.19) it appears that the interchange of the ith and second variables is one such possibility. But this follows the interchange of the ith and the first. But this is, by Theorem 12.2, the interchange of the ith and second followed by a permutation.

From (14.19) we see that the only other admissible interchange is the interchange of the jth and third constraints under the condition

$$f_2(i,j:1,3) > 0. \tag{14.22}$$

This interchange gives the matrix

$$\frac{1}{D_2}\begin{pmatrix} f_1(j:3) & -f_1(i:3) & f_2(i,j:3,2) & f_2(i,j:3,4) \\ -f_1(j:1) & f_1(i:1) & -f_2(i,j:1,2) & -f_2(i,j:1,4) \\ -f_2(j,p:1,3) & f_2(i,p:1,3) & f_3(i,j,p:1,3,2) & F_2(i,j,p:1,3,4) \end{pmatrix} \tag{14.23}$$

with $D_2 = f_2(i, j : 1, 3)$ which by (14.22) is positive. By (14.18) we have $f_2(i, p : 1, 3) < 0$. This configuration appears to have insufficient information to resolve the sign of $f_2(j, p : 1, 3)$. However, if the assumption (14.22) leads to a legitimate simplex step it does impose the additional sign

$$f_2(i, j : 3, 4) > 0. \tag{14.24}$$

In any case, the previous configuration was sufficient to resolve the case of the constraints in three variables. When there are more constraints the additional condition (14.24) may be helpful in analyzing the interaction of various sets of three non-basic constraints combined with the basic constraints.

We remark, also, that if the same constraints i and j solve the maximum problem determining the next simplex step for two steps in a row, the analysis of (14.20) is sufficient to produce either a complete simplex step or to find a redundant constraint. That this be the case when both maximums are achieved by the ith constraint would require the interchange $i - 3$. By (14.19) this is impossible since both the 1, 3 and 3, 3 elements are positive.

Finally, we consider the sign configuration

$$\begin{array}{ccccc} & 1 & 2 & 3 & 4 \\ i & \ominus & - & + & + \\ j & + & \ominus & + & + \\ p & + & + & - & - \end{array} \tag{14.25}$$

The interchange of the ith and first constraints leads to

$$\begin{array}{cccc} - & - & + & + \\ - & - & + & + \\ - & + & + & - \end{array} \tag{14.26}$$

instead of (14.19). Some of these signs are determined as before and the others are consequences of Theorem 10.2. Now we notice that the interchange of the jth constraint with the second is the only admissible simplex interchange. Now to apply the preceding analysis to (14.20), we need only (14.21). This is again a consequence of Theorem 10.2.

15. THE CASE OF SIX CONSTRAINTS IN THREE VARIABLES.

The analysis of the preceding section yields the following Theorem.

Theorem 15.1. *Let us consider a set of Six Constraints in Three Variables which is in Canonical form and with only one constraint not satisfying the basic vertex. If completing a simplex step or finding a redundant constraint or finding the set to be empty requires more than three steps then up to a permutation of the first three columns we may assume the configuration of the non-basic constraints*

$$\begin{array}{cccccc} i & : & \ominus & + & \pm & + \\ j & : & + & \ominus & \pm & + \\ p & : & + & + & - & - \end{array} \tag{15.1}$$

We leave open the question of whether the number of steps can be reduced from three to two by starting with the configuration

$$+ + + - \tag{15.2}$$

for the pth constraint.

References

[1] Dantzig, *Linear Programming and Extensions,* Princeton Univ. Press.

[4] Polya, G., Szego, G., *Problems and Theorems in Analysis,* Springer-Verlag, New York, Heidelberg, Berlin, Berlin, 1972.

[5] Strang, G., *Linear Algebra and its Applications (3rd Ed.),* Harcourt, Brace, Jovanovich, San Diego.

[6] Wu, S. and Coppins, R., *Linear Programming and Extension,* McGraw Hill.

Acknowledgement

I would like to thank Jenny Bourne Wahl for criticizing an earlier version of Sections 1 - 7 of this manuscript.

An Estimate of the Semi-Stable Measure of Small Balls in Banach Spaces

BALRAM S. RAJPUT Department of Mathematics, The University of Tennessee, Knoxville, TN 37923

Abstract. Let $(\mathbb{E}, ||\cdot||)$ be a separable Banach space. Let μ be a symmetric r-semi-stable probability measure of index $0 < \alpha \leq 2$ on $\mathbb{E}$, and let $0 < q < \alpha$. It is proven that if $\int_{\mathbb{E}} ||x||^q d\mu = 1$ then $\mu\{||x|| \leq t\} \leq$ const. $t^{\alpha/2}$, for all $t > 0$, where const. depends only on r, q and α (and not $\mathbb{E}$ or μ). This result compliments similar known results for symmetric Gaussian and α-stable probability measures on $\mathbb{E}$. Two other related results are also proved; these are needed for the proof of the above main result.

1. INTRODUCTION AND PRELIMINARIES

Let $(\mathbb{E}, ||\cdot||)$ be a separable Banach space. Let μ be a symmetric Borel probability measure on $\mathbb{E}$. In a recent paper, M. Lewandowski, M. Ryzner, and T. Zak (1992) showed that, if μ is α-stable, satisfying $\int_{\mathbb{E}} ||x||^q d\mu = 1$ with $0 < q < \alpha$, then $\mu\{||x|| \leq t\} \leq$ const. t, where const. depends only on α and q (and not $\mathbb{E}$ or μ). In the case when μ is centered Gaussian, a similar result is proved earlier by S. Szarek (1991) and also by X. Fernique and by J. Sawa; Sawa requires in addition that $\mathbb{E}$ be a Hilbert space. (For a discussion and references of the Fernique and Sawa contribution, we refer the reader to Lewandowski, Ryzner, and Zak (1992)).

This research is partially supported by the University of Tennessee Science Alliance, a State of Tennessee Center of Excellence.

The main effort of this paper is aimed at proving a version of the above result of Lewandowski, Ryzner, and Zak (1992) for the larger class of semi-stable probability measures. Specifically, we prove the result stated in the abstract. The proof of Lewandowski, Ryzner, and Zak (1992) in the stable case is based on the fact that every $\mathbb{E}$-valued symmetric α-stable random variable is conditionally Gaussian and on the well known Anderson Inequality for Gaussian measures. Since a semi-stable random variable in general is not conditionally Gaussian (Rosinski (1991), p.32), the methods used in Lewandowski, Ryzner, and Zak (1992) do not apply in the more general semi-stable case; a similar situation seems to prevail with regard to the methods of proof used by Szarek and Sawa. Our proof, like the one due to Fernique in the Gaussian case (see Lewandowski, Ryzner, and Zak (1992)), is based on the well known Kantor Inequality. In the Gaussian case ($\alpha = 2$), this approach yields the same upper bound for $\mu\{||x|| \leq t\}$ as obtained in Lewandowski, Ryzner, and Zak (1992) in the stable case (namely, const. t). In the proper semi-stable case, on the other hand, this approach yields the upper bound for $\mu\{||x|| \leq t\}$ as const. $t^{\alpha/2}$. which, in the interesting case, i.e., when t is close to 0, is worse than const. t . (For more on this point see Concluding Remark).

For our proof of the main result, in addition to the Kantor Inequality, we also need an estimate for the lower bound of the tail of symmetric semi-stable probability measures on $\mathbb{E}$; this is obtained in Lemma 1. This lower bound is obtained by using the Paley-Zygmund Inequality and another result which provides a comparison between moments of a semi-stable probability measure and a related F-norm (Proposition 1).

Throughout, r and α will denote real numbers satisfying $0 < r < 1$ and $0 < \alpha < 2$; and the notation $r - SS(\alpha)$ will mean " r-semi-stable index α". Further, throughout $\mathbb{E}$ will denote a real separable Banach space. By a measure on $\mathbb{E}$, we shall always mean that it is defined on its Borel σ-algebra. For the sake of brevity, we refer the reader to Chung, Rajput, and Tortrat (1982) and Rajput and Rama-Murthy (1987) for the definition and properties of $\mathbb{E}$-valued $r - SS(\alpha)$ random variables and $r - SS(\alpha)$ probability measures on $\mathbb{E}$. A fact regarding these which will be important for us is the following: Let X be a symmetric $\mathbb{E}$-valued random variable and let $\mu = \mathcal{L}(X)$, the law of X; then X is an $r - SS(\alpha)$ random variable (equivalently, μ is an $r - SS(\alpha)$ probability measure) $\Leftrightarrow$ μ is infinitely divisible and $\mu^{r^n} = r^{r/\alpha} \cdot \mu$, for all $n = \pm 1, \pm 2, \cdots$, where $a \cdot \mu \equiv \mathcal{L}(aX)$ for a real number a and $\mu^s, s > 0$, denotes the sth root of μ (see Chung, Rajput, and Tortrat (1982) and Rajput and Rama-Murthy (1987)). Note also that if μ is centered Gaussian then μ is $r - SS(2)$ measure, for all $0 < r < 1$.

Before we end this Section, we introduce a few more notation: Let $0 < p < q < \alpha$, then we set

$$C(r, \alpha, p, q) \equiv \left(\frac{\alpha}{\alpha - q}\right)^{\frac{1}{q}} \left(\frac{2^{\frac{1}{\alpha}+\frac{2}{p}}}{r^{\frac{1}{\alpha}}}\right).$$

Let $C \equiv C(r, \alpha, \frac{q}{2}, q)$, then we put

$$K(r, \alpha, q) \equiv \left(\frac{3}{2}\right) \left(\frac{1}{\sqrt{r(1-r)}}\right) \left(\frac{(2C)^{\frac{\alpha+q}{2}}}{2^{\frac{q}{2}} - 1}\right);$$

we note that both $C(r, \alpha, p, q)$ and $K(r, \alpha, q)$ are greater than 1. For a non-negative random variable ξ and $p > 0$, we shall use the notation $\Lambda_p(\xi)$ for $\sup_{t>0} t\,(P\{\xi > t\})^{1/p}$; and, for an $\mathbb{E}$- valued random variable X and $q > 0$, we shall use the notation $||X||_q$ for $(E||X||^q)^{\frac{1}{q}}$.

2. STATEMENTS AND PROOFS OF RESULTS

The main result of this paper is the following:

THEOREM 1. Let μ be a symmetric $r-SS(\alpha)$ probability measure on $\mathbb{E}$ and let $0 < q < \alpha$ and $K \equiv K(r, \alpha, q)$. If $\int_{\mathbb{E}} ||x||^q d\mu = 1$, then $\mu\{||x|| \leq t\} \leq K\ t^{\alpha/2}$, for all $t > 0$.

As noted in Section 1, the proof of the theorem depends on the following Lemma. The proof of the Lemma in turn depends on Proposition 1. The counterpart of the Proposition and that of its Corollary (Corollary 1) for the stable case are well known and have played a useful role in the study of stable measures in Banach spaces (see Araujo and Gine (1980), Linde (1982), Linde (1986)). We state and prove these results and the Lemma first. We have stated these in terms of $\mathbb{E}$-valued $r - SS(\alpha)$ random variables (rather than in terms of $r - SS(\alpha)$ probability measures on $\mathbb{E}$) because we found it notationally convenient. We note that even though we have assumed throughout that $0 < \alpha < 2$, the above and the following results and their proofs are also valid in the case when μ is symmetric Gaussian (the case $r - SS(2)$).

PROPOSITION 1. Let $0 < q < \alpha$, and let X be an $\mathbb{E}$-valued symmetric $r - SS(\alpha)$ random variable, then we have

$$\left(\frac{\alpha - q}{\alpha}\right)^{\frac{1}{q}} ||X||_q \leq \Lambda_\alpha(||X||) \leq \left(\frac{2^{\frac{1}{\alpha}+\frac{2}{q}}}{r^{\frac{1}{\alpha}}}\right) ||X||_q.$$

COROLLARY 1. Let X be an $\mathbb{E}$-valued symmetric $r - SS(\alpha)$ random variable and let $0 < p < q < \alpha$, then we have

$$||X||_q \leq C(r, \alpha, p, q) ||X||_p.$$

LEMMA 1. Let $0 < q < \alpha$ and $\varepsilon > 0$, and set $C = C(r, \alpha, \frac{q}{2}, q)$. Then, for any $\mathbb{E}$-valued symmetric $r - SS(\alpha)$ random variable X satisfying $||X||_q > C\varepsilon$, we have

$$P\{||X|| > \varepsilon\} \geq \left(\frac{1}{C^{\frac{q}{2}}} - \frac{\varepsilon^{\frac{q}{2}}}{\sqrt{E||X||^q}}\right)^2;$$

in particular, if $||X||_q \geq 2C\varepsilon$, then

$$P\{||X|| > \varepsilon\} \geq \left(\frac{1}{C^q}\right)\left(1 - 2^{-\frac{q}{2}}\right)^2 > 0.$$

Proof of Proposition 1: The nontrivial part here is the right inequality. A proof of this in the stable case was given by Lindé (see Lindé (1986), p. 137 and Lindé (1982)). The proof given in Lindé (1982) and Lindé (1986) is based on functional analytic methods and uses certain results of De-Acosta (1977), in particular, the fact that the counterpart of

$\Lambda_\alpha(||X||)$ (in the stable case) is finite. Later another proof of this inequality in the stable case was given by Giné, Marcus, and Zinn (1985). This proof is probabilistic and based on an idea of Pisier (see Giné, Marcus, and Zinn (1985)). We adapt this proof to the present semi-stable case.

For every $n = 1, 2, \ldots$, let $k_n = \left[\frac{1}{r^n}\right]$, the integral part of $1/r^n$. Then $\frac{1}{r^n} = k_n + \varepsilon_n$, $0 \leq \varepsilon_n < 1$. Fix n, and let X_j's be iid random variables with $\mathcal{L}(X_j) = \mu^{r^n}, j = 1, \ldots, k_n$, and let Y_n be independent of X_j's with $\mathcal{L}\{Y_n\} = \mu^{1-r^n k_n}$, where $\mu^s, s > 0$, denotes the sth root of $\mu \equiv \mathcal{L}(X)$. Clearly, we have

$$\begin{aligned}
&P\{\max(||Y_n||, ||X_j||, j = 1, \ldots, k_n) > t\} \\
=&1 - P\{\max(||Y_n||, ||X_j||, j = 1, \ldots, k_n) \leq t\} \\
=&1 - \left(\prod_{j=1}^{k_n} P\{||X_j|| \leq t\}\right) P(||Y_n|| \leq t) \\
\geq&1 - (1 - P\{||X_1|| > t\})^{k_n} \\
=&1 - \left(1 - \mu^{r^n}\{||x|| > t\}\right)^{k_n}. \qquad (1)
\end{aligned}$$

Denote the left side of (1) by $L(t)$ and let $v(t) = \mu^{r^n}\{||x|| > t\} = \mu\{||x|| > r^{-n/\alpha}t\}$. Then (1) becomes $L(t) \geq 1 - (1 - v(t))^{k_n}$; equivalently, $v(t) \leq 1 - (1 - L(t))^{1/k_n}$. Then, using the fact $1 - (1 - x)^{1/m} \leq \frac{1}{m}\left(\frac{x}{1-x}\right), 0 \leq x < 1$, for all $m = 1, 2, \ldots$, one finally gets

$$v(t) \leq \left(\frac{1}{k_n}\right) \frac{L(t)}{1 - L(t)}. \qquad (2)$$

If $t > 0$ is such that $P\{||X|| > t\} < 1/2$, then (2) and Lévy's Inequality (Araujo and Gine (1980), p. 57) yield

$$P\{||X|| > r^{-n/\alpha}t\} \leq \left(\frac{1}{k_n}\right) \frac{2P\{||X|| > t\}}{1 - 2P\{||X|| > t\}}. \qquad (3)$$

Set $t_q = 4^{\frac{1}{q}}||X||q$; then, by Chybeshev's Inequality, we have $P\{||X|| > t_q\} \leq \frac{E||X||^q}{t_q^q} = 1/4$. Thus, by (3), we have

$$k_n P\{||X|| > r^{-n/\alpha}t_q\} \leq \frac{2 \cdot 1/4}{1 - 2 \cdot 1/4} = 1, \qquad (4)$$

for every $n = 1, 2, \ldots$ Set $s_n = t_q/r^{n/\alpha}$; then $\left(\frac{s_n}{t_q}\right)^\alpha = \frac{1}{r^n} = k_n + \varepsilon_n$. Hence, using (4), we get $\left(\frac{s_n}{t_q}\right)^\alpha P\{||X|| > s_n\} = (k_n + \varepsilon_n)P\{||X|| > s_n\} \leq 2$. Therefore, for every $n = 1, 2, \ldots$, we have

$$s_n^\alpha P\{||X|| > s_n\} \leq 2t_q^\alpha. \qquad (5)$$

Now let $s_n < s < s_{n+1}$, then $s^\alpha P\{||X|| > s\} \leq s_{n+1}^\alpha P\{||X|| > s_n\} = \left(\frac{s_n^\alpha}{r}\right) \cdot P\{||X|| > s_n\} \leq (2/r)\, t_q^\alpha$. Thus, we have

$$\sup_{s > s_1} s^\alpha P\{||X|| > s\} \leq (2/r)\, t_q^\alpha.$$

But since, clearly, $s^\alpha P\{||X|| > s\} \leq \frac{2t_q^\alpha}{r}$, if $0 < s \leq s_1 = \frac{t_q}{r^{\frac{1}{\alpha}}}$, we have

$$\sup_{s>0} s^\alpha P\{||X|| > s\} \leq \left(\frac{2}{r}\right) t_q^\alpha = \left(\frac{2}{r}\right) 4^{\alpha/q}||X||_q^\alpha.$$

Showing $\Lambda_\alpha(||X||) \leq \left(\frac{2^{\frac{1}{\alpha}+\frac{2}{q}}}{r^{\frac{1}{\alpha}}}\right) ||X||_q$.

Another proof of this part of the inequality is also possible; this proof uses the analog of the above inequality for the stable case (Lindé, 1986), and a comparison result of the tail probabilities of symmetric stable and semi-stable $\mathbb{E}$-valued random variables due to Rosinski (1987). The proof of this comparision result in turn depends on the theory of single stochastic integrals (Rosinski, 1987). The above proof is direct and does not depend on any of these facts.

The left side inequality is standard; we include a simple proof for completeness. For simplicity of notation, Set $\xi = ||X||$. From above, we have $\Lambda \equiv \lambda_\alpha(\xi) < \infty$; and, clearly, by the definition of $\Lambda, P\{\xi > t\} \leq \min\left\{\frac{\Lambda^\alpha}{t^\alpha}, 1\right\}$. Using this we get $E(\xi^q) = q\int_0^\infty u^{q-1}P(\xi > u)du \leq q\int_0^\infty u^{q-1}\min\{\frac{\Lambda^\alpha}{u^\alpha}, 1\}du = q\int_0^\Lambda u^{q-1}du + q\int_\Lambda^\infty u^{q-1}\left(\frac{\Lambda}{u}\right)^\alpha du$ $= \Lambda^q + \left(\frac{q}{\alpha-q}\right)\Lambda^q = \left(\frac{\alpha}{\alpha-q}\right)\Lambda^q$. This yields $\left(\frac{\alpha-q}{\alpha}\right)^{\frac{1}{q}} ||X||_q \leq \Lambda_\alpha(||X||)$. This completes the proof of the Proposition.

The proof of the Corollary is immediate from the Proposition.

Proof of Lemma 1: Let $\xi = ||X||$ and $\lambda = \frac{\varepsilon^{\frac{q}{2}}}{E\xi^{\frac{q}{2}}}$. Then, since $||X||_{\frac{q}{2}} \geq C^{-1}||X||_q > C^{-1}C\varepsilon$, $0 < \lambda < 1$. Therefore, by the well-known Paley-Zygmund Inequality,

$$\begin{aligned} P\{\xi > \varepsilon\} &= P\left\{\xi^{\frac{q}{2}} > \varepsilon^{\frac{q}{2}}\right\} \\ &= P\left\{\xi^{\frac{q}{2}} > \lambda E\xi^{\frac{q}{2}}\right\} \\ &\geq \frac{(1-\lambda)^2(E\xi^{\frac{q}{2}})^2}{E(\xi^q)} \\ &= \left(\frac{E\xi^{\frac{q}{2}}}{\sqrt{E\xi^q}} - \frac{\varepsilon^{\frac{q}{2}}}{\sqrt{E\xi^q}}\right)^2 \\ &\geq \left(\frac{1}{C^{\frac{q}{2}}} - \frac{\varepsilon^{\frac{q}{2}}}{\sqrt{E\xi^q}}\right)^2, \end{aligned}$$

Since $\frac{E\xi^{\frac{q}{2}}}{E\xi^q} \geq \frac{1}{C^{\frac{q}{2}}}$. This proves the first inequality; the second is now immediate from the first.

Proof of Theorem 1: As in the definition of K, Set $C = C(r, \alpha, \frac{q}{2}, q)$ and let $Y = 2CX$ where X is an $\mathbb{E}$-valued random variable with $\mathcal{L}(X) = \mu$. Let n and m be any positive integers satisfying $0 < r^n m \leq 1$. Let $Y_1, \ldots, Y_m$ be iid's with $\mathcal{L}\{Y_1\} = \nu^{r^n} = r^{n/\alpha}.\nu$, where, as before $\nu^s, s > 0$, is the sth roof $\nu \equiv \mathcal{L}(Y)$; and let Z_m be independent of Y_j's with $\mathcal{L}\{Z_m\} = \nu^{1-r^n m}$. Then using $\left(\nu^{r^n}\right)^{*m} * \nu^{1-r^n m} = \nu^{r^n m} * \nu^{1-r^n m} = \nu$, we have

$\nu = \mathcal{L}(Y) = \mathcal{L}\left(\sum_{i=1}^{m} Y_i + Z_m\right)$. (Here * denotes the usual convolution). Hence, it follows, form Kantor's Inequality (Araujo and Giné (1980), p. 136), that

$$\begin{aligned} &P\{||Y|| \le r^{n/\alpha}\} \\ &= P\{||\sum_{i=1}^{m} Y_i + Z_m|| \le r^{n/\alpha}\} \\ &\le \left(\frac{3}{2}\right)\left[\sum_{i=1}^{m} P\{||Y_i|| > r^{n/\alpha}\} + P\{||Z_m|| > r^{n/\alpha}\right]^{-1/2} \\ &\le \left(\frac{3}{2}\right)\left[mP\{||Y_1|| > r^{n/\alpha}\}\right]^{-1/2} \\ &= \left(\frac{3}{2}\right)[mP\{||Y|| > 1\}]^{-1/2}; \end{aligned} \tag{6}$$

in the last step we have used the fact that $\nu^{r^n} = r^{n/\alpha} \cdot \nu$. Taking $\varepsilon = 1$, in the second inequality of Lemma 1, we have

$$P\{||Y|| > 1\}^{-1/2} \le \frac{C^{\frac{q}{2}}}{\left(1 - 2^{-\frac{q}{2}}\right)},$$

provided $||Y||_q \ge 2C$. Therefore, (6) yields

$$P\{||Y|| \le r^{n/\alpha}\} \le Dm^{-1/2}, \text{ provided } ||Y||_q \ge 2C, \tag{7}$$

where $D = \left(\frac{3}{2}\right)\left(\frac{C^{\frac{q}{2}}}{1-2^{-\frac{q}{2}}}\right) = \left(\frac{3}{2}\right)\frac{(2C)^{\frac{q}{2}}}{(2^{\frac{q}{2}}-1)}$.

Let k_n be the integral part of $1/r^n$, then $\frac{1}{r^n} = k_n + \varepsilon_n$, $0 \le \varepsilon_n < 1$. Taking $m = k_n$ in (7), we get

$$P\{||Y|| \le r^{n/\alpha}\} \le Dk_n^{-1/2} = D\left(\frac{1}{r^n} - \varepsilon_n\right)^{-1/2} = Dr^{n/2}(1 - r^n\varepsilon_n)^{-1/2}.$$

Thus, since $(1-r^n\varepsilon_n) \ge 1-r$ and n was arbitrary, we have $P\{||Y|| \le r^{n/\alpha}\} \le D(r^{n/\alpha})^{\alpha/2}(1-r)^{-1/2}$, for all $n = 1, 2, \ldots$. Now let t be any positive real number satisfying $r^{\frac{n+1}{\alpha}} < t < r^{\frac{n}{\alpha}}$; then the preceeding inequality yields that

$$\begin{aligned} P\{||Y|| \le t\} \le &P\{||Y|| \le r^{n/\alpha}\} \le \frac{D\left(r^{n/\alpha}\right)^{\alpha/2}}{\sqrt{1-r}} \\ &= \frac{D\left(r^{\frac{n+1}{\alpha}}\right)^{\alpha/2}}{\sqrt{r(1-r)}} \le \frac{D\, t^{\alpha/2}}{\sqrt{r(1-r)}}, \end{aligned}$$

if $||Y||_q \ge 2C$. We have thus proven that if $0 < t \le r^{\frac{1}{\alpha}}$ and $||Y||_q \ge 2C$, then $P\{||Y|| \le t\} \le \left(\frac{D}{\sqrt{r(1-r)}}\right)t^{\alpha/2}$. If $t > r^{\frac{1}{\alpha}}$, then $\frac{D\, t^{\alpha/2}}{\sqrt{r(1-r)}} > \frac{D\left(r^{\frac{1}{\alpha}}\right)^{\alpha/2}}{\sqrt{r(1-r)}} = \frac{D}{\sqrt{1-r}} > 1$; thus, the preceeding probability inequality is valid for all $t > 0$. Therefore, recalling that $Y = 2CX$ and observing that $K(r,\alpha,q) = \frac{2^{\alpha/2}C^{\alpha/2}D}{\sqrt{r(1-r)}}$, we have $P\{||X|| \leq t\} \leq K\ t^{\alpha/2}$, provided $||X||_q \geq 1$. This completes the proof.

CONCLUDING REMARK

Let $\mathbb{E} = \mathbb{R}$, the real line; then it is easy to show that, for any symmetric $r - SS(\alpha)$ probability measure μ on $\mathbb{R}$ satisfying $\int_{\mathbb{R}} |x|^q d\mu = 1$ with $0 < q < \alpha, \mu\{|x| \leq t\} \leq$ const. t where the const. depends only on r, α (and not on μ). To see this, we proceed as follows: Let σ be the (finite) symmetric spectral measure (on $\triangle \equiv \{r^{\frac{1}{\alpha}} < |s| \leq 1\}$) of the given measure μ, then $\varphi(y)$, the characteristic function of μ, is given by

$$\varphi(y) = \exp\ -\left\{\int_{\triangle} |ys|^{\alpha} k_{\alpha}(|ys|)\sigma(ds)\right\},$$

where $k_\alpha(t) = |t|^{-\alpha}\sum_{n=-\infty}^{\infty} r^{-n}(1-\cos r^{n/\alpha}t), t \neq 0, k_\alpha(0) = 0$ (Rajput and Rama-Murthy, 1987). Now, if f denotes the probability density function of μ, then, using the fact $k_\alpha(t) \geq d_0(r,\alpha), t \neq 0$ (the constant d_0 depending only on r and α) (Rajput and Rama-Murthy, 1987), we have, for $t > 0$,

$$\begin{aligned}
\mu\{|x| \leq t\} =& 2\int_0^t f(x)dx \\
=& \int_0^t \left\{\frac{4}{2\pi}\int_0^\infty \cos(xy)\varphi(y)dy\right\} dx \\
\leq& \left(\frac{2t}{\pi}\right)\int_0^\infty \varphi(y)dy \\
\leq& \left(\frac{2t}{\pi}\right)\int_0^\infty \exp\ -\left\{d_0|y|^\alpha \int_\triangle |s|^\alpha \sigma(ds)\right\} dy \\
=& \left(\frac{2t}{\pi}\right)\left(\frac{\Gamma(\frac{1}{\alpha})}{\alpha d_0^{1/\alpha}}\right)\left(\frac{1}{(\int_\triangle |s|^\alpha\sigma(ds))^{\frac{1}{\alpha}}}\right).
\end{aligned}$$

Now recalling the fact that $(\int_\triangle |s|^\alpha \sigma(ds))^{\frac{1}{\alpha}} \geq d_1(r,\alpha,q).(\int_{\mathbb{R}} |x|^q d\mu)^{\frac{1}{q}}$ (where d_1 depends only on r, α and q) (Lindé, 1986), we get $\mu\{|x| \leq t\} \leq$ const.$(r,\alpha,q)\, t$, provided $\int_{\mathbb{R}} |x|^q d\mu = 1$, where const.$(r,\alpha,q) = \frac{2\Gamma(\frac{1}{\alpha})}{\pi\alpha d_0^{\frac{1}{\alpha}} d_1}$. This fact notwithstanding, the question whether $t^{\alpha/2}$ can be replaced by t in the statement of Theorem 1 remains open. The bet here seems to be that the answer to this is affirmative!

REFERENCES

1 A. Araujo and E. Giné (1980). The CLT for Real and Banach Valued Random Variables, J. Wiley New York.

2 D. M. Chung, B. S. Rajput, and A. Tortrat (1982). Semi-stable laws on topological vector spaces, Z. Wahrsch. verw. Geb, 60: 209- 218.

3 A. De-Acosta (1977). Asymptotic behavior of stable measures, Ann. of Probab., 5: 494-499.

4 E. Giné, M. B. Marcus, and J. Zinn (1985). A version of Chevet's Theorem for stable processes, J. Functional Anal., 63: 47-73.

5 M. Lewandowski, M. Ryzner, and T. Zak (1992). Stable measure of a small ball, Proc. Amer. Math. Soc., 2: 489-494.

6 W. Lindé (1982). Operators generating stable measures on Banach spaces, Z. Wahrsch. verw. Geb., 60: 171-184.

7 W. Lindé (1986). Probability in Banach Spaces, J. Wiley, New York.

8 B. S. Rajput and K. Rama-Murthy (1987). Spectral representations of semi-stable processes, and semi-stable laws on Banach spaces, J. Multi. Anal., 21: 141-159.

9 J. Rosinski (1987). Bilinear random integrals, Dissertations Mathematicae, CCLIX.

10 J. Rosinski (1991). On a class of infinitely divisible processes as mixtures of Gaussian processes, Stable Processes and Related Topics (S. Cambanis, et al), Birkhauser, Boston, 27-41.

11 S. Szarek (1991). Condition numbers of random matrices, J. of Complexity, 7: 131-149.

Nonsquare Constants of Orlicz Spaces

ZHONGDAU REN Department of Mathematics, University of California, Riverside, CA 92521

Dedicated to Professor M. M. Rao on the occasion of his 65th birthday.

Abstract. Estimation of nonsquare constants, in the sense of James, of Orlicz spaces is given. Clarkson's inequalities for L^p space have been generalized for Orlicz space by using M. M. Rao's interpolation theorem. The exact values of nonsquare constants of a class of reflexive Orlicz spaces are also obtained by using a new quantitative index of N-functions and the inequalities of Clarkson type for Orlicz spaces.

1993 Mathematical Subject Classification: 46B30.

1 Introduction

Let X be a Banach space and let $S(X) = \{x \in X : \|x\| = 1\}$ be the unit sphere of X. In 1964, James[9] called X uniformly nonsquare if there exists a $\delta > 0$ such that for any $x, y \in S(X)$, either $\|\frac{1}{2}(x+y)\| \leq 1-\delta$ or $\|\frac{1}{2}(x-y)\| \leq 1-\delta$. In 1990, Gao and Lau[4] introduced the following.

Definition 1.1 The parameter $J(X)$ of a Banach space X, which will be called nonsquare constant in the sense of James in this paper, is defined by

$$J(X) = \sup\{\min(\|x+y\|, \|x-y\|) : x, y \in S(X)\}. \tag{1}$$

Gao and Lau[4] proved that X is uniformly nonsquare in the sense of James if and only if $J(X) < 2$.

Remark 1.2(See [4]) Schäffer[16] called X uniformly nonsquare if there exists an $a > 1$ such that $\max(\|x+y\|, \|x-y\|) \geq a$ for any $x, y \in S(X)$. Nonsquare constant $g(X)$ of a Banach space X, in the sense of Schäffer, is defined by

$$g(X) = \inf\{\max(\|x+y\|, \|x-y\|) : x, y \in S(X)\}.$$

If $\dim X \geq 2$, then $1 \leq g(X) \leq \sqrt{2} \leq J(X) \leq 2$ and $g(X)J(X) = 2$. Therefore, $1 < g(X)$ if and only if $J(X) < 2$, i. e., X is uniformly nonsquare in the sense of Schäffer if and only if X is uniformly nonsquare in the sense of James(see also Gao and Lau[5]).

In this paper, we only deal with $J(X)$ when X is an Orlicz space.

Let

$$\Phi(u) = \int_0^{|u|} \phi(t)dt \qquad \text{and} \qquad \Psi(v) = \int_0^{|v|} \psi(s)ds$$

be a pair of complementary N-functions, i.e., $\phi(t) \nearrow \infty$ as $t \nearrow \infty$. The Orlicz function space $L^\Phi(\Omega)$ on $\Omega = [0,1]$ or $[0,\infty)$ is defined to be the set $\{x : x$ is Lebesgue measurable on Ω and $\rho_\Phi(\lambda x) = \int_\Omega \Phi[\lambda|x(t)|]dt < \infty$ for some $\lambda > 0\}$. The gauge norm and the Orlicz norm are defined by

$$\|x\|_{(\Phi)} = \inf\left\{c > 0 : \rho_\Phi(\frac{x}{c}) \leq 1\right\}$$

and

$$\|x\|_\Phi = \sup\left\{\int_\Omega |x(t)y(t)|dt : \rho_\Psi(y) \leq 1\right\}.$$

The norms are equivalent: $\|x\|_{(\Phi)} \leq \|x\|_\Phi \leq 2\|x\|_{(\Phi)}$. The closed separable subspace $M^\Phi(\Omega)$ of $L^\Phi(\Omega)$ is defined to be the set $\{x \in L^\Phi(\Omega) : \rho_\Phi(\lambda x) < \infty$ for all $\lambda > 0\}$. In the same way we define the Orlicz sequence space ℓ^Φ and its closed separable subspace h^Φ. An important parameter for analysis in an Orlicz space is the rate of growth of the underling N-function. An N-function $\Phi(u)$ is said to satisfy the Δ_2-condition for large u (for small u or for all $u \geq 0$), in symbol $\Phi \in \Delta_2(\infty)(\Phi \in \Delta_2(0)$ or $\Phi \in \Delta_2)$, if there exist $u_0 > 0$ and $K > 2$ such that $\Phi(2u) \leq K\Phi(u)$ for $u \geq u_0$ (for $0 \leq u \leq u_0$ or for $u \geq 0$). An N-function $\Phi(u)$ is said to satisfy the ∇_2-condition for large u, in symbol $\Phi \in \nabla_2(\infty)$, if there exist $u_0 > 0$ and $a > 1$ such that $\Phi(u) \leq \frac{1}{2a}\Phi(au)$ for $u \geq u_0$. Similarly we define $\Phi \in \nabla_2(0)$ and $\Phi \in \nabla_2$. The basic facts on Orlicz spaces can be found in [10], [11] and [14]. For instance, $L^\Phi[0,1]$ ($L^\Phi[0,\infty)$ or ℓ^Φ) is separable if and only if $\Phi \in \Delta_2(\infty)$ ($\Phi \in \Delta_2$ or $\Phi \in \Delta_2(0)$); $L^\Phi[0,1]$ ($L^\Phi[0,\infty)$ or ℓ^Φ) is reflexive if and only if $\Phi \in \Delta_2(\infty) \bigcap \nabla_2(\infty)$ ($\Phi \in \Delta_2 \bigcap \nabla_2$ or $\Phi \in \Delta_2(0) \bigcap \nabla_2(0)$).

A new quantitative index of $\Phi(u)$ is provided by the following six constants:

$$\alpha_\Phi = \liminf_{u\to\infty} \frac{\Phi^{-1}(u)}{\Phi^{-1}(2u)}, \qquad \beta_\Phi = \limsup_{u\to\infty} \frac{\Phi^{-1}(u)}{\Phi^{-1}(2u)}, \tag{2}$$

$$\alpha_\Phi^0 = \liminf_{u\to 0} \frac{\Phi^{-1}(u)}{\Phi^{-1}(2u)}, \qquad \beta_\Phi^0 = \limsup_{u\to 0} \frac{\Phi^{-1}(u)}{\Phi^{-1}(2u)}, \tag{3}$$

and

$$\bar{\alpha}_\Phi = \inf\left\{\frac{\Phi^{-1}(u)}{\Phi^{-1}(2u)} : 0 < u < \infty\right\}, \qquad \bar{\beta}_\Phi = \sup\left\{\frac{\Phi^{-1}(u)}{\Phi^{-1}(2u)} : 0 < u < \infty\right\}. \tag{4}$$

The following result will play the leading role in this paper.

Theorem 1.3 (i) $\Phi \notin \Delta_2(\infty) \Leftrightarrow \beta_\Phi = 1, \Phi \notin \nabla_2(\infty) \Leftrightarrow \alpha_\Phi = \frac{1}{2}$; (ii) $\Phi \notin \Delta_2(0) \Leftrightarrow \beta_\Phi^0 = 1, \Phi \notin \nabla_2(0) \Leftrightarrow \alpha_\Phi^0 = \frac{1}{2}$; (iii) $\Phi \notin \Delta_2 \Leftrightarrow \bar{\beta}_\Phi = 1, \Phi \notin \nabla_2 \Leftrightarrow \bar{\alpha}_\Phi = \frac{1}{2}$.

The proof of Theorem 1.3 can be found in [14, p.23] and [15].

Another quantitative index of Φ is well known and is provided by the following six constants:

$$A_\Phi = \liminf_{t\to\infty} \frac{t\phi(t)}{\Phi(t)}, \qquad B_\Phi = \limsup_{t\to\infty} \frac{t\phi(t)}{\Phi(t)}, \tag{5}$$

$$A_\Phi^0 = \liminf_{t\to 0} \frac{t\phi(t)}{\Phi(t)}, \qquad B_\Phi^0 = \limsup_{t\to 0} \frac{t\phi(t)}{\Phi(t)}, \tag{6}$$

and

$$\bar{A}_\Phi = \inf\left\{\frac{t\phi(t)}{\Phi(t)} : 0 < t < \infty\right\}, \qquad \bar{B}_\Phi = \sup\left\{\frac{t\phi(t)}{\Phi(t)} : 0 < t < \infty\right\}. \tag{7}$$

It is also known that $\Phi \notin \Delta_2(\infty) \Leftrightarrow B_\Phi = \infty$, $\Phi \notin \nabla_2(\infty) \Leftrightarrow A_\Phi = 1$, $\Phi \notin \Delta_2(0) \Leftrightarrow B_\Phi^0 = \infty$, $\Phi \notin \nabla_2(0) \Leftrightarrow A_\Phi^0 = 1$, $\Phi \notin \Delta_2 \Leftrightarrow \bar{B}_\Phi = \infty$ and $\Phi \notin \nabla_2 \Leftrightarrow \bar{A}_\Phi = 1$. Furthermore, we have the following.

Proposition 1.4 Let Φ and Ψ be a pair of complementary N-functions. Then

$$\frac{1}{A_\Phi} + \frac{1}{B_\Psi} = \frac{1}{A_\Phi^0} + \frac{1}{B_\Psi^0} = \frac{1}{\bar{A}_\Phi} + \frac{1}{\bar{B}_\Psi} = 1. \tag{8}$$

Proposition 1.5 Let $\Phi(u)$ be an N-function. Then

$$2^{-\frac{1}{A_\Phi}} \le \alpha_\Phi \le \beta_\Phi \le 2^{-\frac{1}{B_\Phi}}, \tag{9}$$

$$2^{-\frac{1}{A_\Phi^0}} \le \alpha_\Phi^0 \le \beta_\Phi^0 \le 2^{-\frac{1}{B_\Phi^0}}, \tag{10}$$

and

$$2^{-\frac{1}{\bar{A}_\Psi}} \le \bar{\alpha}_\Phi \le \bar{\beta}_\Phi \le 2^{-\frac{1}{\bar{B}_\Phi}}. \tag{11}$$

The proofs of Propositions 1.4 and 1.5 can be found in [14, p.27], [11] and [15].

2 Lower Bounds of $J(L^{(\Phi)}(\Omega))$ and $J(L^\Phi(\Omega))$

We first consider the Orlicz space $L^{(\Phi)}(\Omega) = (L^\Phi(\Omega), \|\cdot\|_{(\Phi)})$ with $\|\cdot\|_{(\Phi)}$ being the gauge norm defined by an N-function Φ on $\Omega = [0,1]$, or $[0,\infty)$ with the usual Lebesgue measure μ, and the Orlicz sequence space $\ell^{(\Phi)}$.

Theorem 2.1 Let Φ be an N-function. Then, nonsquare constants of $L^{(\Phi)}[0,1]$, $L^{(\Phi)}[0,\infty)$ and $\ell^{(\Phi)}$, in the sense of James, satisfy respectively

$$\max\left(\frac{1}{\alpha_\Phi}, 2\beta_\Phi\right) \le J(L^{(\Phi)}[0,1]), \tag{12}$$

$$\max\left(\frac{1}{\bar{\alpha}_\Phi}, 2\bar{\beta}_\Phi\right) \le J(L^{(\Phi)}[0,\infty)) \tag{13}$$

and

$$\max\left(\frac{1}{\alpha^0_\Phi}, 2\beta^0_\Phi\right) \le J(\ell^{(\Phi)}). \tag{14}$$

Proof To prove (12), we first show

$$\frac{1}{\alpha_\Phi} \le J(L^{(\Phi)}[0,1]). \tag{15}$$

By (2), there exist $1 < u_i \nearrow \infty$ such that $\lim_{i\to\infty} \frac{\Phi^{-1}(u_i)}{\Phi^{-1}(2u_i)} = \alpha_\Phi$. For any given $\epsilon > 0$, there is an $i_0 \ge 1$ such that for $i \ge i_0$

$$\frac{\Phi^{-1}(u_i)}{\Phi^{-1}(2u_i)} < \alpha_\Phi + \epsilon.$$

For simplicity, we set $u_0 = u_{i_0}$. Choose G_1 and G_2 in $[0,1]$ such that $G_1 \bigcap G_2 = \emptyset$ and $\mu(G_1) = \mu(G_2) = \frac{1}{2u_0}$. Put

$$x(t) = \Phi^{-1}(2u_0)\chi_{G_1}(t) \quad \text{and} \quad y(t) = \Phi^{-1}(2u_0)\chi_{G_2}(t),$$

where χ_{G_1} is the characteristic function of G_1. Note that

$$\|\chi_{G_1}\|_{(\Phi)} = \frac{1}{\Phi^{-1}(\frac{1}{\mu(G_1)})}.$$

We have $\|x\|_{(\Phi)} = \|y\|_{(\Phi)} = 1$ and

$$\|x-y\|_{(\Phi)} = \|x+y\|_{(\Phi)} = \frac{\Phi^{-1}(2u_0)}{\Phi^{-1}(u_0)} > \frac{1}{\alpha_\Phi + \epsilon}.$$

Since ϵ is arbitrary, we obtain (15) by (1). Next we prove

$$2\beta_\Phi \le J(L^{(\Phi)}[0,1]). \tag{16}$$

For any given $\epsilon > 0$, by (2), there exists a $v_0 > 1$ such that

$$\frac{\Phi^{-1}(v_0)}{\Phi^{-1}(2v_0)} > \beta_\Phi - \frac{\epsilon}{2}.$$

Choose E_1 and E_2 in $[0,1]$ such that $E_1 \bigcap E_2 = \emptyset$ and $\mu(E_1) = \mu(E_2) = \frac{1}{2v_0}$. Put

$$x(t) = \Phi^{-1}(v_0)[\chi_{E_1}(t) + \chi_{E_2}(t)] \quad \text{and} \quad y(t) = \Phi^{-1}(v_0)[\chi_{E_1}(t) - \chi_{E_2}(t)].$$

Then $\|x\|_{(\Phi)} = \|y\|_{(\Phi)} = 1$ and

$$\|x-y\|_{(\Phi)} = \|x+y\|_{(\Phi)} = \frac{2\Phi^{-1}(v_0)}{\Phi^{-1}(2v_0)} > 2\beta_\Phi - \epsilon.$$

Since ϵ is arbitrary, we get (16) by (1). Hence, (12) follows from (15) and (16).

The proof of (13) is similar to that of (12).

Finally, we prove (14). We first show

$$\frac{1}{\alpha_\Phi^0} \leq J(\ell^{(\Phi)}). \tag{17}$$

For any given $1 > \epsilon > 0$, by (3), there exists $0 < u_0 < \frac{\epsilon}{2}$ such that $[\Phi^{-1}(u_0)/\Phi^{-1}(2u_0)] < \alpha_\Phi^0 + \epsilon$ or, equivalently,

$$\Phi[(\alpha_\Phi^0 + \epsilon)\Phi^{-1}(2u_0)] > u_0.$$

Let $k_0 = [\frac{1}{2u_0}]$ be the integer part of $\frac{1}{2u_0}$. Then $k_0 \leq \frac{1}{2u_0} < k_0 + 1$. Choose $c \geq 0$ such that $2k_0u_0 + \Phi(c) = 1$. Put

$$x = (\overbrace{\Phi^{-1}(2u_0), \cdots, \Phi^{-1}(2u_0)}^{k_0}, c, 0, 0, \cdots)$$

and

$$y = (\overbrace{0, \cdots, 0}^{k_0}, 0, \overbrace{\Phi^{-1}(2u_0), \cdots, \Phi^{-1}(2u_0)}^{k_0}, c, 0, 0, \cdots).$$

Then, we have $\rho_\Phi(x) = \rho_\Phi(y) = 1$, $\|x\|_{(\Phi)} = \|y\|_{(\Phi)} = 1$ and

$$\begin{aligned}
\rho_\Phi\left[\frac{(\alpha_\Phi^0 + \epsilon)(x-y)}{1-\epsilon}\right] &= \rho_\Phi\left[\frac{(\alpha_\Phi^0 + \epsilon)(x+y)}{1-\epsilon}\right] \\
&> \frac{1}{1-\epsilon}\rho_\Phi[(\alpha_\Phi^0 + \epsilon)(x+y)] \\
&= \frac{1}{1-\epsilon}\{2k_0\Phi[(\alpha_\Phi^0 + \epsilon)\Phi^{-1}(2u_0)] + 2\Phi[c(\alpha_\Phi^0 + \epsilon)]\} \\
&> \frac{2k_0u_0}{1-\epsilon} > \frac{1-2u_0}{1-\epsilon} > 1.
\end{aligned}$$

Therefore,

$$\min(\|x-y\|_{(\Phi)}, \|x+y\|_{(\Phi)}) \geq \frac{1-\epsilon}{\alpha_\Phi^0 + \epsilon}.$$

We have proved (17) by (1), since ϵ is arbitrary. Secondly, we prove

$$2\beta_\Phi^0 \leq J(\ell^{(\Phi)}). \tag{18}$$

For any given $1 > \epsilon > 0$, by (3), there exists $0 < v_0 < \frac{\epsilon}{2}$ such that $[\Phi^{-1}(v_0)/\Phi^{-1}(2v_0)] > \beta_\Phi^0 - \frac{\epsilon}{2}$ or, equivalently

$$\Phi\left[\frac{2\Phi^{-1}(v_0)}{2\beta_\Phi^0 - \epsilon}\right] > 2v_0.$$

Let $k_0 = [\frac{1}{2v_0}]$. Then $k_0 \leq \frac{1}{2v_0} < k_0 + 1$. Choose $t \geq 0$ such that $2k_0v_0 + \Phi(t) = 1$. Put

$$x = (\overbrace{\Phi^{-1}(v_0), \cdots, \Phi^{-1}(v_0)}^{k_0}, \overbrace{\Phi^{-1}(v_0), \cdots, \Phi^{-1}(v_0)}^{k_0}, t, 0, 0, \cdots)$$

and

$$y = (\overbrace{\Phi^{-1}(v_0), \cdots, \Phi^{-1}(v_0)}^{k_0}, \overbrace{-\Phi^{-1}(v_0), \cdots, -\Phi^{-1}(v_0)}^{k_0}, 0, t, 0, \cdots).$$

Then $\|x\|_{(\Phi)} = \|y\|_{(\Phi)} = 1$ since $\rho_\Phi(x) = \rho_\Phi(y) = 1$ and

$$\begin{aligned} \rho_\Phi\left[\frac{x-y}{(1-\epsilon)(2\beta_\Phi^0 - \epsilon)}\right] &= \rho_\Phi\left[\frac{x+y}{(1-\epsilon)(2\beta_\Phi^0 - \epsilon)}\right] \\ &> \frac{1}{1-\epsilon}\rho_\Phi\left[\frac{x+y}{2\beta_\Phi^0 - \epsilon}\right] \\ &= \frac{1}{1-\epsilon}\left\{k_0\Phi\left[\frac{2\Phi^{-1}(v_0)}{2\beta_\Phi^0 - \epsilon}\right] + 2\Phi\left(\frac{t}{2\beta_\Phi^0 - \epsilon}\right)\right\} \\ &> \frac{2k_0v_0}{1-\epsilon} > \frac{1-2v_0}{1-\epsilon} > 1. \end{aligned}$$

Therefore,

$$\min(\|x-y\|_{(\Phi)}, \|x+y\|_{(\Phi)}) \geq (1-\epsilon)(2\beta_\Phi^0 - \epsilon).$$

Since ϵ is arbitrary, we obtain (18). Finally, (14) follows from (17) and (18).□

Next we deal with another three classical Orlicz spaces equipped with Orlicz norm.

Theorem 2.2 Let Φ be an N-function. Then nonsquare constants of $L^\Phi[0,1] = (L^\Phi[0,1], \|\cdot\|_\Phi)$, $L^\Phi[0,\infty)$ and ℓ^Φ, in the sense of James, satisfy respectively

$$\max\left(2\beta_\Psi, \frac{1}{\alpha_\Psi}\right) \leq J(L^\Phi[0,1]), \tag{19}$$

$$\max\left(2\bar{\beta}_\Psi, \frac{1}{\bar{\alpha}_\Psi}\right) \leq J(L^\Phi[0,\infty)) \tag{20}$$

and

$$\max\left(2\beta_\Psi^0, \frac{1}{\alpha_\Psi^0}\right) \leq J(\ell^\Phi), \tag{21}$$

where Ψ is the complementary N-function to Φ.

Proof We omit the proof of (19) and turn to prove (20). We first show

$$2\bar{\beta}_\Psi \leq J(L^\Phi[0,\infty)). \tag{22}$$

By the definition of $\bar{\beta}_\Psi$ in (4), for any given $1 > \epsilon > 0$ there exists $0 < v_0 < \infty$ such that

$$\frac{\Psi^{-1}(v_0)}{\Psi^{-1}(2v_0)} > \bar{\beta}_\Psi - \frac{\epsilon}{2}.$$

Choose G_1 and G_2 in $[0,\infty)$ such that $G_1 \bigcap G_2 = \emptyset$ and $\mu(G_1) = \mu(G_2) = \frac{1}{2v_0}$. Put

$$x(t) = \frac{2v_0}{\Psi^{-1}(2v_0)}\chi_{G_1}(t) \quad \text{and} \quad y(t) = \frac{2v_0}{\Psi^{-1}(2v_0)}\chi_{G_2}(t).$$

Note that

$$\|\chi_{G_1}\|_\Phi = \mu(G_1)\Psi^{-1}(\frac{1}{\mu(G_1)}).$$

Therefore, one has $\|x\|_\Phi = \|y\|_\Phi = 1$ and

$$\|x-y\|_\Phi = \|x+y\|_\Phi = \frac{2\Psi^{-1}(v_0)}{\Psi^{-1}(2v_0)} > 2\bar{\beta}_\Psi - \epsilon.$$

Since ϵ is arbitrary, we obtain (22). Next we prove

$$\frac{1}{\bar{\alpha}_\Psi} \leq J(L^\Phi[0,\infty)). \tag{23}$$

For any given $\epsilon > 0$, there is a $u_0 > 0$ such that

$$\frac{\Psi^{-1}(u_0)}{\Psi^{-1}(2u_0)} < \bar{\alpha}_\Psi + \epsilon.$$

Choose E_1 and E_2 in $[0,\infty)$ such that $E_1 \bigcap E_2 = \emptyset$ and $\mu(E_1) = \mu(E_2) = \frac{1}{2u_0}$. Put

$$x(t) = \frac{u_0}{\Psi^{-1}(u_0)}[\chi_{E_1}(t) + \chi_{E_2}(t)] \quad \text{and} \quad y(t) = \frac{u_0}{\Psi^{-1}(u_0)}[\chi_{E_1}(t) - \chi_{E_2}(t)].$$

One has $\|x\|_\Phi = \|y\|_\Phi = 1$ and

$$\|x-y\|_\Phi = \|x+y\|_\Phi = \frac{\Psi^{-1}(2u_0)}{\Psi^{-1}(u_0)} > \frac{1}{\bar{\alpha}_\Psi + \epsilon}.$$

Since ϵ is arbitrary, we obtain (23). Hence, (20) follows from (22) and (23).

To prove (21), we first show

$$2\beta_\Psi^0 \leq J(\ell^\Phi). \tag{24}$$

For given $1 > \epsilon > 0$, there exist $v_n \searrow 0$ such that for all $n \geq 1$

$$\frac{\Psi^{-1}(v_n)}{\Psi^{-1}(2v_n)} > \beta_\Psi^0 - \frac{\epsilon}{2}.$$

We may assume $2v_n \leq 1$ for all $n \geq 1$. Let $k_n = [\frac{1}{2v_n}]$. Then

$$\frac{1}{k_n+1} < 2v_n \leq \frac{1}{k_n}.$$

Since $\frac{\Psi^{-1}(v)}{v} \nearrow \infty$ as $v \searrow 0$, we have

$$2(k_n+1)\Psi^{-1}\left[\frac{1}{2(k_n+1)}\right] > \frac{\Psi^{-1}(v_n)}{v_n} \geq 2k_n\Psi^{-1}\left(\frac{1}{2k_n}\right)$$

and

$$(k_n+1)\Psi^{-1}\left(\frac{1}{k_n+1}\right) > \frac{\Psi^{-1}(2v_n)}{2v_n} \geq k_n\Psi^{-1}\left(\frac{1}{k_n}\right).$$

Put

$$b_n = \left[k_n \Psi^{-1}(\frac{1}{k_n})\right]^{-1}$$

and

$$c_n = 2(k_n+1)\Psi^{-1}\left[\frac{1}{2(k_n+1)}\right] - 2k_n\Psi^{-1}\left(\frac{1}{2k_n}\right).$$

Then $b_n \searrow 0$ and

$$c_n < 2(k_n+1)\Psi^{-1}\left(\frac{1}{2k_n}\right) - 2k_n\Psi^{-1}\left(\frac{1}{2k_n}\right) = 2\Psi^{-1}\left(\frac{1}{2k_n}\right) \to 0$$

as $n \to \infty$. Choose $n_0 \geq 1$ such that $b_n c_n < \epsilon$ for all $n \geq n_0$. For simplicity, we set $v_0 = v_{n_0}$, $k_0 = k_{n_0}$, $c_0 = c_{n_0}$ and $b_0 = b_{n_0}$. Put

$$x = (\overbrace{b_0, b_0, \cdots, b_0}^{k_0}, 0, 0, \cdots, 0, \cdots)$$

and

$$y = (\overbrace{0, 0, \cdots, 0}^{k_0}, \overbrace{b_0, b_0, \cdots, b_0}^{k_0}, 0, 0, \cdots).$$

We have $\|x\|_\Phi = \|y\|_\Phi = b_0 k_0 \Psi^{-1}(\frac{1}{k_0}) = 1$ and

$$\begin{aligned}
\|x-y\|_\Phi &= \|x+y\|_\Phi \\
&= b_0 2k_0 \Psi^{-1}\left(\frac{1}{2k_0}\right) \\
&= b_0 \left\{2(k_0+1)\Psi^{-1}\left[\frac{1}{2(k_0+1)}\right] - c_0\right\} \\
&> b_0 \left[\frac{\Psi^{-1}(v_0)}{v_0} - c_0\right] \\
&> b_0 \left[\frac{1}{v_0}\left(\beta_\Psi^0 - \frac{\epsilon}{2}\right)\Psi^{-1}(2v_0) - c_0\right] \\
&\geq 2\left(\beta_\Psi^0 - \frac{\epsilon}{2}\right) b_0 k_0 \Psi^{-1}\left(\frac{1}{k_0}\right) - b_0 c_0 \\
&= 2\left(\beta_\Psi^0 - \frac{\epsilon}{2}\right) - b_0 c_0 \\
&> 2(\beta_\Psi^0 - \epsilon).
\end{aligned}$$

Since ϵ is arbitrary, we obtain (24). Finally, we prove

$$\frac{1}{\alpha_\Psi^0} \leq J(\ell^\Phi). \tag{25}$$

For any given $1 > \epsilon > 0$, there exist $\frac{1}{2} > u_n \searrow 0$ such that for all $n \geq 1$

$$\frac{\Psi^{-1}(u_n)}{\Psi^{-1}(2u_n)} < \alpha_\Psi^0 + \epsilon.$$

Put $k_n = [\frac{1}{2u_n}]$, $t_n = [2k_n\Psi^{-1}(\frac{1}{2k_n})]^{-1}$ and

$$s_n = (k_n+1)\Psi^{-1}\left(\frac{1}{k_n+1}\right) - k_n\Psi^{-1}\left(\frac{1}{k_n}\right).$$

Since $t_n \searrow 0$ and $s_n \searrow 0$ as $n \to \infty$, there is an $n_0 \geq 1$ such that $t_n s_n < \frac{\epsilon}{4}$ for all $n \geq n_0$ and so, for $n \geq n_0$

$$2t_n s_n < \frac{\epsilon}{2} < \frac{\epsilon}{1+\epsilon} \leq \frac{\epsilon}{\alpha_\Psi^0 + \epsilon}.$$

Let us set $u_0 = u_{n_0}$, $t_0 = t_{n_0}$ and $s_0 = s_{n_0}$ and define

$$x = (\overbrace{t_0, t_0, \cdots, t_0}^{k_0}, \overbrace{t_0, t_0, \cdots, t_0}^{k_0}, 0, 0, \cdots)$$

and

$$y = (\overbrace{t_0, t_0, \cdots, t_0}^{k_0}, \overbrace{-t_0, -t_0, \cdots, -t_0}^{k_0}, 0, 0, \cdots).$$

Then, we have $\|x\|_\Phi = \|y\|_\Phi = t_0 2k_0\Psi^{-1}(\frac{1}{2k_0}) = 1$ and

$$\begin{aligned}
\|x-y\|_\Phi &= \|x+y\|_\Phi \\
&= 2t_0k_0\Psi^{-1}\left(\frac{1}{k_0}\right) \\
&= 2t_0\left[(k_0+1)\Psi^{-1}\left(\frac{1}{k_0+1}\right) - s_0\right] \\
&> 2t_0\left[\frac{\Psi^{-1}(2u_0)}{2u_0} - s_0\right] \\
&> 2t_0\left[\frac{\Psi^{-1}(u_0)}{2u_0(\alpha_\Psi^0+\epsilon)} - s_0\right] \\
&\geq \frac{t_0 2k_0}{\alpha_\Psi^0+\epsilon}\Psi^{-1}\left(\frac{1}{2k_0}\right) - 2t_0 s_0 \\
&> \frac{1-\epsilon}{\alpha_\Psi^0+\epsilon}.
\end{aligned}$$

Since ϵ is arbitrary, we have proved (25). Thus, (21) follows from (24) and (25). □

Some examples will be given in Section 4.

Remark 2.3 James[9] proved that every uniformly nonsquare Banach space is reflexive. For the above six classical Orlicz spaces, this can be easily proved. For instance, by Theorem 1.3 and Theorem 2.1 we have that

$$\Phi \notin \Delta_2(\infty)\bigcap\nabla_2(\infty) \Longrightarrow \max\left(\frac{1}{\alpha_\Phi}, 2\beta_\Phi\right) = 2 \Longrightarrow J(L^{(\Phi)}[0,1]) = 2.$$

Moreover, Chen[1], Hudzik[7] and Wang and Chen[17] proved that uniform nonsquareness coincides with reflexivity for these Orlicz spaces(see also [6]). Some relations between nonsquare constants and other geometric coefficients of Banach space can be found in [5, Theorem 5.4] and [18, Theorem 3.2].

3 A Generalization of Clarkson's Inequalities

Clarkson[2] is the first mathematician to study geometry of Banach space. His results, called Clarkson's inequalities in these later days, deal only with L^p spaces(see also Corollary 3.4 in this section). In 1966, Rao[12] first obtained Riesz-Thorin type interpolation theorem between Orlicz spaces equipped with Orlicz norm(see also [14, p. 226]). In 1972, Cleaver[3] generalized Rao's interpolation theorem for ℓ^p-product of Orlicz spaces(see also [14, p. 240]). In 1985, the author proved that these theorems are still valid for Orlicz spaces equipped with gauge norm(see [14, p.226, p.256]). In this section, by using Rao's theorem with its generalization, we generalize Clarkson's inequalities for the case of Orlicz spaces. The main result of this section is Theorm 3.2, which will be used in Section 4. Let us start with the following.

Lemma 3.1 Let Φ be an N-function. Suppose that $\Phi_0(u) = u^2$, $0 < s \leq 1$ and $\Phi_s(u)$ is defined to be the inverse of

$$\Phi_s^{-1}(u) = [\Phi^{-1}(u)]^{1-s}[\Phi_0^{-1}(u)]^s. \tag{26}$$

Then $\Phi_s \in \Delta_2 \bigcap \nabla_2$.

Proof Since $\frac{1}{2} \leq \bar{\alpha}_\Phi \leq \bar{\beta}_\Phi \leq 1$, $0 < s \leq 1$ and for $u > 0$

$$\frac{\Phi_s^{-1}(u)}{\Phi_s^{-1}(2u)} = \left[\frac{\Phi^{-1}(u)}{\Phi^{-1}(2u)}\right]^{1-s}\left[\frac{\sqrt{u}}{\sqrt{2u}}\right]^s, \tag{27}$$

we have

$$\bar{\beta}_{\Phi_s} = (\bar{\beta}_\Phi)^{1-s}\left(\frac{1}{\sqrt{2}}\right)^s \leq \left(\frac{1}{\sqrt{2}}\right)^s < 1$$

and

$$\bar{\alpha}_{\Phi_s} = (\bar{\alpha}_\Phi)^{1-s}\left(\frac{1}{\sqrt{2}}\right)^s \geq \frac{(\sqrt{2})^s}{2} > \frac{1}{2}.$$

Therefore, the conclusion follows from Theorem 1.3(iii). □

Theorem 3.2 Let Φ be an N-function and Φ_s be the inverse of (26). Suppose that Ω is one of $[0, 1]$, $[0, \infty)$ with Lebesgue measure and N with counting measure. If $0 < s \leq 1$, then we have for any $x, y \in L^{(\Phi_s)}(\Omega)$

$$\left[\|x+y\|_{(\Phi_s)}^{\frac{2}{s}} + \|x-y\|_{(\Phi_s)}^{\frac{2}{s}}\right]^{\frac{s}{2}} \leq 2^{\frac{s}{2}}\left[\|x\|_{(\Phi_s)}^{\frac{2}{2-s}} + \|y\|_{(\Phi_s)}^{\frac{2}{2-s}}\right]^{\frac{2-s}{2}}. \tag{28}$$

Similarly, we have for any $x, y \in L^{\Phi_s}(\Omega)$

$$\left[\|x+y\|_{\Phi_s}^{\frac{2}{s}} + \|x-y\|_{\Phi_s}^{\frac{2}{s}}\right]^{\frac{s}{2}} \leq 2^{\frac{s}{2}}\left[\|x\|_{\Phi_s}^{\frac{2}{2-s}} + \|y\|_{\Phi_s}^{\frac{2}{2-s}}\right]^{\frac{2-s}{2}}. \tag{29}$$

Proof Let $\vec{\Phi}_1 = (\Phi, \Phi)$ and $1 \leq r_1 \leq \infty$.We define

$$X[(\vec{\Phi}_1), r_1] = \{(x, y) : x, y \in M^{(\Phi)}(\Omega); \|(x, y)\|_{(\vec{\Phi}_1), r_1} < \infty\},$$

where

$$\|(x,y)\|_{(\vec{\Phi}_1),r_1} = \begin{cases} \left[\|x\|_{(\Phi)}^{r_1} + \|y\|_{(\Phi)}^{r_1}\right]^{\frac{1}{r_1}}, & \text{if } 1 \le r_1 < \infty \\ \max(\|x\|_{(\Phi)}, \|y\|_{(\Phi)}), & \text{if } r_1 = \infty. \end{cases}$$

Similarly, we can define $X[(\vec{Q}_1), t_1]$, $X[(\vec{\Phi}_2), r_2]$ and $X[(\vec{Q}_2), t_2]$ for $\vec{Q}_1 = (\Phi, \Phi)$ and $\vec{\Phi}_2 = \vec{Q}_2 = (\Phi_0, \Phi_0)$. Now let us choose $r_1 = 1$, $t_1 = \infty$ and $r_2 = t_2 = 2$, and define a linear operator $T : X[(\vec{\Phi}_k), r_k] \to X[(\vec{Q}_k), t_k]$ for $k = 1, 2$ by $T(x,y) = (x+y, x-y)$. Then, one has

$$\begin{aligned} \|T(x,y)\|_{(\vec{Q}_1),t_1} &= \max(\|x+y\|_{(\Phi)}, \|x-y\|_{(\Phi)}) \\ &\le \|x\|_{(\Phi)} + \|y\|_{(\Phi)} \\ &= C_1\|(x,y)\|_{(\vec{\Phi}_1),r_1} \end{aligned}$$

and

$$\begin{aligned} \|T(x,y)\|_{(\vec{Q}_2),t_2} &= \left[\|x+y\|_{(\Phi_0)}^{t_2} + \|x-y\|_{(\Phi_0)}^{t_2}\right]^{\frac{1}{t_2}} \\ &= \left[\|x+y\|_2^2 + \|x-y\|_2^2\right]^{\frac{1}{2}} \\ &= \sqrt{2}\left[\|x\|_2^2 + \|y\|_2^2\right]^{\frac{1}{2}} \\ &= C_2\|(x,y)\|_{(\vec{\Phi}_2),r_2}, \end{aligned}$$

where $C_1 = 1$ and $C_2 = \sqrt{2}$. Let r_s and t_s be determined by

$$\frac{1}{r_s} = \frac{1-s}{r_1} + \frac{s}{r_2} \quad \text{and} \quad \frac{1}{t_s} = \frac{1-s}{t_1} + \frac{s}{t_2}.$$

Then

$$r_s = \frac{2}{2-s} \quad \text{and} \quad t_s = \frac{2}{s}. \tag{30}$$

In view of Rao's interpolation theorem and Cleaver's generalization(see also [14, pp.236-239]), we have $T \in \{X[(\vec{\Phi}_s), r_s] \to X[(\vec{Q}_s), t_s]\}$ and, by $C_1^{1-s}C_2^s = 2^{\frac{s}{2}}$,

$$\|T(x,y)\|_{(\vec{Q}_s),t_s} \le 2^{\frac{s}{2}}\|(x,y)\|_{(\vec{\Phi}_s),r_s}, \tag{31}$$

where $\vec{\Phi}_s = \vec{Q}_s = (\Phi_s, \Phi_s)$ with Φ_s being the inverse of (26) for $0 < s \le 1$. Therefore, (30) implies that $X[(\vec{\Phi}_s), r_s] = \{(x,y) : x, y \in M^{(\Phi_s)}(\Omega)\}$ equipped with norm

$$\|(x,y)\|_{(\vec{\Phi}_s),r_s} = \left[\|x\|_{(\Phi_s)}^{\frac{2}{2-s}} + \|y\|_{(\Phi_s)}^{\frac{2}{2-s}}\right]^{\frac{2-s}{2}} \tag{32}$$

and that

$$\|T(x,y)\|_{(\vec{Q}_s),t_s} = \left[\|x+y\|_{(\Phi_s)}^{\frac{2}{s}} + \|x-y\|_{(\Phi_s)}^{\frac{2}{s}}\right]^{\frac{s}{2}} \tag{33}$$

It follows from (31), (32) and (33) that (28) holds for any $x, y \in M^{(\Phi_s)}(\Omega)$. By Lemma 3.1, we have $L^{(\Phi_s)}(\Omega) = M^{(\Phi_s)}(\Omega)$ which completes the proof of (28).

For Orlicz norm we define

$$Y[\vec{\Phi}_1, r_1] = \{(x, y) : x, y \in M^{\Phi}(\Omega); \|(x, y)\|_{\vec{\Phi}_1, r_1} < \infty\},$$

where

$$\|(x, y)\|_{\vec{\Phi}_1, r_1} = \begin{cases} [\|x\|_{\Phi}^{r_1} + \|y\|_{\Phi}^{r_1}]^{\frac{1}{r_1}}, & \text{if } 1 \leq r_1 < \infty \\ \max(\|x\|_{\Phi}, \|y\|_{\Phi}), & \text{if } r_1 = \infty. \end{cases}$$

Hence, (29) holds by similar arguments. □

Recall that the modulus of convexity and the modulus of smoothness of a Banach space X are $\delta(X, \epsilon)$ defined on $[0, 2]$ and $\varrho(X, \tau)$ on $[0, \infty)$ respectively by

$$\delta(X, \epsilon) = \inf\left\{1 - \frac{1}{2}\|x + y\| : x, y \in S(X), \|x - y\| = \epsilon\right\}$$

and

$$\varrho(X, \tau) = \sup\left\{\frac{1}{2}(\|x + y\| + \|x - y\|) - 1 : x \in S(X), \|y\| = \tau\right\}.$$

We say that X is uniformly convex if $\delta(X, \epsilon) > 0$ for every $2 \geq \epsilon > 0$ and that X is uniformly smooth if $\lim_{\tau \to 0}[\varrho(X, \tau)/\tau] = 0$.

Corollary 3.3 Let Φ be an N-function and let Φ_s be the inverse of (26). Suppose that $0 < s \leq 1$ and that

$$X_s \in \{L^{(\Phi_s)}[0, 1], L^{(\Phi_s)}[0, \infty), \ell^{(\Phi_s)}, L^{\Phi_s}[0, 1], L^{\Phi_s}[0, \infty), \ell^{\Phi_s}\}.$$

Then, X_s is uniformly convex and uniformly smooth. More precicely, one has

$$\delta(X_s, \epsilon) \geq 1 - \frac{1}{2}\left(2^{\frac{2}{s}} - \epsilon^{\frac{2}{s}}\right)^{\frac{s}{2}} \tag{34}$$

and

$$\varrho(X_s, \tau) \leq \left(1 + \tau^{\frac{2}{2-s}}\right)^{\frac{2-s}{2}} - 1. \tag{35}$$

Proof We first deal with gauge norm. If $x, y \in S(X_s)$ and $\|x - y\|_{(\Phi_s)} \geq \epsilon$, one has from (28)

$$\left(\|x + y\|_{(\Phi_s)}^{\frac{2}{s}} + \epsilon^{\frac{2}{s}}\right)^{\frac{s}{2}} \leq 2$$

or, equivalentely,

$$1 - \frac{1}{2}\|x + y\|_{(\Phi_s)} \geq 1 - \frac{1}{2}\left(2^{\frac{2}{s}} - \epsilon^{\frac{2}{s}}\right)^{\frac{s}{2}},$$

which implies (34). Therefore, $\delta(X_s, \epsilon) > 0$ if $0 < \epsilon \leq 2$, i. e., X_s is uniformly convex. On the other hand, if $\|x\|_{(\Phi_s)} = 1$ and $\|y\|_{(\Phi_s)} = \tau$, again by (28) we have

$$\begin{aligned} \frac{1}{2}\left(\|x + y\|_{(\Phi_s)} + \|x - y\|_{(\Phi_s)}\right) &\leq \left[\frac{1}{2}\left(\|x + y\|_{(\Phi_s)}^{\frac{2}{s}} + \|x - y\|_{(\Phi_s)}^{\frac{2}{s}}\right)\right]^{\frac{s}{2}} \\ &\leq \left(1 + \tau^{\frac{2}{2-s}}\right)^{\frac{2-s}{2}}, \end{aligned}$$

which shows (35). Therefore,

$$\lim_{\tau\to 0}\frac{\varrho(X_s,\tau)}{\tau}\le \lim_{\tau\to 0}\frac{1}{\tau}\left[\left(1+\tau^{\frac{2}{2-s}}\right)^{\frac{2-s}{2}}-1\right]=0,$$

i. e., X_s is uniformly smooth.

Using (29), we can show that (34) and (35) are still valid for another three Orlicz spaces equipped with Orlicz norm. □

From this result we can deduce the following.

Corollary 3.4 (Clarkson's inequalities) Suppose that $1 < p < \infty$, $\frac{1}{p}+\frac{1}{q}=1$ and $x, y \in L^p(\Omega)$, where Ω is as in Theorem 3.2. Then, one has for $1 < p \le 2$

$$\left[\|x+y\|_p^q+\|x-y\|_p^q\right]^{\frac{1}{q}}\le 2^{\frac{1}{q}}\left[\|x\|_p^p+\|y\|_p^p\right]^{\frac{1}{p}} \tag{36}$$

and for $2 \le p < \infty$

$$\left[\|x+y\|_p^p+\|x-y\|_p^p\right]^{\frac{1}{p}}\le 2^{\frac{1}{p}}\left[\|x\|_p^q+\|y\|_p^q\right]^{\frac{1}{q}}. \tag{37}$$

Proof If $1 < p \le 2$, we choose $1 < a < p \le 2$. Putting $\Phi(u) = |u|^a$, $\Phi_0(u) = u^2$ and $s = \frac{2(p-a)}{p(2-a)}$, we have $0 < s \le 1$ and for $u \ge 0$

$$\Phi_s^{-1}(u)=u^{\frac{1-s}{a}+\frac{s}{2}}=u^{\frac{1}{p}},$$

i. e., $\Phi_s(u) = |u|^p$. Since $L^{(\Phi_s)}(\Omega) = L^p(\Omega)$, $\|\cdot\|_{(\Phi_s)} = \|\cdot\|_p$ and

$$\lim_{a\searrow 1}\frac{2}{s}=\frac{p}{p-1}=q,\quad \lim_{a\searrow 1}\frac{2-s}{2}=\frac{1}{p}, \tag{38}$$

we obtain (36) by (28). If $2 \le p < \infty$, we choose $2 \le p < b < \infty$. Letting $\Phi(u) = |u|^b$ and $s = \frac{2(b-p)}{p(b-2)}$, again we have $0 < s \le 1$ and $\Phi_s(u) = |u|^p$. Since

$$\lim_{b\nearrow\infty}\frac{2}{s}=p \quad\text{and}\quad \lim_{b\nearrow\infty}\frac{2-s}{2}=\frac{p-1}{p}=\frac{1}{q}, \tag{39}$$

we get (37), again by (28). □

Remark 3.5 In view of (34), (35), (38) and (39), one has

$$\delta(L^p(\Omega),\epsilon)\ge\begin{cases}1-\frac{1}{2}(2^q-\epsilon^q)^{\frac{1}{q}}, & \text{if } 1<p\le 2\\ 1-\frac{1}{2}(2^p-\epsilon^p)^{\frac{1}{p}}, & \text{if } 2\le p<\infty\end{cases}$$

and

$$\varrho(L^p(\Omega),\tau)\le\begin{cases}(1+\tau^p)^{\frac{1}{p}}-1, & \text{if } 1<p\le 2\\ (1+\tau^q)^{\frac{1}{q}}-1, & \text{if } 2\le p<\infty,\end{cases}$$

which imply that $L^p(\Omega)$ is uniformly convex and uniformly smooth. Of course, the above two inequalities can be directly induced from (36) and (37). It should be noted that the modulus of convexity of some special Orlicz spaces was discussed by Rao[13, pp.307-308] and Hudzik[8], independently(see also [14, pp.289-303]).

4 Main Theorems

Now we can estimate upper bounds of nonsquare constants of some Orlicz spaces by using Clarkson type inequalities (28) and (29).

Theorem 4.1 Let Φ be an N-function and Φ_s be the inverse of (26). If $0 < s \leq 1$, then nonsquare constants of $L^{(\Phi_s)}(\Omega)$ and $L^{\Phi_s}(\Omega)$, in the sense of James, satisfy respectively

$$J(L^{(\Phi_s)}(\Omega)) \leq 2^{1-\frac{s}{2}} \tag{40}$$

and

$$J(L^{\Phi_s}(\Omega)) \leq 2^{1-\frac{s}{2}}, \tag{41}$$

where Ω is as in Theorem 3.2.

Proof By (28), we have for any $x, y \in S(L^{(\Phi_s)}(\Omega))$

$$2\left[\min(\|x+y\|_{(\Phi_s)}, \|x-y\|_{(\Phi_s)})\right]^{\frac{2}{s}} \leq \|x+y\|_{(\Phi_s)}^{\frac{2}{s}} + \|x-y\|_{(\Phi_s)}^{\frac{2}{s}} \leq 2^{\frac{2}{s}}$$

or

$$\min(\|x+y\|_{(\Phi_s)}, \|x-y\|_{(\Phi_s)}) \leq 2^{1-\frac{s}{2}},$$

which implies (40). Similarly, we can prove (41) by using (29). □

Example 4.2 If $1 < p < \infty$ and $\Phi(u) = |u|^{2p} + 2|u|^p$, then $\Phi^{-1}(u) = (\sqrt{u+1} - 1)^{\frac{1}{p}}$ for $u \geq 0$ and for $0 < s \leq 1$

$$\Phi_s^{-1}(u) = (\sqrt{u+1} - 1)^{\frac{1-s}{p}} u^{\frac{s}{2}}.$$

It is easily seen that

$$\alpha_{\Phi_s} = \beta_{\Phi_s} = \lim_{u\to\infty} \frac{\Phi_s^{-1}(u)}{\Phi_s^{-1}(2u)} = \left(\frac{1}{2}\right)^{\frac{1-s}{2p}+\frac{s}{2}},$$

$$\alpha_{\Phi_s}^0 = \beta_{\Phi_s}^0 = \lim_{u\to 0} \frac{\Phi_s^{-1}(u)}{\Phi_s^{-1}(2u)} = \left(\frac{1}{2}\right)^{\frac{1-s}{p}+\frac{s}{2}}$$

and that $\bar{\alpha}_{\Phi_s} = \alpha_{\Phi_s}^0$ and $\bar{\beta}_{\Phi_s} = \beta_{\Phi_s}$. Therefore, we have from (40) and Theorem 2.1 that

$$2^{1-\frac{s}{2}} \geq J(L^{(\Phi_s)}[0,1]) \geq 2^{1-(\frac{1-s}{2p}+\frac{s}{2})},$$

$$2^{1-\frac{s}{2}} \geq J(\ell^{(\Phi_s)}) \geq \begin{cases} 2^{\frac{1-s}{p}+\frac{s}{2}}, & \text{if } 1 < p \leq 2 \\ 2^{1-(\frac{1-s}{p}+\frac{s}{2})}, & \text{if } 2 \leq p < \infty \end{cases}$$

and

$$2^{1-\frac{s}{2}} \geq J(L^{(\Phi_s)}[0,\infty)) \geq \begin{cases} 2^{\frac{1-s}{p}+\frac{s}{2}}, & \text{if } 1 < p \leq \frac{3}{2} \\ 2^{1-(\frac{1-s}{2p}+\frac{s}{2})}, & \text{if } \frac{3}{2} \leq p < \infty. \end{cases}$$

Example 4.3(see[4]) If $1 < p < \infty$ and $L^p(\Omega) \in \{L^p[0,1], L^p[0,\infty), \ell^p\}$, then

$$J(L^p(\Omega)) = \max\left(2^{\frac{1}{p}}, 2^{1-\frac{1}{p}}\right). \tag{42}$$

In fact, putting $\Phi_p(u) = |u|^p$, we have $\alpha_{\Phi_p} = \beta_{\Phi_p} = \alpha^0_{\Phi_p} = \beta^0_{\Phi_p} = \bar{\alpha}_{\Phi_p} = \bar{\beta}_{\Phi_p} = 2^{-\frac{1}{p}}$ and so, by Theorem 2.1,

$$\max\left(2^{\frac{1}{p}}, 2^{1-\frac{1}{p}}\right) \le J(L^p(\Omega)).$$

The opposite inequality of the above follows from (40), (38) and (39). Thus, (42) holds.

Combining Theorem 2.1 and Theorem 4.1 with Theorem 1.3, we can find the exact values of nonsquare constants in the sense of James of a class of reflexive Orlicz spaces equipped with gauge norm. The first main theorem of this paper is as follows.

Theorem 4.4 Let Φ be an N-function and Φ_s be the inverse of (26) with $0 < s \le 1$.

(i) If $\Phi \notin \Delta_2(\infty) \bigcap \nabla_2(\infty)$, then

$$J(L^{(\Phi_s)}[0,1]) = 2^{1-\frac{s}{2}}. \tag{43}$$

(ii) If $\Phi \notin \Delta_2(0) \bigcap \nabla_2(0)$, then

$$J(\ell^{(\Phi_s)}) = 2^{1-\frac{s}{2}}. \tag{44}$$

(iii) If $\Phi \notin \Delta_2 \bigcap \nabla_2$, then

$$J(L^{(\Phi_s)}[0,\infty)) = 2^{1-\frac{s}{2}}. \tag{45}$$

Proof (i) In virtue of (12) and (40), one has

$$\max\left(\frac{1}{\alpha_{\Phi_s}}, 2\beta_{\Phi_s}\right) \le J(L^{(\Phi_s)}[0,1]) \le 2^{1-\frac{s}{2}}. \tag{46}$$

If $\Phi \notin \Delta_2(\infty)$, then $\frac{1}{2} \le \alpha_\Phi \le \beta_\Phi = 1$ by Theorem 1.3(i). Therefore, we have from (27)

$$\alpha_{\Phi_s} = (\alpha_\Phi)^{1-s}\left(\frac{1}{2}\right)^{\frac{s}{2}} \ge \left(\frac{1}{2}\right)^{1-\frac{s}{2}},$$

$$2\beta_{\Phi_s} = 2(\beta_\Phi)^{1-s}\left(\frac{1}{2}\right)^{\frac{s}{2}} = 2^{1-\frac{s}{2}}$$

and so,

$$\max\left(\frac{1}{\alpha_{\Phi_s}}, 2\beta_{\Phi_s}\right) = 2^{1-\frac{s}{2}}. \tag{47}$$

If $\Phi \notin \nabla_2(\infty)$, then $\frac{1}{2} = \alpha_\Phi \le \beta_\Phi \le 1$ by Theorem 1.3(i). Since

$$\frac{1}{\alpha_{\Phi_s}} = 2^{1-\frac{s}{2}} \ge 2\beta_{\Phi_s},$$

again, (47) holds. Finally, (43) follows from (46) and (47).

(ii) and (iii) are similar to (i). $\Box$

To find the exact values of nonsquare constants of a class of reflexive Orlicz spaces equipped with Orlicz norm we need the following two lemmas.

Lemma 4.5 Let Φ be an N-function and let Φ_s be the inverse of (26). Then

$$\frac{1}{A_{\Phi_s}} = \frac{1-s}{A_\Phi} + \frac{s}{2}, \quad \frac{1}{B_{\Phi_s}} = \frac{1-s}{B_\Phi} + \frac{s}{2}, \tag{48}$$

$$\frac{1}{A^0_{\Phi_s}} = \frac{1-s}{A^0_\Phi} + \frac{s}{2}, \quad \frac{1}{B^0_{\Phi_s}} = \frac{1-s}{B^0_\Phi} + \frac{s}{2} \tag{49}$$

and

$$\frac{1}{\bar{A}_{\Phi_s}} = \frac{1-s}{\bar{A}_\Phi} + \frac{s}{2}, \quad \frac{1}{\bar{B}_{\Phi_s}} = \frac{1-s}{\bar{B}_\Phi} + \frac{s}{2}. \tag{50}$$

Proof Note that $\Phi_s'(t) = \phi_s(t)$ a.e on $(0, \infty)$. Putting $t = \Phi_s^{-1}(u)$ for $0 < u < \infty$ one has

$$\begin{aligned} \frac{\Phi_s(t)}{t\Phi_s'(t)} &= \frac{u[\Phi_s^{-1}(u)]'}{\Phi_s^{-1}(u)} \\ &= \frac{(1-s)[\Phi^{-1}(u)]^{-s}[\Phi^{-1}(u)]'u^{1+\frac{s}{2}} + [\Phi^{-1}(u)]^{1-s}\frac{s}{2}u^{\frac{s}{2}}}{[\Phi^{-1}(u)]^{1-s}u^{\frac{s}{2}}} \\ &= \frac{(1-s)u[\Phi^{-1}(u)]'}{\Phi^{-1}(u)} + \frac{s}{2}. \end{aligned} \tag{51}$$

Therefore,

$$\frac{1}{A_{\Phi_s}} = \limsup_{t\to\infty} \frac{\Phi_s(t)}{t\Phi_s'(t)} = (1-s)\limsup_{u\to\infty} \frac{u[\Phi^{-1}(u)]'}{\Phi^{-1}(u)} + \frac{s}{2} = \frac{1-s}{A_\Phi} + \frac{s}{2}.$$

Similarly, we can prove the others in (48), (49) and (50) by using (51).□

Lemma 4.6 Let Φ, Ψ be a pair of complementary N-functions. Suppose that

$$C_\Phi = \lim_{t\to\infty} \frac{t\phi(t)}{\Phi(t)} \tag{52}$$

exists. Then

(i) $\gamma_\Phi = \lim_{u\to\infty} \frac{\Phi^{-1}(u)}{\Phi^{-1}(2u)}$ exists and $\gamma_\Phi = 2^{-\frac{1}{C_\Phi}}$;

(ii) $C_\Psi = \lim_{t\to\infty} \frac{t\psi(t)}{\Psi(t)}$ exists and $\gamma_\Psi = \lim_{v\to\infty} \frac{\Psi^{-1}(v)}{\Psi^{-1}(2v)} = 2^{-\frac{1}{C_\Psi}}$;

(iii) $2\gamma_\Phi\gamma_\Psi = 1$.

Similarly, if $C^0_\Phi = \lim_{t\to 0} \frac{t\phi(t)}{\Phi(t)}$ exists, then $\gamma^0_\Phi = \lim_{u\to 0} \frac{\Phi^{-1}(u)}{\Phi^{-1}(2u)} = 2^{-\frac{1}{C^0_\Phi}}$, $C^0_\Psi = \lim_{t\to 0} \frac{t\psi(t)}{\Psi(t)}$ exists, $\gamma^0_\Psi = 2^{-\frac{1}{C^0_\Psi}}$ and $2\gamma^0_\Phi\gamma^0_\Psi = 1$.

Proof The assertions follow from Propositions 1.4 and 1.5.□

The second main theorem of this paper is as follows.

Theorem 4.7 Let Φ, Φ_s and s be as in Theorem 4.4.

(i) If $\Phi \notin \Delta_2(\infty) \cap \nabla_2(\infty)$ and C_Φ defined by (52) exists, then

$$J(L^{\Phi_s}[0,1]) = 2^{1-\frac{s}{2}}. \tag{53}$$

(ii) If $\Phi \notin \Delta_2(0) \bigcap \nabla_2(0)$ and C_Φ^0 exists, then

$$J(\ell^{\Phi_s}) = 2^{1-\frac{s}{2}}. \tag{54}$$

(iii) If $\Phi \notin \Delta_2 \bigcap \nabla_2$ and C_Φ exists in the case that $\Phi \notin \Delta_2(\infty) \bigcap \nabla_2(\infty)$ or C_Φ^0 exists in the case that $\Phi \notin \Delta_2(0) \bigcap \nabla_2(0)$, then

$$J(L^{\Phi_s}[0,\infty)) = 2^{1-\frac{s}{2}}. \tag{55}$$

Proof (i) By (19) and (41), we have

$$\max\left(2\beta_{\Psi_s^+}, \frac{1}{\alpha_{\Psi_s^+}}\right) \le J(L^{\Phi_s}[0,1]) \le 2^{1-\frac{s}{2}}, \tag{56}$$

where Ψ_s^+ is the complementary N-function to Φ_s. The conditions given in (i) imply that $C_\Phi = \infty$ or $C_\Phi = 1$.

In the case that $C_\Phi = \infty$, we have $\frac{1}{C_{\Phi_s}} = \frac{1-s}{C_\Phi} + \frac{s}{2} = \frac{s}{2}$ by (48), $\frac{1}{C_{\Psi_s^+}} = 1 - \frac{1}{C_{\Phi_s}} = 1 - \frac{s}{2}$ by (8) and so, in view of Lemma 4.6,

$$\alpha_{\Psi_s^+} = \beta_{\Psi_s^+} = \gamma_{\Psi_s^+} = 2^{-\frac{1}{C_{\Psi_s^+}}} = 2^{\frac{s}{2}-1}.$$

Therefore,

$$\max\left(2\beta_{\Psi_s^+}, \frac{1}{\alpha_{\Psi_s^+}}\right) = 2^{1-\frac{s}{2}}. \tag{57}$$

In the case that $C_\Phi = 1$, one has $\frac{1}{C_{\Phi_s}} = 1 - \frac{s}{2}$, $\frac{1}{C_{\Psi_s^+}} = \frac{s}{2}$ and $\alpha_{\Psi_s^+} = \beta_{\Psi_s^+} = \gamma_{\Psi_s^+} = 2^{-\frac{s}{2}}$. Again (57) holds. Hence, (53) follows from (56) and (57).

(ii) The conditions given in (ii) imply that

$$\max\left(2\beta^0_{\Psi_s^+}, \frac{1}{\alpha^0_{\Psi_s^+}}\right) = 2^{1-\frac{s}{2}}. \tag{58}$$

Therefore, (54) follows from (21), (41) and (58).

(iii) By (20) and (41), we have

$$\max\left(2\bar{\beta}_{\Psi_s^+}, \frac{1}{\bar{\alpha}_{\Psi_s^+}}\right) \le J(L^{\Phi_s}[0,\infty)) \le 2^{1-\frac{s}{2}}. \tag{59}$$

Note that $\Phi \notin \Delta_2 \bigcap \nabla_2$ if and only if $\Phi \notin \Delta_2(\infty) \bigcap \nabla_2(\infty)$ or $\Phi \notin \Delta_2(0) \bigcap \nabla_2(0)$ and that

$$\max\left(2\beta_{\Psi_s^+}, \frac{1}{\alpha_{\Psi_s^+}}, 2\beta^0_{\Psi_s^+}, \frac{1}{\alpha^0_{\Psi_s^+}}\right) \le \max\left(2\bar{\beta}_{\Psi_s^+}, \frac{1}{\bar{\alpha}_{\Psi_s^+}}\right), \tag{60}$$

since $\max(\beta_{\Psi_s^+}, \beta^0_{\Psi_s^+}) \le \bar{\beta}_{\Psi_s^+}$ and $\bar{\alpha}_{\Psi_s^+} \le \min(\alpha_{\Phi_s^+}, \alpha^0_{\Psi_s^+})$ always hold. Finally, (55) follows from (59), (60), (57) and (58). □

Example 4.8 Let $1 < p < \infty$ and let $M(u)$ be the inverse of

$$M^{-1}(u) = [\ln(1+u)]^{\frac{1}{2p}} u^{\frac{1}{4}}, \quad u \geq 0.$$

Then

$$J(L^{(M)}[0,1]) = J(L^{(M)}[0,\infty)) = 2^{\frac{3}{4}}, \tag{61}$$

$$J(L^{M}[0,1]) = J(L^{M}[0,\infty)) = 2^{\frac{3}{4}} \tag{62}$$

and

$$2^{\frac{3}{4}} \geq \{J(\ell^{(M)}, J(\ell^{M})\} \geq \begin{cases} 2^{\frac{1}{4}+\frac{1}{2p}}, & \text{if } 1 < p \leq 2 \\ 2^{\frac{3}{4}-\frac{1}{2p}}, & \text{if } 2 \leq p < \infty. \end{cases} \tag{63}$$

In fact, putting $\Phi(u) = e^{|u|^p} - 1$ in Theorem 4.4 we have $\Phi^{-1}(u) = [\ln(1+u)]^{\frac{1}{p}}$ for $u \geq 0$ and

$$\Phi_s^{-1}(u) = [\ln(1+u)]^{\frac{1-s}{p}} u^{\frac{s}{2}}.$$

Therefore, $M(u) = \Phi_s(u)|_{s=\frac{1}{2}}$. Since

$$\lim_{u\to\infty} \frac{\Phi^{-1}(u)}{\Phi^{-1}(2u)} = 1 \quad \text{and} \quad \lim_{u\to 0} \frac{\Phi^{-1}(u)}{\Phi^{-1}(2u)} = 2^{-\frac{1}{p}},$$

it follows from Theorem 1.3 that $\Phi \notin \Delta_2(\infty)$, $\Phi \notin \Delta_2$ and $\Phi \in \Delta_2(0) \bigcap \nabla_2(0)$. Therefore, (61) follows from (43) and (45). On the other hand, we have $C_\Phi = \infty$. In virtue of Theorem 4.7(i) and (iii), we obtain (62).

Finally, we prove (63). Set $N(v) = \Psi_s^+(v)|_{s=\frac{1}{2}}$. Noticing that $C_\Phi^0 = p$, we have $\frac{1}{C_M^0} = \frac{1}{2p} + \frac{1}{4}$ by Lemma 4.5. It follows from Lemma 4.6 that

$$\alpha_M^0 = \beta_M^0 = 2^{-(\frac{1}{4}+\frac{1}{2p})}$$

and

$$\alpha_N^0 = \beta_N^0 = 2^{\frac{1}{2p}-\frac{3}{4}}.$$

Therefore, (63) follows from (14), (21), (40) and (41).

References

[1] S. T. Chen, Non-squareness of Orlicz spaces(in Chinese), Chinese Ann. Math., **6**A(1985), 619-624.(**MR** 87j: 46064)

[2] J. A. Clarkson, Uniformly convex spaces, Trans. Amer. Math. Soc., **40**(1936), 396-414.

[3] C. E. Cleaver, On the extension of Lipschitz-Hölder maps on Orlicz spaces, Studia Math., **42**(1972), 195-204.

[4] J. Gao and K. S. Lau, On the geometry of spheres in normed linear spaces, J. Austral. Math. Soc., **48**A(1990), 101-112.

[5] J. Gao and K. S. Lau, On two classes of Banach spaces with uniform normal structure, Studia Math., **99**(1991), 41-56.

[6] R. Grzaslewicz, H. Hudzik and W. Orlicz, Uniform non-$\ell_n^{(1)}$ property in some normed spaces, Bull. Pol. Acad. Soc. Math., **34**(1986), 161-171.

[7] H. Hudzik, Uniformly non-$\ell_n^{(1)}$ Orlicz spaces with Luxemburg norm, Studia Math., **81**(1985), 41-54.

[8] H. Hudzik, An estimation of the modulus of convexity in a class of Orlicz spaces, Math. Japon., **32**(1987), 227-237.

[9] R. C. James, Uniformly nonsquare spaces, Annals of Math., **80**(1964), 542-550.

[10] M. A. Krasnosel'skii and Ya. B. Rutickii, *Convex Functions and Orlicz Spaces,* P. Noordhoff Ltd. Groningen, 1961.

[11] J. Lindenstrauss and L. Tzafriri, *Classical Banach Spaces,* I and II, Springer, Berlin, 1977 and 1979.

[12] M. M. Rao, Interpolation, ergodicity and martigales, J. Math. Mech., **16**(1966), 543-567.

[13] M. M. Rao, *Measure Theory and Integration,* Wiley-Interscience, New York, 1987.

[14] M. M. Rao and Z. D. Ren, *Theory of Orlicz Spaces,* Marcel Dekker, New York, 1991.

[15] Z. D. Ren, Packing in Orlicz function spaces, Ph. D. Dissertation(Advisor: N. E. Gretsky), University of California, Riverside, 1994.

[16] J. J. Schäffer, *Geometry of Spheres in Normed Spaces,* Marcel Dekker, New York and Basel, 1976.

[17] Y. W. Wang and S. T. Chen, Non-squareness, B-convexity and flatness of Orlicz spaces, Comment. Math. Prace Mat., **28**(1988), 155-165.

[18] H. K. Xu, Measure of noncompactness and normal type structure in Banach spaces, Panamer. Math. J., **3**(1993), No.2, 17-34.

Recursive Multiple Wiener Integral Expansion for Nonlinear Filtering of Diffusion Processes

SERGEY LOTOTSKY[†] Center for Applied Mathematical Sciences, University of Southern California, Los Angeles, CA 90089-1113

BORIS L. ROZOVSKII[‡] Center for Applied Mathematical Sciences, University of Southern California, Los Angeles, CA 90089-1113

Abstract

A recursive in time Wiener chaos representation of the optimal nonlinear filter is derived for a time homogeneous diffusion model with uncorrelated noises. The existing representations are either not recursive or require a prior computation of the unnormalized filtering density, which is time consuming. An algorithm is developed for computing a recursive approximation of the filter, and the error of the approximation is obtained. When the parameters of the model are known in advance, the on-line speed of the algorithm can be increased by performing part of the computations off line.

Key Words: nonlinear filtering, Wiener chaos, recursive filters.

1 INTRODUCTION

In a typical filtering model, a non-anticipative functional $f_t(x)$ of the unobserved signal process $(x(t))_{t\geq 0}$ is estimated from the observations $y(s),\ s \leq t$. The best mean square estimate is known to be the conditional expectation $\mathbf{E}[f_t(x)|y(s),\ s \leq t]$, called the optimal filter. When the observation noise is additive, the Kallianpur-Striebel formula (Kallianpur (1980), Liptser and Shiryayev (1992)) provides the representation of the optimal filter as follows:

$$\mathbf{E}[f_t(x)|y(s),\ s \leq t] = \frac{\phi_t[f]}{\phi_t[1]},$$

[†]This work was partially supported by ONR Grant #N00014-95-1-0229.

[‡]This work was partially supported by ONR Grant #N00014-95-1-0229 and ARO Grant DAAH 04-95-1-0164.

where $\phi_t[\cdot]$ is a functional called *the unnormalized optimal filter.* In the particular case $f_t(x) = f(x(t))$, there are two approaches to computing $\phi_t[f]$.

In the first approach (Lo and Ng (1983), Mikulevicius and Rozovskii (1995), Ocone (1983)), the functional $\phi_t[f]$ is expanded in a series of multiple integrals with respect to the observation process. This approach can be used to obtain representations of general functionals, but these representations are not recursive in time. In fact, there is no closed form differential equation satisfied by $\phi_t[f]$.

In the second approach (Kallianpur (1980), Liptser and Shiryayev (1992), Rozovskii (1990)), it is proved that, under certain regularity assumptions, the functional $\phi_t[f]$ can be written as

$$\phi_t[f] = \int f(x)u(t,x)dx \tag{1.1}$$

for some function $u(t,x)$, called *the unnormalized filtering density.* Even though the computation of $u(t,x)$ can be organized recursively in time, and there are many numerical algorithms to do this (Budhiraja and Kallianpur (1995), Elliott and Glowinski (1989), Florchinger and LeGland (1991), Ito (1996), Lototsky et al. (1996), etc.), these algorithms are time consuming because they involve evaluation of $u(t,x)$ at many spatial points. Moreover, computation of $\phi_t[f]$ using this approach requires subsequent evaluation of the integral (1.1).

The objective of the current work is to develop a recursive in time algorithm for computing $\phi_t[f]$ without computing $u(t,x)$. The analysis is based on the multiple integral representation of the unnormalized filtering density (Lototsky et al. (1996), Mikulevicius and Rozovskii (1995), Ocone, (1983)) with subsequent Fourier series expansion in the spatial domain. For simplicity, in this paper we consider a one-dimensional diffusion model with uncorrelated noises. In the proposed algorithm, the computations involving the parameters of the model can be done separately from those involving the observation process. If the parameters of the model are known in advance, this separation can substantially increase the on-line speed of the algorithm.

2 REPRESENTATION OF THE UNNORMALIZED OPTIMAL FILTER

Let $(\Omega, \mathcal{F}, \mathrm{P})$ be a complete probability space, on which standard one-dimensional Wiener processes $(V(t))_{t\geq 0}$ and $(W(t))_{t\geq 0}$ are given. Random processes $(x(t))_{t\geq 0}$ and $(y(t))_{t\geq 0}$ are defined by the equations

$$\begin{aligned} x(t) &= x_0 + \int_0^t b(x(s))ds + \int_0^t \sigma(x(s))dV(s), \\ y(t) &= \int_0^t h(x(s))ds + W(t). \end{aligned} \tag{2.1}$$

In applications, $x(t)$ represents the unobserved state process subject to estimation from the observations $y(s),\ s \leq t$. The σ - algebra generated by $y(s),\ s \leq t$, will be denoted by $\mathcal{F}_t^y$.

The following is assumed about the model (2.1):

(**A1**) The Wiener processes $(V(t))_{t\geq 0}$ and $(W(t))_{t\geq 0}$ are independent of x_0 and of each other;

(**A2**) The functions $b(x)$, $\sigma(x)$, and $h(x)$ are infinitely differentiable and bounded with all the derivatives;

(**A3**) The random variable x_0 has a density $p(x)$, $x \in \mathbf{R}$, so that the function $p = p(x)$ is infinitely differentiable and, together with all the derivatives, decays at infinity faster than any power of x.

Let $f = f(x)$ be a measurable function such that

$$|f(x)| \leq L(1 + |x|^{k_0}) \tag{2.2}$$

for some $k_0 \geq 0$ and $L > 0$. Assumptions (**A2**) and (**A3**) imply that $\mathbf{E}|f(x(t))|^2 < \infty$ for all $t \geq 0$ (Liptser and Shiryayev, 1992). Suppose that $T > 0$ is fixed. It is known (Kallianpur (1980), Liptser and Shiryayev (1992)) that the best mean square estimate of $f(x(t))$ given $y(s)$, $s \leq t \leq T$, is $\hat{f}(x(t)) = \mathbf{E}[f(x(t))|\mathcal{F}_t^y]$, and this estimate can be written by the Kallianpur-Striebel formula as follows:

$$\hat{f}(x(t)) = \frac{\tilde{\mathbf{E}}[f(x(t))\rho(t)|\mathcal{F}_t^y]}{\tilde{\mathbf{E}}[\rho(t)|\mathcal{F}_t^y]}, \tag{2.3}$$

where

$$\rho(t) = \exp\Big\{\int_0^t h(x(s))dy(s) - \frac{1}{2}\int_0^t |h(x(s))|^2 ds\Big\},$$

and $\tilde{\mathbf{E}}$ is the expectation with respect to measure $\tilde{\mathrm{P}}(\bullet) := \int_\bullet (\rho(T))^{-1} d\mathrm{P}$. Moreover, under measure $\tilde{\mathrm{P}}$, the observation process $(y(t))_{0\leq t\leq T}$ is a Wiener process independent of $(x(t))_{0\leq t\leq T}$.

The conditional expectation $\tilde{\mathbf{E}}[f(x(t))\rho(t)|\mathcal{F}_t^y]$ will be referred to as the unnormalized optimal filter and will be denoted by $\phi_t[f]$. In this section, a recursive in time representation of $\phi_t[f]$ is derived for an arbitrary function f satisfying (2.2).

It is known (Rozovskii, 1990) that, under assumptions (**A1**) – (**A3**), there exists a random field $u(t,x)$, $t \geq 0$, $x \in \mathbf{R}$, called the unnormalized filtering density, such that

$$\phi_t[f] = \int_{\mathbf{R}} u(t,x) f(x) dx. \tag{2.4}$$

Denote by $P_t\varphi(x)$ the solution of the equation

$$\begin{aligned} \frac{\partial v(t,x)}{\partial t} &= \frac{1}{2}\frac{\partial^2(\sigma^2(x)v(t,x))}{\partial x^2} - \frac{\partial(b(x)v(t,x))}{\partial x}, \quad t > 0; \\ v(0,x) &= \varphi(x), \end{aligned}$$

and consider $0 = t_0 < t_1 < \ldots < t_M = T$, a partition of $[0,T]$ with steps $\Delta_i = t_i - t_{i-1}$ (this partition will be fixed hereafter). The following theorem gives a recursive representation of the unnormalized filtering density at the points of the partition.

THEOREM 2.1. (cf. Mikulevicius and Rozovskii (1995), Ocone (1983)) Under assumptions (**A1**) - (**A3**),

$$\begin{aligned} &u(t_0,x) = p(x), \\ &u(t_i,x) = P_t u(t_{i-1},\cdot)(x) + \\ &\sum_{k\geq 1}\int_0^{\Delta_i}\int_0^{s_k}\cdots\int_0^{s_2} P_{t-s_k}h\ldots hP_{s_1}u(t_{i-1},\cdot)(x)dy^{(i)}(s_1)\ldots dy^{(i)}(s_k), \ \mathrm{P}-\text{a.s.} \end{aligned} \tag{2.5}$$

for $i = 1,\ldots,M$, where $y^{(i)}(t) = y(t+t_{i-1}) - y(t_{i-1})$, $0 \leq t \leq \Delta_i$.

Proof. This follows from Theorem 2.3 in Lototsky et al. (1996) and Theorem 3.1 in Ito (1951).

To simplify the further presentation, the following notations are introduced. For an $\mathcal{F}^y_{t_{i-1}}$ - measurable function $g = g(x, \omega)$ and $0 \leq t \leq \Delta_i$,

$$\begin{aligned} &F_0^{(i)}(t,g)(x) := P_t g(x), \\ &F_k^{(i)}(t,g)(x) := \int_0^t \int_0^{s_k} \ldots \int_0^{s_2} P_{t-s_k} h \ldots h P_{s_1} g(x) dy^{(i)}(s_1) \ldots dy^{(i)}(s_k), \ k \geq 1. \end{aligned} \tag{2.6}$$

With these notations, (2.5) becomes

$$u(t_i, x) = \sum_{k \geq 0} F_k^{(i)}(\Delta_i, u(t_{i-1}, \cdot))(x), \ i = 1, \ldots, M. \tag{2.7}$$

It is known (Ladyzhenskaia et al. (1968), Rozovskii, (1990)) that [1]

$$||P_t \cdot ||_0 \leq e^{Ct} || \cdot ||_0. \tag{2.8}$$

Then it follows by induction that for every $t \in [0, \Delta_i]$, $i = 1, \ldots, M$, and $k \geq 0$, the operator $g \mapsto F_k^{(i)}(t, g)$ is linear and bounded from $L_2(\Omega, \tilde{\mathrm{P}}; L_2(\mathbf{R}))$ to itself and $\tilde{\mathbf{E}}||F_k^{(i)}(t,g)||_0^2 \leq e^{Ct}[(Ct)^k/k!]\tilde{\mathbf{E}}||g||_0^2$. This, in particular, implies that $u(t_i, \cdot) \in L_2(\mathbf{R})$ P- and $\tilde{\mathrm{P}}$- a.s.

In the following theorem, the unnormalized filtering density is expanded with respect to an orthonormal basis in $L_2(\mathbf{R})$. With a special choice of the basis, this expansion will be used later to construct the recursive representation of the unnormalized optimal filter.

THEOREM 2.2. If $\{e_n\}_{n \geq 0}$ is an orthonormal basis in $L_2(\mathbf{R})$ and random variables $\psi_n(i)$, $n \geq 0$, $i = 0, \ldots, M$, are defined recursively by

$$\begin{aligned} &\psi_n(0) = (p, e_n)_0, \\ &\psi_n(i) = \sum_{k \geq 0} \Big(\sum_{l \geq 0} (F_k^{(i)}(\Delta_i, e_l), e_n)_0 \psi_l(i-1) \Big), \ i = 1, \ldots, M, \end{aligned} \tag{2.9}$$

then

$$u(t_i, \cdot) = \sum_{n \geq 0} \psi_n(i) e_n, \ \mathrm{P - a.s.} \tag{2.10}$$

Proof. The proof is carried out by induction. Representation (2.10) is obvious for $i = 0$. If it is assumed for some $i - 1 \geq 0$, then (2.7) and the continuity of $F_k^{(i)}(\Delta_i, \cdot)$ imply that

$$\begin{aligned} &\psi_n(i) := (u(t_i, \cdot), e_n)_0 = \sum_{k \geq 0} (F_k^{(i)}(\Delta_i, u(t_{i-1}, \cdot)), e_n)_0 = \\ &\sum_{k \geq 0} \Big(\sum_{l \geq 0} (F_k^{(i)}(\Delta_i, e_l), e_n)_0 \psi_l(i-1) \Big), \end{aligned}$$

and (2.10) follows with $\psi_n(i)$ given by (2.9).

REMARK. Direct computations show that the infinite sums in (2.9) can be interchanged even though the double sum need not converge absolutely. The absolute convergence holds

---

[1] $|| \cdot ||_0$ and $(\cdot, \cdot)_0$ are the norm and the inner product in $L_2(\mathbf{R})$. The value of the constant C depends only on the parameters of the model and is usually different in different places.

if $\sum_n \sqrt{\tilde{\mathbf{E}}|\psi_n(i)|^2} < \infty$, which is the case when $\{e_n\}$ is the Hermite polynomial basis (2.11). For practical computations, both infinite sums in (2.9) must be truncated. These truncations are studied in Section 3.

To get a representation of $\phi_t[f]$, it now seems natural, according to (2.4), to multiply both sides of (2.10) by $f(x)$ and integrate, but this cannot be done in general because (2.10) is an equality in $L_2(\mathbf{R})$ and f need not be square integrable. The difficulty is resolved by choosing a special basis $\{e_n\}$ so that integral $\int_{\mathbf{R}} f(x)e_n(x)dx$ can be defined for every function f satisfying (2.2).

Specifically, let $\{e_n\}$ be the Hermite basis in $L_2(\mathbf{R})$ (Gottlieb and Orszag (1977), Hille and Phillips (1957)):

$$e_n(x) = \frac{1}{\sqrt{2^n \pi^{1/2} n!}} e^{-x^2/2} H_n(x), \tag{2.11}$$

where $H_n(x)$ is the nth Hermite polynomial defined by

$$H_n(x) = (-1)^n e^{x^2} \frac{d^n}{dx^n} e^{-x^2}, \quad n \geq 0.$$

Then the following result is valid.

THEOREM 2.3. If assumptions (**A1**) – (**A3**) and (2.2) hold and e_n is defined by (2.11), then

$$\phi_{t_i}[f] = \sum_{n \geq 0} f_n \psi_n(i), \ \mathrm{P} - \text{a.s.}, \tag{2.12}$$

where $f_n = \int_{\mathbf{R}} f(x)e_n(x)dx$ and $\psi_n(i)$ is given by (2.9).

Proof. Condition (2.2) and fast decay of $e_n(x)$ at infinity imply that f_n is well defined for all n. Then (2.12) will follow from (2.4) and (2.10) if the series $\sum_{n \geq 0} f_n \psi_n(i)$ is P - a.s. absolutely convergent for all $i = 0, \ldots, M$. Since measures P and $\tilde{\mathrm{P}}$ are equivalent, it suffices to show that

$$\sum_{n \geq 0} |f_n| \, \tilde{\mathbf{E}}|\psi_n(i)| < \infty. \tag{2.13}$$

Arguments similar to those in Hille and Phillips (1957), paragraph (21.3.3), show that

$$\int_{\mathbf{R}} |x|^{k_0} |e_n(x)| dx \leq C n^{(2k_0+1)/4},$$

which implies that

$$|f_k| \leq C n^{(2k_0+1)/4}. \tag{2.14}$$

On the other hand, it follows from the proof of Theorem 2.6 in Lototsky et al. (1996) that for every integer γ there exists a constant $C(\gamma)$ such that

$$\tilde{\mathbf{E}}|\psi_n(i)| \leq \sqrt{\tilde{\mathbf{E}}|\psi_n(i)|^2} \leq \frac{C(\gamma)}{n^\gamma}. \tag{2.15}$$

Taking γ sufficiently large and combining (2.14) and (2.15) results in (2.13).

REMARK. It is known (Hille and Phillips (1957), paragraph (21.3.2)) that $\sup_x |e_n(x)| \leq C n^{-1/12}$. Together with (2.15), this inequality implies that, for the Hermite basis, the series in (2.10) converges uniformly in $x \in \mathbf{R}$, P - a.s.

3 RECURSIVE APPROXIMATION OF THE UNNORMALIZED OPTIMAL FILTER

It was already mentioned that the infinite sums in (2.9) must be approximated by truncating the number of terms, if the formula is to be used for practical computations. Multiple integrals in (2.6) must also be approximated. The effects of these approximations are studied below.

For simplicity, it is assumed that the partition of $[0,T]$ is uniform ($\Delta_i = \Delta$ for all $i = 1,\dots,M$). With obvious modifications, the results remain valid for an arbitrary partition.

Given a positive integer κ, define random variables $\psi_{n,\kappa}(i)$, $n = 0,\dots,\kappa$, $i = 0,\dots,M$, by

$$\begin{aligned}
&\psi_{n,\kappa}(0) = (p, e_n)_0,\\
&\psi_{n,\kappa}(i) = \sum_{l=0}^{\kappa} \Big((P_\Delta e_l, e_n)_0 + (P_\Delta h e_l, e_n)_0[y(t_i) - y(t_{i-1})]+\\
&(1/2)(P_\Delta h^2 e_l, e_n)_0[(y(t_i) - y(t_{i-1}))^2 - \Delta]\Big)\psi_{n,\kappa}(i-1),\ i = 1,\dots,M.
\end{aligned} \tag{3.1}$$

Then the corresponding approximations to $u(t_i, x)$ and $\phi_{t_i}[f]$ are

$$\begin{aligned}
u_\kappa(t_i, x) &= \sum_{n=0}^{\kappa} \psi_{n,\kappa}(i) e_n(x),\\
\phi_{t_i,\kappa}[f] &= \sum_{n=0}^{\kappa} \psi_{n,\kappa}(i) f_n.
\end{aligned} \tag{3.2}$$

The errors of these approximations are given in the following theorem.

THEOREM 3.1. If assumptions (**A1**) – (**A3**) and (2.2) hold and the basis $\{e_n\}$ is chosen according to (2.11) then

$$\max_{1\le i\le M} \sqrt{\tilde{\mathbf{E}}\|u_\kappa(t_i,\cdot) - u(t_i,\cdot)\|_0^2} \le C\Delta + \frac{C(\gamma)}{\kappa^{\gamma-1/2}\Delta}, \tag{3.3}$$

$$\max_{1\le i\le M} \sqrt{\tilde{\mathbf{E}}|\phi_{t_i,\kappa}[f] - \phi_{t_i}[f]|^2} \le C\Delta + \frac{C(\gamma)}{\kappa^{\gamma-1/2}\Delta}. \tag{3.4}$$

Proof. To simplify the presentation, the notation $|||\cdot|||_0 := \sqrt{\tilde{\mathbf{E}}\|\cdot\|_0^2}$ will be used. All constants are denoted by C.

1. Proof of (3.3) is carried out in three steps.

Step 1. Define

$$\begin{aligned}
u^1(t_0, x) &:= p(x),\\
u^1(t_i, x) &:= \sum_{k=0}^{2} F_k^{(i)}(\Delta, u(t_{i-1},\cdot))(x).
\end{aligned}$$

It is proved in Lototsky et al. (1996), Theorem 2.4, that

$$\max_{0\le i\le M} |||u(t_i,\cdot) - u^1(t_i,\cdot)|||_0 \le C\Delta. \tag{3.5}$$

Step 2. Define

$$\begin{aligned}
&\bar{F}_0^{(i)}(\Delta, g)(x) := P_\Delta g(x),\\
&\bar{F}_1^{(i)}(\Delta, g)(x) := [y(t_i) - y(t_{i-1})] P_\Delta h g(x),\\
&\bar{F}_2^{(i)}(\Delta, g)(x) := (1/2)[(y(t_i) - y(t_{i-1}))^2 - \Delta] P_\Delta h^2 g(x),
\end{aligned}$$

and then by induction

$$\bar{u}^1(t_0, x) := p(x),$$
$$\bar{u}^1(t_i, x) = \sum_{k=0}^{2} \bar{F}_k^{(i)}(\Delta, \bar{u}^1(t_{i-1}, \cdot))(x).$$

Since $y(t) - y(t_{i-1})$, $t > t_{i-1}$, is independent of $\mathcal{F}_{t_{i-1}}^y$ under measure $\tilde{\mathrm{P}}$, it follows that

$$|||\bar{u}^1(t_i, \cdot) - u^1(t_i, \cdot)|||_0^2 = |||P_\Delta(\bar{u}^1(t_{i-1}, \cdot) - u^1(t_{i-1}, \cdot))|||_0^2 + \\ |||\sum_{k=0}^{2} \bar{F}_k^{(i)}(\Delta, \bar{u}^1(t_{i-1}, \cdot)) - F_k^{(i)} u^1(t_{i-1}, \cdot))|||_0^2. \tag{3.6}$$

Next,

$$|||\sum_{k=0}^{2} \bar{F}_k^{(i)}(\Delta, \bar{u}^1(t_{i-1}, \cdot)) - F_k^{(i)}(\Delta, u^1(t_{i-1}, \cdot))|||_0^2 \leq \\ 4 \sum_{k=1}^{2} \Big(|||\bar{F}_k^{(i)}(\Delta, \bar{u}^1(t_{i-1}, \cdot) - u^1(t_{i-1}, \cdot))|||_0^2 + \\ |||\bar{F}_k^{(i)}(\Delta, u^1(t_{i-1}, \cdot)) - F_k^{(i)}(\Delta, u^1(t_{i-1}, \cdot))|||_0^2 \Big). \tag{3.7}$$

It follows from (2.8) and the definition of $\bar{F}_k^{(i)}$ that

$$4|||\bar{F}_k^{(i)}(\Delta, \bar{u}^1(t_{i-1}, \cdot) - u^1(t_{i-1}, \cdot))|||_0^2 \leq \\ \frac{(C\Delta)^k}{k!} e^{C\Delta} |||\bar{u}^1(t_{i-1}, \cdot) - u^1(t_{i-1}, \cdot)|||_0^2, \ k = 1, 2. \tag{3.8}$$

The Taylor formula and the definition of P_t imply

$$||P_{\Delta - s} h P_s g - P_\Delta h g||_0^2 \leq C s^2 ||g||_{\mathbf{H}^2}^2,$$

where $||\cdot||_{\mathbf{H}^a}$ is the norm in the corresponding Sobolev space. Then

$$|||\bar{F}_1^{(i)}(\Delta, u^1(t_{i-1}, \cdot)) - F_1^{(i)}(\Delta, u^1(t_{i-1}, \cdot))|||_0^2 \leq C\tilde{\mathbf{E}} ||u^1(t_{i-1}, \cdot)||_{\mathbf{H}^2}^2 \Delta^3; \\ |||\bar{F}_2^{(i)}(\Delta, u^1(t_{i-1}, \cdot)) - F_2^{(i)}(\Delta, u^1(t_{i-1}, \cdot))|||_0^2 \leq C\tilde{\mathbf{E}} ||u^1(t_{i-1}, \cdot)||_{\mathbf{H}^2}^2 \Delta^4. \tag{3.9}$$

Finally, the continuity of operator P_t in $\mathbf{H}^2(\mathbf{R})$ (Ladyzhenskaia et al. (1968), Rozovskii, (1990)) implies

$$\tilde{\mathbf{E}} ||u^1(t_{i-1}, \cdot)||_{\mathbf{H}^2}^2 \leq e^{CT} ||p||_{\mathbf{H}^2}^2 \leq C. \tag{3.10}$$

Combining inequalities (3.6) – (3.10) results in

$$|||\bar{u}^1(t_i, \cdot) - u^1(t_i, \cdot)|||_0^2 \leq e^{C\Delta} |||\bar{u}^1(t_{i-1}, \cdot) - u^1(t_{i-1}, \cdot)|||_0^2 + C\Delta^3$$

(at least for sufficiently small Δ), which, by the Gronwall Lemma, implies

$$\max_{0 \leq i \leq M} |||\bar{u}^1(t_i, \cdot) - u^1(t_i, \cdot)|||_0^2 \leq C\Delta^2. \tag{3.11}$$

Step 3. The same arguments as in the proof of Theorem 2.6 in Lototsky et al. (1996) show that

$$|||\bar{u}^1(t_i, \cdot) - u_\kappa(t_i, \cdot)|||_0 \leq \frac{C(\gamma)}{\kappa^{\gamma - 1/2} \Delta}. \tag{3.12}$$

Combining (3.5), (3.11), (3.12), and the triangle inequality results in (3.3).

2. The natural way of proving (3.4) is to use (3.3) and the Cauchy inequality. To deal with the technical difficulty that $f \notin L_2(\mathbf{R})$, the following spaces are introduced (Rozovskii (1990), Sec. 4.3): for $r \in \mathbf{R}$, $L_2(\mathbf{R}, r) = \{\varphi : \int_{\mathbf{R}} \varphi^2(x)(1+x^2)^r dx < \infty\}$. The weighted Sobolev spaces $\mathbf{H}^n(\mathbf{R}, r)$ are defined in a similar way. Then $L_2(\mathbf{R}, r)$ is a Hilbert space with inner product

$$(\varphi_1, \varphi_2)_r := \int_{\mathbf{R}} \varphi_1(x)\varphi_2(x)(1+x^2)^r dx$$

and the corresponding norm $||\cdot||_r$. If $\varphi_1 \in L_2(\mathbf{R}, r)$ and $\varphi_2 \in L_2(\mathbf{R}, -r)$, then $\int_{\mathbf{R}}\varphi_1(x)\varphi_2(x)dx$ is well defined and will be denoted by $(\varphi_1, \varphi_2)_0$. Condition (2.2) implies that $f \in L_2(\mathbf{R}, -r)$ for all $r > k_0 + 1/2$. On the other hand, assumptions **(A2)** and **(A3)** imply that $u(t, \cdot) \in L_2(\mathbf{R}, r)$ for all $r \in \mathbf{R}$ (Rozovskii, 1990, Theorem 4.3.2), and the same is true for $u^1(t_i, \cdot)$, $\bar{u}^1(t_i, \cdot)$, and $u_\kappa(t_i, \cdot)$.

Fix an even integer $r > k_0 + 1/2$ and define $\beta(x) := (1+x^2)^{r/2}$. Notation $|||\cdot|||_r := \sqrt{\tilde{\mathbf{E}}||\cdot||_r^2}$ will also be used.

By the Cauchy inequality,

$$\begin{aligned}\sqrt{\tilde{\mathbf{E}}|\phi_{t_i,\kappa}[f] - \phi_{t_i}[f]|^2} &\equiv \sqrt{\tilde{\mathbf{E}}(u_\kappa(t_i,\cdot) - u(t_i,\cdot), f)_0^2} \leq \\ \sqrt{||f||_{-r}^2 |||u_\kappa(t_i,\cdot) - u(t_i,\cdot)|||_r^2} &\leq ||f||_{-r}(|||u(t_i,\cdot) - u^1(t_i,\cdot)|||_r + \\ |||u^1(t_i,\cdot) - \bar{u}^1(t_i,\cdot)|||_r &+ |||\bar{u}^1(t_i,\cdot) - u_\kappa(t_i,\cdot)|||_r).\end{aligned} \tag{3.13}$$

Since the operator P_t is linear bounded from $\mathbf{H}^n(\mathbf{R}, r)$ to itself (Ladyzhenskaia et al. (1968), Rozovskii (1990)), the arguments of steps 1 and 2 can be repeated to conclude that

$$|||u(t_i,\cdot) - u^1(t_i,\cdot)|||_r + |||u^1(t_i,\cdot) - \bar{u}^1(t_i,\cdot)|||_r \leq C\Delta. \tag{3.14}$$

Next, it follows from the proof of Theorem 2.6 in Lototsky et al. (1996) that for every positive integer γ there exists $C(\gamma)$ such that for all $i = 0, \ldots, M$

$$\tilde{\mathbf{E}}(\bar{u}^1(t_i,\cdot), e_n)_0^2 \leq \frac{C(\gamma)}{n^{2\gamma+r}}. \tag{3.15}$$

Similarly, by (3.12), there is $C(\gamma)$ so that

$$|||\bar{u}^1(t_i,\cdot) - u_\kappa(t_i,\cdot)|||_0^2 \leq \frac{C(\gamma)}{\kappa^{2\gamma+r-1}\Delta^2}. \tag{3.16}$$

On the other hand, repeated application of the relations $e_n' = (\sqrt{n}e_{n-1} - \sqrt{n+1}e_{n+1})/\sqrt{2}$ and $-e_n'' + (1+x^2)e_n = 2(n+1)e_n$ shows that

$$(g, e_n)_{r/2}^2 \leq C \sum_{m=n-r/2}^{n+r/2} m^r (g, e_m)_0^2$$

(if $m < 0$, the corresponding term in the sum is set to be zero), and consequently

$$\sum_{n\geq 0}(g, e_n)_{r/2}^2 \leq C \sum_{n\geq 0} n^r (g, e_n)_0^2.$$

Combining the last inequality with the identities

$$||g||_r^2 = ||g\beta||_0^2 = \sum_n (g\beta, e_n)_0^2 = \sum_n (g, e_n)_{r/2}^2$$

results in

$$|||\bar{u}^1(t_i,\cdot) - u_\kappa(t_i,\cdot)|||_r^2 = \sum_{n\geq 0} \tilde{\mathbf{E}}(\bar{u}^1(t_i,\cdot) - u_\kappa(t_i,\cdot), e_n)_{r/2}^2 \leq$$
$$C\sum_{n\geq 0} n^r \tilde{\mathbf{E}}(\bar{u}^1(t_i,\cdot) - u_\kappa(t_i,\cdot), e_n)_0^2 = C\sum_{n=0}^{\kappa} n^r \tilde{\mathbf{E}}(\bar{u}^1(t_i,\cdot) - u_\kappa(t_i,\cdot), e_n)_0^2 +$$
$$C\sum_{n>\kappa} n^r \tilde{\mathbf{E}}(\bar{u}^1(t_i,\cdot), e_n)_0^2.$$

Now, (3.15) and (3.16) imply

$$|||\bar{u}^1(t_i,\cdot) - u_\kappa(t_i,\cdot)|||_r^2 \leq C\kappa^r \sum_{n=1}^{\kappa} \tilde{\mathbf{E}}(\bar{u}^1(t_i,\cdot) - u_\kappa(t_i,\cdot), e_n)_0^2 +$$
$$C(\gamma)\sum_{n>\kappa} n^{-2\gamma} \leq \kappa^r |||\bar{u}^1(t_i,\cdot) - u_\kappa(t_i,\cdot)|||_0^2 + \frac{C(\gamma)}{\kappa^{2\gamma-1}} \leq \frac{C(\gamma)}{\kappa^{2\gamma-1}\Delta^2}.$$

Together with (3.13) and (3.14), the last inequality implies (3.4).

REMARK. The constants in (3.3) and (3.4) are determined by the bounds on the functions b, σ, h, and p and their derivatives and by the length T of the time interval. The constants in (3.4) also depend on L and k_0 from (2.2).

The error bounds in (3.3) and (3.4) involve two asymptotic parameters: Δ (the size of the partition of the time interval) and κ (the number of the spatial basis functions). With the appropriate choice of these parameters, the errors can be made arbitrarily small.

In Lototsky et al. (1996), the multiple integrals (2.6) were approximated using the Cameron-Martin version of the Wiener chaos decomposition. The analysis was carried out only for the unnormalized filtering density, but the results can be extended to the unnormalized optimal filter $\phi_t[f]$ in the same way as it is done in the present work. The overall error of approximation from Lototsky et al. (1996) has the same order in Δ and κ as (3.3), but the approximation formulas are more complicated.

Formulas (3.1) and (3.2) provide an effective numerical algorithm for computing both the unnormalized filtering density $u(t,x)$ and the unnormalized optimal filter $\phi_t[f]$ independently of each other. If the ultimate goal is an estimate of $f(x(t_i))$ (e.g. estimation of moments of $x(t_i)$), it can be achieved with a given precision recursively in time without computing $u(t_i,x)$ as an intermediate step. This approach looks especially promising if the parameters of the model (i.e. functions b, σ, h and the initial density p) are known in advance. In this case, the values of $(P_\Delta e_l, e_n)_0$, $(P_\Delta h e_l, e_n)_0$, $(1/2)(P_\Delta h^2 e_l, e_n)_0$, and $f_n = (f, e_n)_0$, $n, l = 1, \ldots, \kappa$, can be pre-computed and stored. When the observations become available, the coefficients $\psi_n(i)$ are computed according to (3.1) and then $\phi_{t_i,\kappa}[f]$ is computed according to (3.2). As a result, the algorithm avoids performing on line the time consuming operations of solving partial differential equations and computing integrals. Moreover, only increments of the observation process are required at each step of the algorithm.

REFERENCES

Budhiraja, A. and Kallianpur, G. (1995). Approximations to the Solution of the Zakai Equations using Multiple Wiener and Stratonovich Integral Expansions, Technical Report 447, Center for Stochastic Processes, University of North Carolina, Chapel Hill, NC 27599-3260.

Elliott, R. J. and Glowinski, R. (1989). Approximations to solutions of the Zakai filtering equation, *Stoch. Anal. Appl., 7(2)*:145–168.

Florchinger, P. and LeGland, F. (1991). Time discretization of the Zakai equation for diffusion processes observed in correlated noise, *Stoch. and Stoch. Rep., 35(4)*:233–256.

Gottlieb, D. and Orszag, S. A. (1977). *Numerical Analysis of Spectral Methods: Theory and Applications*, CBMS-NSF Regional Conference, Series in Applied Mathematics, Vol.26.

Hille, E. and Phillips, R. S. (1957). *Functional Analysis and Semigroups*, Amer. Math. Soc. Colloq. Publ., Vol. XXXI.

Ito, K. (1951). Multiple Wiener integral, *J. Math. Soc. Japan, 3*:157–169.

Ito, K. (1996). Approximation of the Zakai equation for nonlinear filtering, *SIAM J. Cont. Opt.* (to appear).

Kallianpur, G. (1980). *Stochastic Filtering Theory*, Springer.

Ladyzhenskaia, O. A., Solonikov, V. A., and Ural'tseva, N. N. (1968). *Linear and quasi-linear equations of parabolic type*, American Mathematical Society, Providence, Rhode Island.

Liptser, R. S. and Shiryayev, A. N. (1992). *Statistics of Random Processes*, Springer.

Lo, J. T.-H. and Ng, S.-K. (1983). Optimal orthogonal expansion for estimation I: Signal in white Gaussian noise, *Nonlinear Stochastic Problems* (Bucy, R. and Moura, J., ed.), D. Reidel Publ. Company, pp. 291–309.

Lototsky, S., Mikulevicius, R., and Rozovskii, B. L. (1996). Nonlinear Filtering Revisited: A Spectral Approach, *SIAM Journal on Control and Optimization*, to appear.

Mikulevicius, R. and Rozovskii, B. L. (1995). Fourier-Hermite Expansion for Nonlinear Filtering, Festschrift in honor of A. N. Shiryayev.

Ocone, D. (1983). Multiple integral expansions for nonlinear filtering, *Stochastics, 10*:1–30.

Rozovskii, B. L. (1990). *Stochastic Evolution Systems*, Kluwer Academic Publishers.

A Berry–Esseen Type Estimate for Hilbert Space Valued U-Statistics and On Bootstrapping Von Mises Statistics

V.V. SAZONOV Steklov Mathematical Institute, Moscow, and Hong Kong University of Science and Technology

This paper consists of two parts related to each other only by employing a common approach. This approach consists in using the technique developed for the proof of Berry-Esseen type estimates and Edgeworth type expansions for Hilbert space valued independent random variables. A number of researches contributed to this area and a rather complete account of the related work up to 1990 can be found in the survey paper by Bentkus et al. (1990). Here we will mention only papers by Götze (1979), Yurinskii (1982), and Sazonov, Ulyanov and Zalesskii (1988, 1991), which are most closely related to the present work.

First consider Hilbert space valued U-statistics. Let $X_1, \dots, X_n$ be independent identically distributed (i.i.d.) random variables with values in a measurable space $(\mathbf{X}, \mathcal{X})$. Denote P the distribution of X_1: $P(X_1 \in A)$, $A \in \mathcal{X}$. Let Φ be a map defined on $(\mathbf{X} \times \mathbf{X}, \mathcal{X} \times \mathcal{X})$ with values in a separable Hilbert space H such that $\Phi(x_1, x_2) = \Phi(x_2, x_1)$, x_1, $x_2 \in \mathbf{X}$. The inner product and norm in H will be denoted $(.,.)$ and $||.||$ respectively. Assume that $E\Phi(X_1, X_2) = 0$ (this assumption is not essential and is made for simplicity) and $E||\Phi(X_1, X_2)|| < \infty$. The U-statistic with kernel Φ corresponding to the sequence

$X_1, \ldots, X_n$ is defined as

$$U_n = \binom{n}{2}^{-1} \sum_{1 \le i < j \le n} \Phi(X_i, X_j).$$

The Hoeffding decomposition represents U_n as

$$U_n = 2n^{-1} \sum_{j=1}^{n} g_1(X_1) + 2n^{-1}(n-1)^{-1} \sum_{1 \le i < j \le n} g_2(X_i, X_j),$$

where

$$g_1(x) = E\Phi(x, X_1),$$
$$g_2(x, y) = \Phi(x, y) - g_1(x) - g_2(y).$$

Without loss of generality we may assume that $g_1(x)$ is defined everywhere. We will also assume that $0 < \sigma^2 = E||g_1(X_1)||^2 < \infty$. Clearly by the Fubini theorem

$$Eg_1(X_1) = E\Phi(X_1, X_2) = 0.$$

Denote V the covariance operator of $g_1(X_1)$:

$$(Vx, y) = E(x, g_1(X_1))(y, g_1(X_1))$$

and $\sigma_1^2 \geq \sigma_2^2 \ldots$ be the eigenvalues of V. Finally let Y be a Gaussian H-valued random variable with mean zero and covariance operator $\sigma^{-2}V$.

By the central limit theorem in Hilbert space, if

$$E||g_1(X_1)||^2 < \infty, \;\; E||g_2(X_1, X_2)||^{4/3} < \infty,$$

then

$$\Delta_n(a) = \sup_{r \geq 0} |P(||2^{-1}n^{1/2}\sigma^{-1}U_n - a|| \leq r) - P(||Y - a|| \leq r)| \to 0 \text{ as } n \to \infty.$$

(see Borovskikh (1986), Borovskikh and Koroljuk (1990)). We are interested in estimating $\Delta_n(a)$ in terms of P and n. The first estimate of this type for real valued U-statistics was obtained by Serfling (1973) and since then a number of gradual improvements were published (for details see Lee (1990), Bentkus, Götze and Zitikis (1992)). The best at present estimate was proved by Koroljuk and Borovskikh (1986). It states that

$$\sup_{r \geq 0} |P(2^{-1}n^{1/2}\sigma^{-1}U_n \leq r) - P(Y \leq r)| \leq c(\sigma^{-1}E|g_1(X_1)|^3 + \sigma^{-5/3}E|g_2(x_1, x_2)|^{5/3})n^{-1/2},$$

and it turns out that $5/3$ here can not be replaced by any smaller number (see Bentkus, Götze, Zitikis (1992)).

For H-valued X_i it was shown in Borovskikh (1986), Borovskikh and Koroljuk (1990, 1991, 1994), that

$$\Delta_n(0) = o(n^{-1/2}).$$

In Puri and Sazonov (1991) an estimate of $\Delta_n(a)$ with an explicit dependence on moments $E||g_1(X_1)||^3$, $E||g_2(X_1, X_2)||^2$ and on a was obtained, however this estimate is only of order $O(n^{-1/3})$. Our new result is the following estimate of $\Delta_n(a)$, obtained jointly with Yu. Borovskikh and Madan Puri, which has an explicit dependence on moments and is of order $O(n^{-1/2})$:

$$\Delta_n(a) \le c(1+||a||^3)\sigma^4(\prod_{j=1}^{9}\sigma_j^{-1})\sigma_9^{-1}(E||\Phi(X_1,X_2)||^3)^2 n^{-1/2}. \tag{1}$$

The main tools in the proof of this estimate are an extended version of the Götze lemma on the estimation of the characteristic function of the square of the norm of a sum of independent Hilbert space valued random variables (see Götze (1979), Yurinskii (1982) and especially Sazonov, Ulyanov and Zalesskii (1988), Lemma 11), and moment inequalities for Hilbert space valued U-statistics (see Koroljuk and Borovskikh (1994)). A rather long proof of (1) will be published elsewhere.

Next consider the bootstrapping of von Mises ω^2-statistics. Let $X_1, X_2, \dots$ be i.i.d. real random variables defined on a probability space $(\Omega, \mathcal{B}, P)$. Assume that the distribution function $F(x) = P(X_1 \le x)$ is continuous and denote by $F_n(x) = n^{-1}\sum_{j=1}^{n} 1_{\{X_j \le x\}}$ the empirical distribution function corresponding to $X_1, \dots X_n$, $n = 1, 2, \dots$ (here and in what follow 1_B denotes the indicator function of a set B, i.e. $1_B(b) = 1$ if $b \in B$ and $1_B(B) = 0$ if $b \notin B$). The von Mises ω^2-statistics are defined as

$$\omega_n^2 = n\int_{-\infty}^{\infty}(F_n(x) - F(x))^2 dF(x).$$

To define the bootstrapped version of ω_n^2 a resample sequence $X_1^*, \dots, X_n^*$ of i.i.d. random variables with distribution function F_n is introduced. The variables $X_1^*, \dots, X_n^*$ may be defined on $(\Omega, \mathcal{B}, P)$, and in this case they are assumed to be independent of $X_1, \dots, X_n$ or on an another probability space $(\Omega^*, \mathcal{B}, P^*)$. We prefer to suppose them to be defined on Ω^*. If F_n^* is the empirical distribution function corresponding to $X_1^*, \dots, X_n^*$, the bootstrapped version of ω_n^2 is

$$(\omega_n^2)^* = n\int_{-\infty}^{\infty}(F_n^*(x) - F_n(x))^2 dF_n(x).$$

We will also consider generalized Bayesian bootstrap versions of ω_n^2. Let $U_1, U_2, \dots$ be a sequence of real i.i.d. random variables defined on a probability space $(\Omega', \mathcal{B}', P')$, such that $P'(U_1 = 0) = 0$, $E'U_1 = 1$,

$$E' e^{tU_1} < \infty \quad \text{for some} \quad t > 0. \tag{2}$$

Condition (2) is equivalent to the existence of $g > 0$, $T > 0$ such that

$$E' e^{t(U-1)} \le e^{-gt^2/2} \quad \text{for all} \quad t : |t| \le T \tag{2'}$$

(see Petrov (1975), Ch. 3, Lemma 5). The generalized Bayesian bootstrap version of ω_n^2 is defined as

$$(\omega_n^2)^B = n\sigma^{-2}\int_{-\infty}^{\infty}(F_n^B(x) - F_n(x))^2 dF_n(x),$$

where $\sigma^2 = E'\,(U_1 - 1)^2$, $F_n^B(x) = \sum_{j=1}^n \Delta_j 1_{\{X_j \le x\}}$ and

$$\Delta_j = U_j(\sum_{j=1}^{n} U_j)^{-1}, \qquad j = 1, \dots, n.$$

Our aim is to estimate

$$\delta_n(\omega) = \sup_{y \ge 0} |P(\omega_n^2 \le y) - P^*((\omega_n^2)^* \le y)|$$

and

$$\Delta_n(\omega) = \sup_{y \ge 0} |P(\omega_n^2 \le y) - P'((\omega_n^2)^B \le y)|.$$

As far as we know δ_n, Δ_n have not been estimated before. The following results have been obtained jointly with Albert Lo and will be published with full proofs elsewhere.

1. With probability 1 for any $\epsilon > 0$

$$\delta_n(\omega) \le c(\epsilon) n^{-1+\epsilon}, \qquad n = 1, 2, \dots, \tag{3}$$

where $c(\epsilon)$ depends only on ϵ.

2. With probability 1

$$\Delta_n(\omega) \le c(\beta, g, T)(\log n)^{3/2} n^{-1/2}, \qquad n = 1, 2, \dots, \tag{4}$$

where $c(\beta, g, T)$ depends only on $\beta = \sigma^{-4} E'\,(U_1 - 1)^4$ and on g, T from $(2')$ (one can write down a $c(\beta, g, T)$ with an explicit dependence on β, g, T).

It follows from the proofs that (3) and (4) are true for all ω such that all $X_1(\omega), X_2(\omega), \dots$ are different (the set of such ω has P probability 1).

To prove (3) we write

$$\delta_n(\omega) \le \delta_{n1} + \delta_{n2}(\omega) + \delta_{n3},$$

where

$$\delta_{n1} = \sup_{y \ge 0} |P(\omega_n^2 \le y) - G(y)|, \quad \delta_{n2} = \sup_{y \ge 0} |P^*((\omega_n^2)^* \le y) - G_n(y)|,$$

$$\delta_{n3} = \sup_{y \ge 0} |G(y) - G_n(y)|;$$

here $G(y)$ may be represented as $G(y) = P(|Y| \le y)$, Y being a Hilbert space valued Gaussian random variable with $EY = 0$ and the covariance operator V having eigenvalues $(\pi j)^{-2}$, $j = 1, 2, \dots$; $G_n(y)$ is defined below, see (5) and what follows it.

It is well known that $\delta_{n1} \le cn^{-1}$, $n = 1, 2, \dots$ (see e.g. Bentkus et al. (1990), Götze (1979)). To estimate $\delta_{n2}(\omega)$ we observe that $(\omega_n^2)^*$ can be represented as

$$(\omega_n^2)^* = \|n^{-1/2} \sum_{l=1}^{n} Y_{nl}\|^2,$$

where Y_{nl} for each $n = 1, 2, \dots$, $l = 1, \dots, n$ is a $L_2[0,1]$-valued random variable defined by

$$Y_{nl}(t) = 1_{\{Z_l^* \le k/n\}} - k/n \quad \text{if} \quad (k-1)/n < t \le k/n, \quad k = 1, \dots, n,$$

where $Z_l^* = m_n(X_l^*)$ and m_n is a map defined on $X_k(\omega)$, $k = 1, \dots, n$ by $m_n(X_{(k)}(\omega)) = k/n$, $X_{(k)}(\omega)$, $k = 1, \dots, n$, being the values of the order statistics corresponding to $X_1, \dots, X_n$ ($X_{(k)}(\omega)$, $k = 1, \dots, n$, are assumed to be different and they are different with probability 1). Then we apply a Berry-Esseen type estimate (Corollary 7 in Sazonov, Ulyanov and Zalesskii (1988)) to the sequence Y_{nl}, $l = 1, \dots, n$ of $L_2[0,1]$ i.i.d. random variables to obtain

$$\sup_{y \geq 0} |P^*(n^{-1/2}|\sum_{l=1}^{n} Y_{nl}| \leq y) - G_n(y)| \leq c(\epsilon) n^{-1+\epsilon}, \tag{5}$$

where $G_n(y) = P(|N_n| \leq y)$ and N_n is a Gaussian $L_2[0,1]$-valued random variable with $EN_n = 0$ and having the same covariance operator V_n as Y_{n1}. The eigenvalues of V_n are

$$w_{nk} = \begin{cases} (4n^2 \sin^2(\pi k/(2n)))^{-1}, & k = 1, \dots, n-1 \\ 0, & k \geq n. \end{cases}$$

Finally, using explicit expression for eigenvalues of V and V_n one can prove that $\delta_{n3} \leq cn^{-1}$ and this leads to the estimate of $\delta_n(\omega)$ stated above.

The proof of (4) goes basically along the same lines. The main differences are: in proving (4) we have to use a Berry-Esseen type estimate for non identically distributed independent Hilbert space valued summands (see Ulyanov (1987)) and instead of $\|n^{-1/2} \sum_{i=1}^{n} Y_{nl}\|$ we have to deal with the norm of a similar normalized sum multiplied by the random factor $n/|\sum_{j=1}^{n} U_j|$. Due to the presence of this random factor our approach leads to a lower order estimate (4) as compared to (3).

It is appropriate to note that estimates (3) and (4) have only theoretical value since when the distribution function F is continuous there is no need to bootstrap ω_n^2.

We are grateful to Andrew Rukhin for the information related to the operator V_n, in particular for the reference to Rutherford (1951), which permitted to identify precisely the eigenvalues of V_n.

References

Bentkus, V., Götze, F., Paulaskas, V. and Rackauskas, V. (1990). The accuracy of Gaussian approximation in Banach spaces, SFB 343 "Diskrete Strukturen in der Mathematik", Preprint 90-100, Univeristät Bielefeld.

Bentkus, V., Götze, F. and Zitikis, R. (1992). Lower estimates of the convergence rate for *U*-statistics, SFB 343 "Diskrete Strukturen in der Mathematik", Preprint 92-075, Universität Bielefeld.

Borovskikh, Yu. V. (1986). Theory of *U*-statistics in Hilbert space, Preprint No. 86.78, Institute of Mathematics, Ukrain. Acad. of Sci., Kiev, (Russian).

Borovskikh, Yu. V. and Koroljuk, V. S. (1990). *UH*-statistics, *Sov. Math. Dokl., 40*: 432-435.

Götze, F. (1979). Asymptotic expansions for bivariate von Mises functionals, *Z. Wahrsch. Verw. Gebiete, 50*: 333-355.

Koroljuk, V. S. and Borovskikh, Yu. V. (1986). Approximation of non degenerate U-statistics, *Theory Probab. Appl., 30*: 439-450.

Koroljuk, V. S. and Borovskikh, Yu. V. (1991). Rate of convergence in the central limit theorem for UH-statistics, *Theor. Probab. Math. Statist., 43*: 79-85.

Koroljuk, V. S. and Borovskikh, Yu. V. (1994). *Theory of U-statistics*, Kluwer, Dodrecht.

Lee, A. J. (1990). *U-statistics: theory and practice*, M. Dekker, New York.

Petrov, V. V. (1975). *Sums of independent random variables*, Springer-Verlag, Berlin.

Puri, M. L. and Sazonov, V. V. (1991). On Hilbert space valued U-statistics. *Theory Probab. Appl., 36*: 604-605.

Rutherford, D. E. (1951). Some continuant determinants arising in Physics and Chemistry, *Proc. Roy. Soc. Edin. A, 63*: 232-241.

Sazonov, V. V., Ulyanov, V. V. and Zalesskii, A. (1988). Normal approximation in Hilbert space I-II, *Theor. Probab. Appl., 33*: 207-227, 473-487.

Sazonov, V. V., Ulyanov, V. V. and Zalesskii, A. (1991). A precise estimate of the rate of convergence in the central limit theorem in Hilbert space, *Mat. USSR Sbornik, 68*: 453-482.

Serfling, R. J. and Grams, W. E. (1973). Convergence rates for U-statistics and related statistics, *Ann. Statist., 1*: 153-160.

Ulyanov, V. V. (1987). Asymptotic expansions for distributions of sums of independent random variables in H, *Theory Probab. Appl., 31*: 25-39.

Yurinskii, V. V. (1982). On the accuracy of normal approximation of the probability of hitting a ball, *Theory Probab. Appl., 27*: 280-289.

On the Strong Form of the Faber Theorem

BORIS SHEKTMAN, Department of Mathematics, University of South Florida, Tampa, FL 33620

ABSTRACT. Let H_n^∞ be the space of polynomials of degree at most $n-1$ on the unit circle with the uniform norm. We show that the dual spaces $(H_n^\infty)^*$ cannot be embedded uniformly in any $\mathcal{L}_1$-space. This result implies (but is not equivalent to) the famous theorem of Faber.

Let H_n be the space of polynomials of degree at most $n-1$ and let H_n^p be the n-dimensional Banach space H_n considered as a subspace of $L_p(\mathbb{T})$, where $\mathbb{T}$ is the unit circle. We use $d(X, Y)$ to denote the Banach-Mazur distance, $\pi_1(T)$ to denote the absolutely summing norm of an operator T and $\gamma_\infty(T)$ for the L_∞-factorization norm of T. (cf [4]).

Theorem 1. *There exists a constant $C > 0$ such that, if E_n is an n-dimensional subspace of an $\mathcal{L}_1$-space, then*

$$d((H_n^\infty)^*, E_n) \geq C \log\ n.$$

Proof. Let X be an $\mathcal{L}_1$-space. Let $E_n \subset X$ be an n-dimensional subspace. Let

$$T : (H_n^\infty)^* \hookrightarrow X$$

be such that $T(H_n^\infty)^* = E_n$. We want to prove that

$$\|T\|\|T^{-1}\| \geq C \log\ n.$$

Assume without loss of generality that $\|T^{-1}\| = 1$. Then (cf [3]) there are vectors $x_1^*, \ldots, x_n^* \in X^*$ such that

$$T^* x_j^* = e^{ij\theta}$$

and $\|x_j^*\| \leq 1$.

Consider the following operators:

$$J : H_n^\infty \to H_n^1;\ Je^{ik\theta} = e^{ik\theta},$$
$$S : H_n^1 \to \ell_1^{(n)};\ Se^{ik\theta} = \frac{1}{k+1} e_k,$$

where $e_1, \ldots, e_n$ is the canonical basis in $\ell_1^{(n)}$,

$$A : \ell_1^{(n)} \to X^*;\ Ae_k = x_k^*.$$

We illustrate these operators on the diagram

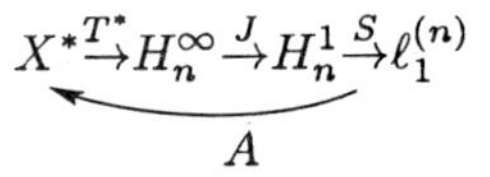

By the Pietsch Factorization Theorem (cf [4])

$$\pi_1(J) = \|J\| = 1.$$

By the Hardy inequality (cf [1]) there exists a constant $c > 0$ independent of n, such that

$$\|S\| \leq c$$

Finally, we have

$$\|A(\Sigma a_j e_j)\| = \|\Sigma a_j x_j^*\| \leq \Sigma |a_j| = \|\Sigma a_j e_j\|$$

and hence

$$\|A\| \leq 1.$$

Since A maps an $\mathcal{L}_1$-space into an $\mathcal{L}_\infty$-space, we have

$$\gamma_\infty(A) = \|A\| = 1.$$

Next observe that

$$tr(ASJT^*) = tr(T^*ASJ) = \sum_{k=0}^{n} \frac{1}{k+1} \geq \log\ n.$$

By trace-duality (cf [4])

$$\log n \leq tr(ASJT^*) \leq \pi_1(SJT^*)\gamma_\infty(A)$$
$$\leq \|S\|\pi_1(J)\|T^*\|\|A\| \leq c\|T\|.$$

Thus

$$\|T\| = \|T^*\| \geq \frac{1}{c} \log\ n. \ \square$$

Theorem 2. *The results of the Theorem 1 remain true if we replace the space* H_n^∞ *by any of the following spaces:*

(1) *span* $[e^{i\lambda_k\theta}]_{k=1}^n \subset C(\mathbb{T})$ *where* λ_k *are arbitrary distinct integers,*
(2) *span* $[\cos j\theta]_{j=1}^n \subset C[0,\pi]$
(3) *span* $[t^k]_{k=1}^n \subset C[a,b]$.

Proof. The proof is the same as that of Theorem 1 if we replace the use of the Hardy inequality by the generalized Hardy inequality (cf [2]) for the trigonometric polynomials and by the Sidon inequality (cf [6]) for the cosine polynomials. The isometry between the algebraic and cosine polynomials proves the remaining case. □

Corollary 1 (Faber Theorem). *Let* $P_n : C(\mathbb{T}) \to H_n$ *be an arbitrary sequence of projections onto the trigonometric polynomials. Then* $\|P_n\| \geq C\log n$.

Proof. Let $P_n : C(\mathbb{T}) \to H_n^\infty$. Then $P_n^* : (H_n^\infty)^* \to C^*(\mathbb{T})$. Moreover, $P_n^*(H_n^\infty)^*$ is an n-dimensional subspace of $C^*(\mathbb{T})$ and $\|(P_n^*)^{-1}\| = 1$. Since $C^*(\mathbb{T})$ is an $\mathcal{L}_1$-space, we conclude

$$C\log n \leq d((H_n^\infty)^*, P_n^*(H_n^\infty)^*) \leq \|P_n^*\|\|(P_n^*)^{-1}\| = \|P_n\|. \ \square$$

Remark 1. The result of Theorem 1 is sharp.

Proof. Let Q_n be the Fourier projections from $C(\mathbb{T})$ onto H_n. Then $\|Q_n\| \sim \log n$ and

$$d((H_n^\infty)^*, Q_n^*(H_n^\infty)^*) = d((H_n^\infty)^*, H_n^1) \leq \|Q_n\|. \ \square$$

We will now mention two examples to conclude that Theorem 1 does not follow from the Faber Theorem.

Example 1. Let $A(\mathbb{T})$ be the disk-algebra. Then, for any sequence of projections P_n from $A(\mathbb{T})$ onto H_n, we have (cf [6]) $\|P_n\| \geq C\log n$. Despite this fact $(H_n^\infty)^*$ is uniformly isomorphic to the subspaces of $A^n(\mathbb{T})$. Indeed it was shown by Bourgain and Pelczynski (cf [5]) that H_n^∞ are uniformly isomorphic to subspaces V_n of $A(\mathbb{T})$ which are uniformly complemented. Using the reasoning of Corollary 1, we get the desired conclusion. □

It may appear that the reason for this example is in the fact that $A(\mathbb{T})$ is not an $\mathcal{L}_\infty$-space and the relative projectional constant of H_n^∞ in $A(\mathbb{T})$ is not isomorphic-invariant.

Therefore

Example 2. Let $V_n \subset C(\mathbb{T})$ be such that

$$d(V_n, \ell_2^{(n)}) = 1.$$

It is well-known (cf [4]) that for any sequence of projections P_n from $C(\mathbb{T})$ onto V_n, we have $\|P_n\| \geq \frac{1}{2}\sqrt{n}$. Yet V_n^* are isometric to $\ell_2^{(n)}$ and hence (by Dvoretzky's Theorem) V_n^* are uniformly isomorphic to subspaces of $C^*(\mathbb{T})$ which is an $\mathcal{L}_1$-space. □

I am grateful to B. Chalmers, V. Gurarie and S. Kisliakov for many suggestions and for the encouragement to write up this note.

References

[1] Hoffman, K., *Banach Spaces of Analytic Functions*, Prentice-Hall, Englewood Cliffs, NJ, 1962.

[2] McGehee, O. C., Pigno, L. and Smith, B., *Hardy's inequality and the L^1 norm of exponential sums*, Ann. Math., **113**((1981), 613-618.

[3] Rudin, W., *Functional Analysis*, McGraw-Hill Inc., 1973.

[4] Tomczak-Jaegerman, N., *Banach-Mazur Distances and Finite-Dimensional Operator Ideals*, Wiley, New-York, 1985.

[5] Wojtaszczyk P., *Banach Spaces for Analysts*, Cambridge Univ. Press, Cambridge, 1991.

[6] Zygmund A., *Trigonometric Series*, Cambridge Univ. Press, Cambridge, 1959.

Nonlinear Filtering Theory for Stochastic Reaction–Diffusion Equations

S. L. HOBBS Naval Command Control, and Ocean Surveillance Center, Code 784, San Diego, CA 92152-6040

S. S. SRITHARAN Naval Command Control, and Ocean Surveillance Center, Code 574, San Diego, CA 92152-6040
Supported by the ONR Mathematical Sciences and Mechanics Divisions; Affiliations: University Of Colorado and SDSU

This paper is dedicated to the occasion of Professor M. M. Rao's sixty fifth birthday

Abstract

We examine the nonlinear filtering problem for the stochastic reaction-diffusion equation with additive noise. We derive the Zakai and Fujisaki-Kallianpur-Kunita filtering equations for the evolution of the un-normalized and normalized conditional expectation, and prove existence and uniqueness theorems for measure valued solutions.

1 INTRODUCTION

The reaction-diffusion equation is an elementary model in a number of physical applications including combustion. The purpose of this article is to study the filtering problem for the stochastic reaction-diffusion equation with bounded nonlinearity and additive, trace class white noise. One of the important applications would be in combustion control with partial observations. In this paper we will derive the Zakai and Fujisaki-Kallianpur-Kunita (FKK) filtering equations for the evolution of the (un-normalized and normalized) conditional expectation (filter), and prove existence and uniqueness theorems for measure valued solutions. The filtering problem for finite dimensional systems has been studied for many years but filtering for systems governed by partial differential equations has only recently begun to be developed [Sri94]. In [Ahm94] a nonlinear filtering result for semilinear equations by an entirely different approach is announced.

The reaction-diffusion equation governs the state of a system $\boldsymbol{u}(t)$ which evolves in time according to a parabolic partial differential equation with a non-linear interaction term. With stochastic forcing by a Wiener process $\boldsymbol{w}(t)$, we have on a complete probability space $(\Omega, \mathcal{F}, \boldsymbol{m})$,

$$d\boldsymbol{u}(t) = [A\boldsymbol{u} + B(\boldsymbol{u})]\, dt + G d\boldsymbol{w}(t) \tag{1}$$

where A is an elliptic partial differential operator, B is a nonlinear function of $\boldsymbol{u}$, and G is a bounded linear operator (which could be the identity). To this stochastic evolution equation we add an observation vector $\boldsymbol{z}(t) \in \boldsymbol{R}^m$ for each time t which evolves according to its own stochastic ordinary differential equation,

$$d\boldsymbol{z}(t) = \boldsymbol{h}(\boldsymbol{u}(t))dt + d\boldsymbol{w}_2(t) \tag{2}$$

where $\boldsymbol{h}(\cdot)$ is an m-vector valued function, $\boldsymbol{w}_2(\cdot)$ is an m-vector Wiener process which is not necessarily independent of $\boldsymbol{w}(\cdot)$. The filtering problem is to estimate some (scalar) function $f(t, \boldsymbol{u}(t))$ of the system state at time t given the set of observations $\{z(s) : 0 < s \leq t\}$ up to time t. It is well known that the best (minimum variance) estimate of this quantity is the conditional expectation

$$\boldsymbol{\Pi}_t(f) = E[f(t, \boldsymbol{u}(t))|\mathcal{F}_t^z] \tag{3}$$

given the σ-field $\mathcal{F}_t^z = \sigma\{\boldsymbol{z}(s) : 0 < s \leq t\}$ generated by the observations.

For the time evolution of $\boldsymbol{\Pi}_t(f)$ we will derive the following equation using the method of [FKK72]:

$$\begin{aligned} d\boldsymbol{\Pi}_t(f) &= \boldsymbol{\Pi}_t(\partial_t f + \mathcal{L} f)dt \\ &+ [\boldsymbol{\Pi}_t(\mathcal{M} f) - \boldsymbol{\Pi}_t(f)\boldsymbol{\Pi}_t(\boldsymbol{h})] \cdot [d\boldsymbol{z} - \boldsymbol{\Pi}_t(\boldsymbol{h})dt] \end{aligned} \tag{4}$$

where

$$\mathcal{L}f = \frac{1}{2}\mathrm{tr}(GQG^*\partial_{uu}f) + < A\boldsymbol{u} + B(\boldsymbol{u}), \partial_u f >, \tag{5}$$

and

$$\mathcal{M}f = G_2^*\partial_u f + f\boldsymbol{h}. \tag{6}$$

Here Q is the covariance operator for the process $\boldsymbol{w}(t)$ and G_2 is a linear operator (discribed below) from the observation space to the state Hilbert space.

The Zakai equation [Zak69]

$$d\boldsymbol{\Theta}_t(f) = \boldsymbol{\Theta}_t(\partial_t f + \mathcal{L}f)dt + \boldsymbol{\Theta}_t(\mathcal{M}f) \cdot d\boldsymbol{z}(t) \tag{7}$$

is satisfied by an *un-normalized* conditional expectation of f given $\mathcal{F}_t^z$, where the two are related by the equation

$$\boldsymbol{\Theta}_t(f) = \boldsymbol{\Pi}_t(f)exp\{\int_0^t \boldsymbol{\Pi}_s(\boldsymbol{h}) \cdot d\boldsymbol{z}(s) - \frac{1}{2}\int_0^t |\boldsymbol{\Pi}_s(\boldsymbol{h})|^2 ds\} \tag{8}$$

and its inverse

$$\boldsymbol{\Pi}_t(f) = \boldsymbol{\Theta}_t(f)/\boldsymbol{\Theta}_t(1). \tag{9}$$

A measure valued solution to the FKK equation gives an evolving conditional probability measure Π_t defined on the Borel subsets of the state space $\boldsymbol{H}$ in which $\boldsymbol{u}(t)$ takes its values, so that the conditional expectation may be computed as

$$\boldsymbol{\Pi}_t(f) = \int_H f(t,\boldsymbol{u})\Pi_t(d\boldsymbol{u}) \tag{10}$$

and an un-normalized measure Θ_t satisfies

$$\boldsymbol{\Theta}_t(f) = \int_H f(t,\boldsymbol{u})\Theta_t(d\boldsymbol{u}) \tag{11}$$

for any f in a restricted class of test functions.

2 ASSUMPTIONS AND NOTATION

We will make the following assumptions.

(A1) *The linear (unbounded) operator $A : D(A) \subset \boldsymbol{H} \to \boldsymbol{H}$ is self-adjoint and generates a strongly continuous semi-group $S(t), t \geq 0$, of bounded linear operators on $\boldsymbol{H}$.*

(A2) *The nonlinear operator* $B : \boldsymbol{H} \to \boldsymbol{H}$ *is defined and continuous on all of* $\boldsymbol{H}$ *and satisfies the following Lipschitz and linear growth conditions: there exists a constant* $C > 0$ *such that* $||B(\boldsymbol{u}) - B(\boldsymbol{v})|| \leq C||\boldsymbol{u} - \boldsymbol{v}||$ *and* $||B(\boldsymbol{u})||^2 \leq C^2(1 + ||\boldsymbol{u}||^2)$ *for all* $\boldsymbol{u}, \boldsymbol{v} \in \boldsymbol{H}$.

(A3) *We assume the Wiener process* $\boldsymbol{w}(t)$ *takes values in its own separable Hilbert space* $\boldsymbol{U}$.

(A4) $G : \boldsymbol{U} \to \boldsymbol{H}$ *is a bounded linear transformation and does not depend on* t *or* $\boldsymbol{u}$.

We will also assume that $\boldsymbol{w}(t)$ has a trace class covariance operator ([DZ92], ch 4.1). That is, $E\boldsymbol{w}(s) \otimes \boldsymbol{w}(t) = (t \wedge s)Q$ where $Q : \boldsymbol{U} \to \boldsymbol{U}$ is a positive definite, self-adjoint, bounded linear operator with finite trace. If we denote by ν_i and $\boldsymbol{v}_i$ the eigenvalues and (complete orthonormal set of) eigenfunctions of Q then the trace of Q is $\sum_{i=1}^{\infty} \nu_i$ and we may write ([DZ92], ch 4.1)

$$\boldsymbol{w}(t) = \sum_{i=1}^{\infty} \sqrt{\nu_i}\beta_i(t)\boldsymbol{v}_i \tag{12}$$

as an expansion of $\boldsymbol{w}(t)$ where the β_i's are independent standard real valued Wiener processes. The convergence is in $\boldsymbol{U}$.

In order to derive our results we will generally restrict the functions f and $\boldsymbol{h}$ to being of class C_b^2 (notation below) but explicit conditions will be given in the theorems below.

For the purpose of expressing (and proving) the Zakai and FKK filtering equations, it is useful to model the noise process $\boldsymbol{w}(t)$ (or more precisely $G\boldsymbol{w}(t)$) as the sum

$$Gd\boldsymbol{w}(t) = G_1 d\boldsymbol{w}_1(t) + G_2 d\boldsymbol{w}_2(t), \tag{13}$$

where $G_2 : \boldsymbol{R}^m \to \boldsymbol{H}$ is a bounded, full rank linear operator. This is possible by setting $G_1 d\boldsymbol{w}_1(t)$ equal to the difference between the first and last terms. The Wiener process $\boldsymbol{w}_1(t)$ takes its values in a separable Hilbert space $\boldsymbol{U}_1$ ($\boldsymbol{U}_1 = \boldsymbol{H}$ is one possibility) and $G_1 : \boldsymbol{U}_1 \to \boldsymbol{H}$ bounded linear. If we use (13) and adjoin a random initial condition, we get

$$d\boldsymbol{u}(t) = [A\boldsymbol{u} + B(\boldsymbol{u})]\, dt + G_1 d\boldsymbol{w}_1(t) + G_2 d\boldsymbol{w}_2(t) \tag{14}$$

$$\boldsymbol{u}(0) = \boldsymbol{u}_0. \tag{15}$$

We will fix throughout $T > 0$ and work on the time interval $[0, T]$. The results of the paper also hold on $[0, \infty)$, except that in this case convergence, *e.g.*, in $C([0,\infty),\boldsymbol{H})$, is in the topology of uniform convergence on compact subsets of $[0,\infty)$.

There will be several filtrations $\mathcal{F}_t$ of increasing σ-fields $\subset \mathcal{F}$. Usually these filtrations are generated by one or more processes with independent increments; for such processes $\boldsymbol{v}(t)$ we will denote by $\mathcal{F}_t^v$ the completion of $\sigma\{\boldsymbol{v}(s) : 0 < s \le t\}$, and we include in every $\mathcal{F}_0^v$ all P-null sets.

3 SOLVABILITY OF THE STOCHASTIC REACTION DIFFUSION EQUATION

We can take the following basic probability space:

$$\Omega = \boldsymbol{H} \times C([0,T];\boldsymbol{U}),$$

$$\mathcal{F} = \mathcal{B}(\boldsymbol{H} \times C([0,T];\boldsymbol{U}))$$

$$\boldsymbol{m} = \boldsymbol{\mu}_0 \times \boldsymbol{\lambda}$$

where $\mathcal{B}(\cdot)$ is the Borel algebra, $\boldsymbol{\mu}_0$ is the distribution of the initial data and $\boldsymbol{\lambda}$ is the Wiener measure.

On the basic probability space $(\Omega,\mathcal{F},\boldsymbol{m})$ given above we will take as our normal filtration $\mathcal{F}_t^{u_0,w}$ the complete σ-fields generated by $\boldsymbol{u}_0$ and $\boldsymbol{w}(\cdot)$ and all P-null sets ([DZ92], ch 3.3 and 7.1). Since $\boldsymbol{u}_0$ is $\mathcal{F}_0^{u_0}$-measurable, and $\mathcal{F}_0^{u_0,w} \in \mathcal{F}$, the solution $\boldsymbol{u}(t)$ of (14) and (15) will be a predictable process with respect to this σ-field.

Definition 1 *For any $\boldsymbol{H}$-valued $\mathcal{F}_0^{u_0,w}$-measurable random variable $\boldsymbol{u}_0$, a $\mathcal{F}_t^{u_0,w}$ predictable $\boldsymbol{H}$-valued process $\boldsymbol{u}(t)$, $0 \le t \le T$ is a* **mild solution** *of (14) and (15) if, for all $t \in [0,T]$*

$$i)\ \ \boldsymbol{m}\{\int_0^t ||\boldsymbol{u}(s)||^2 ds < \infty\} = 1$$

and

$$ii)\ \ \boldsymbol{u}(t) = S(t)\boldsymbol{u}_0 + \int_0^t S(t-s)B(\boldsymbol{u}(s))ds$$

$$+ \int_0^t S(t-s)Gd\boldsymbol{w}(s),\quad \boldsymbol{m}\text{-a.s.}.$$

The condition that the process $\boldsymbol{u}(t)$ be predictable with respect to the filtration $\mathcal{F}_t^{u_0,w}$ is important for it will play a role in our main result. This means ([DZ92], ch 3.3) that $\boldsymbol{u}(t) = \boldsymbol{u}(t, \omega)$ is measurable with respect to $\mathcal{P}_T$ the (completed) σ-field generated by all subsets of $[0, T] \times \Omega$ which have the form $(s, t] \times F$ where $0 \leq s < t \leq T$ and $F \in \mathcal{F}_s^{u_0,w}$.

Theorem 1 *Let assumptions (A1) to (A4) hold and the initial value $\boldsymbol{u}_0$ be an $\boldsymbol{H}$-valued random variable which is independent of $\boldsymbol{w}(\cdot)$, and with $E||\boldsymbol{u}_0||^q < \infty$ for some $q \geq 2$. Then the initial value problem (14) and (15) has a unique (up to equivalence) mild solution $\boldsymbol{u}(t)$. Further, $\boldsymbol{u}(t)$ has a version whose trajectories are continuous a.s., i.e., $\boldsymbol{u}(\cdot) \in C([0, T]; \boldsymbol{H})$, and there exists $C > 0$ (depending on T) such that $sup_{t \in [0,T]} E||\boldsymbol{u}(t)||^q \leq C(1 + E||\boldsymbol{u}_0||^q)$.*

Denoting by $\boldsymbol{X}_q$ the Banach space of $\boldsymbol{H}$-valued predictable processes $\boldsymbol{v}(t)$ such that the norm $(sup_{0 \leq t \leq T} E||\boldsymbol{v}(t)||^q)^{1/q} < \infty$, proof of the existence of $\boldsymbol{u}(t)$ is obtained by taking $\boldsymbol{u}(t)$ as the limit of successive approximations of the mapping $K : \boldsymbol{X}_q \mapsto \boldsymbol{X}_q$ defined by

$$K\boldsymbol{v}(t) = S(t)\boldsymbol{u}_0 + \int_0^t S(t-s)B(\boldsymbol{v}(s))ds + \int_0^t S(t-s)G d\boldsymbol{w}(s). \tag{16}$$

K is a contraction mapping on sufficiently small subintervals of $[0, T]$ ([DZ92], Theorem 7.4 or [Ich82]). We note three things that we will need for Theorems 2 and 3, the main results of this paper: First, although the solution $\boldsymbol{u}(t)$ only satisfies (14) in a mild sense (it need not take values in $D(A)$), it is $\boldsymbol{H}$-valued and not just a distributional solution. Second, as a mild solution, $\boldsymbol{u}(t)$ is predictable and hence adapted to the filtration $\mathcal{F}_t^{u_0,w}$ ([DZ92], ch 7.1). An examination of the proof shows that $\mathcal{F}_t^{u_0,w}$ is the smallest σ-field that can be used for $\mathcal{F}_t$, so $\boldsymbol{u}(t)$ is indeed $\mathcal{F}_t^{u_0,w}$ adapted. Third, $\boldsymbol{u}(t)$ is a *measurable function of* $\boldsymbol{u}_0$ and $\boldsymbol{w}(s)$ for $s \leq t$. For on the subinterval (t_i, t_{i+1}), $\boldsymbol{u}(t)$ is the limit of a sequence of the form $\boldsymbol{u}^j(t) = K\boldsymbol{u}^{j-1}(t)$ where $K\boldsymbol{v}(t)$ is the limit of sums of the form

$$S(t - t_i)\boldsymbol{v}(t_i) + \sum_k S(t - s_k)B(\boldsymbol{v}(s_k))(s_k - s_{k-1})$$

$$+ \sum_k S(t - s_k)G(\boldsymbol{w}(s_k) - \boldsymbol{w}(s_{k-1})) \tag{17}$$

and the sums are taken over a partition $\{s_k\}$ of the subinterval. We see that each term in these sums is clearly measurable with respect to the starting data ($S(t)$ is continuous) and the σ-field generated by $\boldsymbol{w}(t)$ on the relevant subinterval only.

4 DERIVATION OF THE FKK AND ZAKAI EQUATIONS

In this section we will derive evolution equations for the conditional expectation $\boldsymbol{\Pi}_t(f) = E\left[f(\boldsymbol{u}(\cdot,t))|\mathcal{F}_t^z\right]$. Let us define the innovation process $\boldsymbol{Y}(t) = \{Y_1(t), \cdots, Y_n(t)\}$ as,

$$\boldsymbol{Y}(t) = \boldsymbol{z}(t) - \int_0^t \boldsymbol{\Pi}_r(\boldsymbol{h})dr. \tag{18}$$

Lemma 1 *[FKK72] Let $\boldsymbol{u}(\cdot)$ be the solution of (9) (Theorem 1), $\boldsymbol{z}(t)$ be the observation process defined in (10) and $\boldsymbol{h}(\cdot) \in C_b(\boldsymbol{R}^m)$. Then $(\boldsymbol{Y}(t), \mathcal{F}_t^z, \boldsymbol{m})$ is an m-vector standard Wiener process. Moreover, the two sigma fields $\sigma\{Y(r) - Y(s), t \le s < r \le T\}$ and $\mathcal{F}_t^z$ are independent.*

The following martingale representation result due to Fujisaki-Kallianpur-Kunita [FKK72] is the key to the derivation of the nonlinear filtering equation.

Lemma 2 *Every square integrable martingale $(\boldsymbol{M}(t), \mathcal{F}_t^z, \boldsymbol{m})$ is sample continuous and can be represented as a stochastic integral with respect to the innovation process:*

$$\boldsymbol{M}(t) = \mathcal{E}[\boldsymbol{M}(0)] + \int_0^t \boldsymbol{\Phi}(s) \cdot d\boldsymbol{Y}(s), \quad s \in [0,T], \tag{19}$$

where

$$E\int_0^T |\boldsymbol{\Phi}(t)|^2 dt < +\infty \tag{20}$$

and $\boldsymbol{\Phi}(t)$ is jointly measurable in $(0,T) \times \Omega$ and adapted to $\mathcal{F}_t^z$.

Definition 2 *The class of* **cylindrical test functions** $\mathcal{C}_{CY}$ *as follows:*

$$\begin{aligned}\mathcal{C}_{CY} = \{&f(\cdot,\cdot) : [-a,T] \times \boldsymbol{H} \to \boldsymbol{R}; f(t,\boldsymbol{u}) = \phi(t, (\boldsymbol{u},\boldsymbol{e}_1), \cdots, (\boldsymbol{u},\boldsymbol{e}_n)),\\ &\boldsymbol{e}_i \in D(A), i = 1, \cdots, n; \phi \in C_0^\infty((-a,T) \times \boldsymbol{R}^n)\}, a > 0. \end{aligned}\tag{21}$$

We now define

$$\boldsymbol{M}_f(t) := \boldsymbol{\Pi}_t(f(t)) - \boldsymbol{\Pi}_0(f(0)) - \int_0^t \boldsymbol{\Pi}_s(\frac{\partial f}{\partial s}(s) + \mathcal{L}f(s))ds, \tag{22}$$

where $f \in \mathcal{C}_{CY}$ and $\mathcal{L}$ given by (5).

Lemma 3 *For all* $f \in \mathcal{C}_{CY}$*,* $(\boldsymbol{M}_f(t), \mathcal{F}_t^z, \boldsymbol{m})$ *is a square integrable martingale in* $[0, T]$.

This follows from the fact that for the mild solution, combining the results in [DZ92] and [Vio76] we can conclude that, for $f \in \mathcal{C}_{CY}$,

$$\hat{\boldsymbol{M}}_f(t) := f(t, \boldsymbol{u}(t)) - f(0, \boldsymbol{u}(0)) - \int_0^t (\frac{\partial f}{\partial s}(s, \boldsymbol{u}(s)) + \mathcal{L}f(s, \boldsymbol{u}(s)))ds \tag{23}$$

is a square integrable $\mathcal{F}_t^{w,u_0}$-martingale (see also [HSS95] for details).

We now note the following estimate,

$$E\int_0^T |f(\boldsymbol{u}(t)\boldsymbol{h}(\boldsymbol{u}(t))|^2 dt < +\infty, \tag{24}$$

Under the condition (24) we can follow the method in [FKK72] to obtain the explicit form of $\boldsymbol{\Phi}(t)$ in (19) using lemmas (2) and (3):

$$\boldsymbol{\Phi}(t) = \boldsymbol{\Pi}_t(\mathcal{M}f(t)) - \boldsymbol{\Pi}_t(f(t))\boldsymbol{\Pi}_t(\boldsymbol{h}), \tag{25}$$

where $\mathcal{M}$ is defined in (6).

We thus get the **Fujisaki-Kallianpur-Kunita equation** (4) for $f \in \mathcal{C}_{CY}$.

We have due to the boundedness of $\boldsymbol{h}$,

$$E\int_0^T |\boldsymbol{\Pi}_s(\boldsymbol{h})|^2 ds < \infty. \tag{26}$$

Define $\boldsymbol{\Theta}_t(f)$ for $f \in \mathcal{C}_{CY}$ as,

$$\boldsymbol{\Theta}_t(f) = \boldsymbol{\Pi}_t(f)\exp\left\{\int_0^t \boldsymbol{\Pi}_s(\boldsymbol{h}) \cdot d\boldsymbol{z}(s) - \frac{1}{2}\int_0^t |\boldsymbol{\Pi}_s(\boldsymbol{h})|^2 ds\right\}. \tag{27}$$

Then by Ito formula (see [HSS95]) we get the **Zakai equation** (7) for $f \in \mathcal{C}_{CY}$.

5 KOLMOGOROV'S BACKWARD EQUATION

The proof of the uniqueness of measure valued solutions in Theorem 3 will be based on having a unique solution of Kolmogorov's backward equation,

$$\partial_t \Phi(t, \boldsymbol{v}) = \frac{1}{2} tr(GQG^* \partial_{vv} \Phi(t, \boldsymbol{v})) + < A\boldsymbol{v} + B(\boldsymbol{v}) + G_2 \boldsymbol{x}(t), \partial_v \Phi(t, \boldsymbol{v}) >$$

$$+ \boldsymbol{h}(\boldsymbol{v}) \cdot \boldsymbol{x}(t) \Phi(t, \boldsymbol{v}), \quad t > \tau, \boldsymbol{v} \in D(A), \tag{28}$$

and

$$\Phi(\tau, \boldsymbol{v}) = \Psi(\boldsymbol{v}), \quad \boldsymbol{v} \in \boldsymbol{H}. \tag{29}$$

Definition 3 *A* **strict solution** *to (28)-(29) is a function* $\Phi : [\tau, T] \times \boldsymbol{H} \to \boldsymbol{R}$ *such that*

$$(i) \quad \Phi \in C_b([0, T] \times \boldsymbol{H})$$

$$(ii) \quad \Phi(t, \cdot) \in C_b^2(\boldsymbol{H}), \quad \forall t \geq 0,$$

$$(iii) \quad \Phi \in C^1([0, T] \times D(A))$$

and (28) is satisfied for and $\boldsymbol{v} \in D(A)$ *and* $t \geq 0$.

Proposition 1 *[DZ92] Let* $\boldsymbol{h}(\cdot), \Psi(\cdot) \in C_b^2(\boldsymbol{H})$. *Then (28) and (29) has a unique strict solution for* $0 \leq t \leq T$ *and it is given by the Feynman-Kac formula*

$$\Phi(t, \boldsymbol{v}) = E[\Psi(\boldsymbol{u}^x(t, \boldsymbol{v})) \exp\left(\int_\tau^t \boldsymbol{h}(\boldsymbol{u}^x(s, \boldsymbol{v})) \cdot \boldsymbol{x}(s) ds\right)] \tag{30}$$

where $\boldsymbol{u}(t, \boldsymbol{v})$ *is a solution of*

$$d\boldsymbol{u}^x(t) = [A\boldsymbol{u}^x + B(\boldsymbol{u}^x) + G_2 \boldsymbol{x}(t)]\, dt + G d\boldsymbol{w}(t) \tag{31}$$

$$\boldsymbol{u}^x(\tau, \tau, \boldsymbol{v}) = \boldsymbol{v} \in \boldsymbol{H}. \tag{32}$$

6 MEASURE VALUED SOLUTIONS AND SOLVABILITY OF THE FKK AND ZAKAI EQUATIONS

Let $\mathcal{M}(\boldsymbol{H})$ be the vector space of finite σ-additive measures on the Borel σ-field $\mathcal{B}(\boldsymbol{H})$; this is a subspace of the dual of $C_b(\boldsymbol{H})$ and can be given the inherited weak topology. Denote by $\mathcal{M}_+(\boldsymbol{H})$ the subset of positive measures and $\mathcal{P}(\boldsymbol{H})$ the subset of probability measures on this Borel σ-field. In order to define measure valued solutions for the Zakai and FKK equations and show the existence of such solutions we again need the class of cylindrical test functions introduced earlier.

Definition 4 *A $\mathcal{M}_+(\boldsymbol{H})$-valued process Θ_t is called a* **measure valued solution of the Zakai equation on** $[0,T]$ *if the following five conditions hold:*

(i) Θ_t *is* $\mathcal{F}_t^z$ *adapted, i.e.,* Θ_t *is* $\mathcal{F}_t^z$ *measurable for all* $t \in [0,T]$,

$$(ii)\ E\int_0^T \int_H ||\boldsymbol{u}||^q \Theta_t(d\boldsymbol{u})dt < \infty,\ q \geq 2, \tag{33}$$

$$(iii)\ E| < \Theta_t, 1 > |^2 < +\infty,\quad t \in [0,T], \tag{34}$$

$$(iv)\ E\int_0^T | < \Theta_t, 1 > |^2 dt < +\infty, \tag{35}$$

(v) for all $f \in \mathcal{C}_{CY}$ *and* $t \in [0,T]$ *the weak Zakai equation holds*

$$< \Theta_t, f(t) > = < \Theta_0, f(0) > +$$

$$\int_0^t < \Theta_s, \partial_s f(s) + \mathcal{L}f(s)) > ds + \int_0^t < \Theta_s, \mathcal{M}f(s) > \cdot d\boldsymbol{z}(s) \tag{36}$$

$\boldsymbol{m}$*-a.s.*

Definition 5 *A $\mathcal{P}(\boldsymbol{H})$-valued process Π_t is called a* **measure valued solution of the FKK equation on** $[0,T]$ *if the following three conditions hold:*

(i) Π_t *is* $\mathcal{F}_t^z$ *adapted, , i.e.,* Π_t *is* $\mathcal{F}_t^z$ *measurable for all* $t \in [0,T]$,

$$(ii)\ E\int_0^T \int_H ||\boldsymbol{u}||^q \Pi_t(d\boldsymbol{u})dt < \infty,\ q \geq 2, \tag{37}$$

(iii) for all $f \in \mathcal{C}_{CY}$ *and* $t \in [0,T]$ *the weak FKK equation holds*

$$< \Pi_t, f(t) >=< \Pi_0, f(0) > +$$

$$\int_0^t < \Pi_s, (\partial_s f(s) + \mathcal{L}f(s)) > ds +$$

$$\int_0^t < \Pi_s, \mathcal{M}f(s) - f(s)\boldsymbol{h}(s) > \cdot [d\boldsymbol{z}(s) - < \Pi_s, \boldsymbol{h}(s) > ds] \tag{38}$$

$\boldsymbol{m}$*-a.s.*

Point (ii) says that Θ_t and Π_t have at least finite second moments.

In order to prove measure valued solvability for the FKK and Zakai equations, we will need to mention the existence of conditional probability measures; these are the kernels in the following definition [Get75]. A *kernel* from the measurable space $(\Omega, \mathcal{A})$ to the measurable space $(\boldsymbol{H}, \mathcal{B})$ is a real function $\mu(\omega, B)$ defined for each $\omega \in \Omega$ and $B \in \mathcal{B}$ such that $\omega \mapsto \mu(\omega, A)$ is $\mathcal{A}$-measurable for all $B \in \mathcal{B}$ and $B \mapsto \mu(\omega, B)$ is a positive finite measure for all $\omega \in \Omega$. We now come to one of our main results.

Theorem 2 *Assume that the hypotheses of Theorem 1 hold and that* $\boldsymbol{h} \in C_b^2(\boldsymbol{H}; \boldsymbol{R}^m)$. *Assume also that* $\Psi \in C_b^2(\boldsymbol{H})$. *Then there exists a unique measure valued solution* Π_t *of the FKK equation (38) on* $[0,T]$, *and there exists a unique measure valued solution* Θ_t *of the Zakai equation (36) on* $[0,T]$. *Also,* Π_t *and* Θ_t *are related by*

$$\Theta_t(\cdot) = exp\{\int_0^t < \Pi_s, \boldsymbol{h} > \cdot d\boldsymbol{z}(s) - \frac{1}{2}\int_0^t | < \Pi_s, \boldsymbol{h} > |^2 ds\}\Pi_t(\cdot) \tag{39}$$

and its inverse

$$\Pi_t(\cdot) = \Theta_t(\cdot) / < \Theta_t, 1 > . \tag{40}$$

Proof: Since Π_t and Θ_t will be related by the invertible transform (39), (40) it suffices to show existence and uniqueness for only one besides the relation (39),(40). It will be convenient to show existence for Π_t and uniqueness for Θ_t. The key step for the existence of Π_t is the following lemma on the existence of kernels [Get75].

Lemma 4 : *([Get75], Prop 4.1): Let* $\boldsymbol{Y}$ *be homeomorphic to a Borel subset of a compact metric space (*$\boldsymbol{Y}$ *is a* Lusin space *[Get75]), and denote by* $\mathcal{B}_b(\boldsymbol{Y})$ *and* $\mathcal{B}_b(\Omega)$ *respectively bounded Borel functions on* $\boldsymbol{Y}$ *and* Ω. *Suppose that* $T : \mathcal{B}_b(\boldsymbol{Y}) \to \mathcal{B}_b(\Omega)$ *is linear a.e., positive a.e., and satisfies* $0 \le f_n \uparrow f$ *implies* $Tf_n \uparrow Tf$ *for any sequence of functions* $\{f_n\}$ *and* $f \in \mathcal{B}_b(\boldsymbol{Y})$. *Then there exists a bounded kernel* $\mu(\cdot,\cdot)$ *from* $(\Omega, \mathcal{A})$ *to* $(\boldsymbol{Y}, \mathcal{B}(\boldsymbol{Y}))$ *such that* $Tf(\omega) = \int_Y f(\boldsymbol{u})\mu(\omega, d\boldsymbol{u})$, *for all* $f \in \mathcal{B}_b(\boldsymbol{Y})$. *(Equal here is as elements of* $\mathcal{B}_b(\Omega)$.*)*

To prove the theorem we first use the lemma to obtain a kernel which is a candidate for our desired measure. Now every complete, separable metric space (*Polish space*) is a Lusin space (see [Get75] and the reference [3] contained therein, p 201), so the Hilbert space $\boldsymbol{H}$ satisfies the condition of the lemma. At any fixed $t \in [0,T]$, $f(t,\cdot)$ is bounded Borel on $\boldsymbol{H}$.

Now we set $\mathcal{A} = \mathcal{F}_t^z$ as the σ-field on Ω and define T in the lemma by

$$Tf(t,\cdot)(\omega) = E[f(t,\boldsymbol{u}(t))|\mathcal{F}_t^z](\omega).$$

The expectation is with respect to $\boldsymbol{m}$ on Ω, and $\boldsymbol{u}(t) = \boldsymbol{u}(t,\omega)$ is a measurable function on Ω. It is easy to check that this T satisfies the hypothesis of the lemma: linearity, positivity, and 'continuity' for bounded nondecreasing sequences. Thus, there is a kernel (depending on t) Π_t such that

$$Tf(t,\cdot)(\omega) = \int_H f(t,\boldsymbol{u})\Pi_t(\omega, d\boldsymbol{u}).$$

We conclude that

$$\boldsymbol{\Pi}_t(f) = E[f(t,\boldsymbol{u}(t))|\mathcal{F}_t^z] = \int_H f(t,\boldsymbol{u})\Pi_t(\cdot, d\boldsymbol{u}). \tag{41}$$

for all bounded Borel f.

We now check that the kernel Π_t is indeed a measure valued solution of the FKK equation. Point (i) of the definition follows from the lemma, for the definition of 'kernel' implies that μ_t is $\mathcal{F}_t^z$ measurable. (ii) follows from the bound given in Theorem 1 and the Monotone Convergence Theorem. For we can apply

$$E[g_n(\boldsymbol{u})|\mathcal{F}_t^z] = \int_H g_n(\boldsymbol{u})\Pi_t(\cdot, d\boldsymbol{u})$$

to the bounded functions $g_n(\boldsymbol{u}) = ||\boldsymbol{u}||^q \wedge n$, $n \in N$, and then take the limit as $n \uparrow \infty$.

Verifying (38) is the main work in this argument. Our approach is to simply substitute (41) into (4). However, one finds that without further restrictions on $f \in \mathcal{B}_b(\boldsymbol{H})$ the resulting expressions $\partial_t f + \mathcal{L}f$ and $\mathcal{M}f$ are not bounded Borel functions and (38) is not implied from (4) and (41). For this reason we restrict our class of test functions to $f \in \mathcal{C}_{CY}$, and we indicate how to make sense of the terms in (4) through (6).

Now, $< A\boldsymbol{u}, \partial_u f(t,\boldsymbol{u}) >$ will mean $< \boldsymbol{u}, A\partial_u f(t,\boldsymbol{u}) >$, using the self-adjointness of A and noting again that $\forall \boldsymbol{u} \in \boldsymbol{H}$, $\partial_u f(t,\boldsymbol{u}) \in D(A)$ because $\boldsymbol{e}_i \in D(A)$ for $i = 1, 2, ...$ from the definition of $\mathcal{C}_{CY}$. This term is well defined for all $\boldsymbol{u} \in \boldsymbol{H}$ and, as a real valued function of t, it is C^∞ on $[0,T]$ hence it is bounded Borel as required by (38) and (41). Next, $< B(\boldsymbol{u}), \partial_u f(t,\boldsymbol{u}) >$ is well defined by the above comments on $\partial_u f$ and the hypothesis that $D(B) = \boldsymbol{H}$ and B has linear growth (assumption A2).

Finally, $\mathrm{tr}(GQG^*\partial_{uu} f(t,\boldsymbol{u}))$ is well defined for all $\boldsymbol{u} \in \boldsymbol{H}$ and is in fact a C^∞ function of $t \in [0,T]$ (and therefore is bounded Borel). For using (21) and choosing the same orthonormal set $\{\boldsymbol{e}_i\}$ as there ([DZ92], p 416) we have

$$tr(GQG^*\partial_{uu} f(t,\boldsymbol{u})) = \sum_{i=1}^{\infty} < GQG^*\partial_{uu} f\boldsymbol{e}_i, \boldsymbol{e}_i >$$

$$= \sum_{i=1}^{\infty} < GQG^*(\sum_{k=1}^{n}\sum_{\ell=1}^{n} \partial_{k\ell}\phi(...)\boldsymbol{e}_k \otimes \boldsymbol{e}_\ell)\boldsymbol{e}_i, \boldsymbol{e}_i >$$

$$= \sum_{k=1}^{n}\sum_{\ell=1}^{n} \partial_{k\ell}\phi(t, < \boldsymbol{u}, \boldsymbol{e}_1 >, ..., < \boldsymbol{u}, \boldsymbol{e}_n >) < GQG^*\boldsymbol{e}_k, \boldsymbol{e}_\ell > \qquad (42)$$

and this is clearly well defined and C^∞ as a function of $t \in [0,T]$ for every $\boldsymbol{u} \in \boldsymbol{H}$.

The term $G_2^*\partial_u f(t,\boldsymbol{u}))$ arising in $\mathcal{M}f$ is easy to handle: As a function of t it is in $C^\infty([0,T];\boldsymbol{R}^m)$ since G_2^* is bounded and $f \in \mathcal{C}_{CY}$.

Our proof of the uniqueness of Θ_t is adapted from [Sri94] which is an infinite dimensional generalization of a method of Rozovskii [Roz91]. We also point out that an analogous method was used by Vishik and Komech [VK84] for the uniqueness theorem of the direct Kolmogorov equation associated to the stochastic Navier-Stokes equation.

Fix any $\boldsymbol{x} \in C([0,T], \boldsymbol{R}^m)$ and define the following three processes on $[0,T]$:

$$q_t = exp\{\int_0^t \boldsymbol{x}(s)\cdot d\boldsymbol{z}(s) - \frac{1}{2}\int_0^t |\boldsymbol{x}(s)|^2 ds\},$$

$$\rho_t^{-1} = exp\{-\int_0^t \boldsymbol{h}(\boldsymbol{u}^x(s)) \cdot d\boldsymbol{z}(s) + \frac{1}{2}\int_0^t |\boldsymbol{h}(\boldsymbol{u}^x(s))|^2 ds\},$$

$$\gamma_t = q_t \rho_t^{-1}.$$

For any $f \in \mathcal{C}_{CY}$ apply the Ito formula to $< \Theta_t, f(t) > \gamma_t$ to obtain

$$< \Theta_t, f(t) > \gamma_t =< \Theta_0, f(0) >$$

$$+ \int_0^t < \Theta_s, \partial_s f(s) + (\mathcal{L} + \mathcal{M}\boldsymbol{x}) f(s) > \gamma_s ds$$

$$+ \int_0^t \gamma_s [< \Theta_s, f(s) > (\boldsymbol{x}(s) - \boldsymbol{h}(\boldsymbol{u}^x(s))) + < \Theta_s, \mathcal{M} f(s) >] \cdot d\boldsymbol{w}_2(s).$$

The last term is a martingale so

$$E < \Theta_t, f(t) > \gamma_t = E < \Theta_0, f(0) >$$

$$+ E \int_0^t \int < \Theta_s, \partial_s f(s) + (\mathcal{L} + \mathcal{M}\boldsymbol{x}) f(s) > \gamma_s ds. \tag{43}$$

Now, let us take the unique solution $\Phi(t, \boldsymbol{v})$ (see (30)-(31)) of the backward Kolmogorov equation (28) corresponding to the initial data $\Psi(\boldsymbol{v})$. Here $\Psi(\boldsymbol{v})$ is a *cylindrical test function* in $\boldsymbol{v}$. We will consider the smooth approximations $\Phi_{n,\epsilon}(t, \boldsymbol{v}) \in \mathcal{C}_{CY}$ of $\Phi(t, \boldsymbol{v})$ (see [VK84]). In (43) we set

$$f(t, \boldsymbol{v}) = \Phi_{n,\epsilon}(T - t, \boldsymbol{v}) \tag{44}$$

and take the limit $\epsilon \to 0$ and $n \to \infty$. Using the convergence properties of $\Phi_{n,\epsilon}$ and its derivatives we deduce that the integral term on the right hand side of (43) goes to zero and we obtain,

$$E\{< \Theta_T, \Psi > \gamma_T\}$$

$$= E\left[\Psi(\boldsymbol{u}^x(T, 0, \boldsymbol{v})) \exp\left(\int_0^T \boldsymbol{h}(\boldsymbol{u}^x(r, 0, \boldsymbol{v})) \cdot \boldsymbol{x}(r) dr\right)\right], \tag{45}$$

where $\boldsymbol{u}^x$ solves (31). Now, using Girsanov's transformation we get

$$E\left[\Psi(\boldsymbol{u}^x(T, 0, \boldsymbol{v})) \exp\left(\int_0^T \boldsymbol{h}(\boldsymbol{u}^x(r, 0, \boldsymbol{v})) \cdot \boldsymbol{x}(r) dr\right)\right]$$

$$= E\left[\Psi(\boldsymbol{u}(T, 0, \boldsymbol{v})) q_T\right], \tag{46}$$

where $\boldsymbol{u}$ solves (14). To justify this step we need to use finite dimensional approximations of (31) and (14), use Girsanov transform to these finite dimensional diffusion processes and then use the weak convergence of the probability distributions of $\boldsymbol{u}$ and $\boldsymbol{u}^x$ to obtain (46) in the limit [HSS95].

We will now apply the absolutely continuous change of measure $(\Omega, \Sigma, \boldsymbol{m})$ to $(\Omega, \Sigma, \tilde{\boldsymbol{m}})$ defined by

$$\frac{d\tilde{\boldsymbol{m}}}{d\boldsymbol{m}} = \rho_T^{-1}. \tag{47}$$

Then under the new measure we can write (45)-(46) as

$$\tilde{E}\{< \Theta_T, \Psi > q_T\} = \tilde{E}\left[\tilde{E}\left[\Psi(\boldsymbol{u}(T,0,\boldsymbol{v}))\rho_T|\mathcal{F}_T^z\right]q_T\right]. \tag{48}$$

Since processes of the form q_t defined above are dense in $L^2(\Omega, \mathcal{F}_t^z, \tilde{\boldsymbol{m}})$ [Roz90], we conclude that

$$< \Theta_T, \Psi >= \tilde{E}\left[\Psi(\boldsymbol{u}(T,0,\boldsymbol{v}))\rho_T|\mathcal{F}_T^z\right], \quad m\text{-a.s.} \tag{49}$$

Since for an arbitrary measure-valued solution Π_t of the FKK equation, Θ_t defined by (39) satisfies (36), we have thus established the uniqueness of Π_t and Θ_t in the interval $[0, T]$.

References

[Ahm94] N. U. Ahmed. Nonlinear filtering for stochastic differential equations in Hilbert spaces. In W. F. Ames, editor, *14th IMACS World conference on computational and applied mathematics*, pages 5–8, 1994.

[DZ92] G. DaPrato and J. Zabczyk. *Stochastic Equations in Infinite Dimensions.* Cambridge University Press, New York, 1992.

[FKK72] M. Fujisaki, G. Kallianpur, and H. Kunita. Stochastic differential equations for the nonlinear filtering problem. *Osaka J. Math.*, 9:19–40, 1972.

[Get75] R.K. Getoor. On the construction of kernels. In P.A. Meyer, editor, *Seminaire de Probabilités IX. Lecture Notes in Mathematics, vol-465.* Springer-Verlag, 1975.

[HSS95] S. L. Hobbs, G. Sobko, and S. S. Sritharan. Nonlinear filtering theory of stochastic semilinear partial differential equations. To be published, 1995.

[Ich82] A. Ichikawa. Stability of semilinear stochastic evolution equations. *J. Math. Anal. Appl.*, 90:12–44, 1982.

[Roz90] B.L. Rozovskii. Lecture notes on linear stochastic partial differential equations. Lecture Notes 25, Dept. Math., University of North Carolina, 1990.

[Roz91] B.L. Rozovskii. A simple proof of uniqueness for Kushner and Zakai equations. *Stochastic Analysis*, ed. E. Mayer-Wolf and E. Merzbach and A. Schwartz:449–458, 1991.

[Sri94] S. S. Sritharan. Nonlinear filtering of stochastic Navier-Stokes equation. In T. Funaki and W. A. Woyczynski, editors, *Nonlinear Methods on Stochastic Partial Differential Equations: Burgers Turbulence and Hydrodynamic Limit.* Springer-Verlag, 1994.

[Vio76] M. Viot. *Solution faibles D'equations aux derivees partielles stochastique nonlineaires.* These, Universite Pierre et Marie Curie, Paris, 1976.

[VK84] M. J. Vishik and A. I. Komech. On Kolmogorov's equations corresponding to the two dimensional stochastic Navier-Stokes system. *Trans. Moscow Math. Soc.*, pages 1–42, 1984.

[Zak69] M. Zakai. On the optimal filtering of diffusion processes. *Z. Wahrscheinlichkeitstheorie. Verw. Geb.*, 11:230–243, 1969.

An Operator Characterization of Oscillatory Harmonizable Processes

RANDALL J. SWIFT Department of Mathematics,
Western Kentucky University, Bowling Green, Kentucky

Dedicated to Professor M.M. Rao, advisor and friend, on the occasion of his 65th birthday.

1 INTRODUCTION

A class of nonstationary stochastic processes which are encountered in some applications is the class of modulated stationary processes $X(t)$. These processes are obtained when a stationary process $X_0(t)$ is multiplied by some nonrandom modulating function $A(t)$:

$$X(t) = A(t)X_0(t).$$

This class of processes has been investigated by Joyeux (1987) and Priestley (1981). The book by Yaglom (1987) provides a nice treatment of these processes. In particular, if $A(t)$ admits a generalized Fourier transform, the class of oscillatory processes, studied by Priestley (1981) is obtained. In some physical situations, the assumption of stationarity for the process $X_0(t)$ is unrealistic Rao (1982). If this condition is relaxed, and $X_0(t)$ is assumed to be harmonizable and if $A(t)$ admits a generalized Fourier transform, the process $X(t)$ is not oscillatory, but is oscillatory harmonizable.

This paper investigates the properties of oscillatory harmonizable processes. Section 2 recalls the basic theory of harmonizable processes required for the subsequent analysis. Section 3 introduces and develops the class of oscillatory harmonizable processes. In this section, the spectral representation of oscillatory harmonizable processes is obtained. This representation is used to deduce relationships between the oscillatory harmonizable processes and other classes of nonstationary processes. Section 4 obtains an important and useful operator characterization for oscillatory harmonizable processes.

2 PRELIMINARIES

In the following work, there is always an underlying probability space, (Ω, Σ, P), whether this is explicitly stated or not.

DEFINITION 2.1 For $p \geq 1$, define $L_0^p(P)$ to be the set of all complex valued $f \in L^p(\Omega, \Sigma, P)$ such that $E(f) = 0$, where $E(f) = \int_\Omega f(\omega)dP(\omega)$ is the expectation.

In this paper, we will consider second order stochastic processes. More specifically, mappings $X : \mathbb{R} \to L_0^2(P)$.

DEFINITION 2.2 A stochastic process $X : \mathbb{R} \to L_0^2(P)$ is *stationary* (stationary in the wide or Khintchine sense) if its covariance $r(s,t) = E(X(s)\overline{X(t)})$ is continuous and is a function of the difference of its arguments, so that

$$r(s,t) = \tilde{r}(s-t).$$

An equivalent definition of a stationary process is one whose covariance function can be represented as

$$\tilde{r}(\tau) = \int_{\mathbb{R}} e^{i\lambda\tau} dF(\lambda), \tag{1}$$

for a unique non-negative bounded Borel measure $F(\cdot)$. This alternate definition is a consequence of a classical theorem of Bochner's (Gihman and Skorohod, 1974), and motivates the following definition.

DEFINITION 2.3 A stochastic process $X : \mathbb{R} \to L_0^2(P)$ is *weakly harmonizable* if its covariance $r(\cdot,\cdot)$ is expressible as

$$r(s,t) = \int_{\mathbb{R}} \int_{\mathbb{R}} e^{i\lambda s - i\lambda' t} dF(\lambda, \lambda') \tag{2}$$

where $F : \mathbb{R} \times \mathbb{R} \to \mathbb{C}$ is a positive semi-definite bimeasure, hence of finite Fréchet variation.

The integrals in (2) are strict Morse-Transue, (Chang and Rao, 1986). A stochastic process, $X(\cdot)$, is *strongly harmonizable* if the bimeasure $F(\cdot,\cdot)$ in (2) extends to a complex measure and hence is of bounded Vitali variation. In either case, $F(\cdot,\cdot)$ is termed the *spectral bi-measure* (or *spectral measure*) of the harmonizable process.

Comparison of equation (2) with equation (1) shows that when $F(\cdot,\cdot)$ concentrates on the diagonal $\lambda = \lambda'$, both the weak and strong harmonizability concepts reduce to the stationary concept. Harmonizable processes retain the powerful Fourier analytic methods inherent with stationary processes, as seen in Bochner's theorem, (1); but they relax the requirement of stationarity.

The structure and properties of harmonizable processes has been investigated and developed extensively by M.M. Rao and others. The following sources are listed here to provide a partial summary of the literature. The papers by Rao (1978, 1982, 1989, 1991, 1994) provide a basis for the theory. Chang and Rao (1986) develop the necessary bi-measure theory. A study of sample path behavior for harmonizable processes is considered by Swift (1996b). Some results on moving average representations were obtained by Mehlman (1992). The structure of harmonizable isotropic random fields and some applications has been considered by Swift (1994, 1995, 1996a). Second order processes with harmonizable increments has been investigated also by Swift (1996c), The forthcoming book by Kakihara gives a general treatment of multidimensional second order processes which include the harmonizable class.

3 OSCILLATORY HARMONIZABLE PROCESSES

M.B. Priestley (1981), introduced and studied a generalization of the class of stationary processes. This generalization is given by:

DEFINITION 3.1 A stochastic process $X : I\!R \to L_0^2(P)$ is *oscillatory* if it has representation

$$X(t) = \int_R A(t,\lambda)e^{i\lambda t}dZ(\lambda)$$

where $Z(\cdot)$ is a stochastic measure with orthogonal increments and

$$A(t,\lambda) = \int_R e^{itx}H(\lambda, dx)$$

with $H(\cdot, B)$ a Borel function on $I\!R$, $H(\lambda,\cdot)$ a signed measure and $A(t,\lambda)$ having an absolute maximum at $\lambda = 0$ independent of t.

Using this representation the covariance of an oscillatory process is

$$r(s,t) = \int_{I\!R} A(s,\lambda)\overline{A(t,\lambda)}e^{i\lambda(s-t)}d\Phi(\lambda).$$

The idea of definition 2.3 provides the motivation for the following definition:

DEFINITION 3.2 A stochastic process $X : I\!R \to L_0^2(P)$ is *oscillatory weakly harmonizable*, if its covariance has representation

$$r(s,t) = \int_{I\!R}\int_{I\!R} A(s,\lambda)\overline{A(t,\lambda')}e^{i\lambda s - i\lambda' t}dF(\lambda,\lambda')$$

where $F(\ ,\ )$ is a function of bounded Fréchet variation, and

$$A(t,\lambda) = \int_{I\!R} e^{itx}H(\lambda, dx)$$

with $H(\cdot, B)$ a Borel function on $I\!R, H(\lambda,\cdot)$ a signed measure and $A(t,\lambda)$ having an absolute maximum at $\lambda = 0$ independent of t.

Note that if $A(t,\lambda) = 1$, this class coincides with the weakly harmonizable processes. As Priestley's definition provides an extension to the class of stationary processes, definition 3.2 provides an extension to the class of weakly harmonizable processes.

Observe, further, that in this definition, for $F(\cdot,\cdot)$ concentrating on the diagonal, $\lambda = \lambda'$, the oscillatory processes are obtained. Thus the oscillatory harmonizable processes also provide an extension to the class introduced by Priestley, which we will now term *oscillatory stationary* .

Using this definition, it is possible to obtain the spectral representation of an oscillatory harmonizable process $X(\cdot)$.

THEOREM 3.1 *The spectral representation of an oscillatory weakly harmonizable stochastic process is:*

$$X(t) = \int_{I\!R} A(t,\lambda)e^{i\lambda t}dZ(\lambda)$$

where $Z(\cdot)$ *is a stochastic measure satisfying*

$$E(Z(B_1)\overline{Z(B_2)}) = F(B_1, B_2)$$

with $F(\cdot,\cdot)$ *a function of bounded Fréchet variation.*

Proof: Let $X(\cdot)$ be an oscillatory weakly harmonizable process. Then, the covariance $r(\cdot,\cdot)$ has representation

$$r(s,t) = \int_{I\!R}\int_{I\!R} A(s,\lambda)\overline{A(t,\lambda')}e^{i\lambda s - i\lambda' t}dF(\lambda,\lambda').$$

Applying a form of Karhunen's theorem, (Yaglom, 1987, volume 2, pages 33 - 41) gives the spectral representation of $X(\cdot)$ as

$$X(t) = \int_{I\!R} A(t,\lambda)e^{i\lambda t}dZ(\lambda),$$

which is the desired result. □

The following condition on the signed measure H, for oscillatory strongly harmonizable processes show these processes are actually a subclass of the strongly harmonizable processes. A similar result was obtained by R. Joyeux (1987), for the oscillatory stationary processes.

THEOREM 3.2 *If* $X(\cdot)$ *is an oscillatory strongly harmonizable process with*

$$\int_{I\!R} | H(\lambda, dx) | < \infty$$

uniformly in $\lambda \in I\!R$, *then* $X(\cdot)$ *is strongly harmonizable.*

Proof: Let

$$\tilde{Z}(A) = \int_{I\!R} H(\lambda, A-\lambda)dZ(\lambda)$$

where A is a Borel set of $I\!R$ and

$$A - \lambda = \{x - \lambda : x \in A\}.$$

$\tilde{Z}(\cdot)$ is a stochastic measure since $H(\lambda,\cdot)$ is a signed measure, and uniformly bounded by K.

Now set

$$\tilde{X}(t) = \int_{I\!R} e^{i\lambda t}d\tilde{Z}(\lambda).$$

Claim: $\tilde{X}(\cdot)$ is a strongly harmonizable process.

If one lets $\tilde{Z}(A,B) = E(\tilde{Z}(A)\overline{\tilde{Z}(B)})$ A, B Borel sets of $I\!R$, it must be shown that

$$\int_{I\!R}\int_{I\!R} | \tilde{F}(d\omega, d\omega') | < \infty.$$

Now

$$\begin{aligned} E(\tilde{Z}(d\omega)\overline{\tilde{Z}(d\omega')}) &= \int_{I\!R}\int_{I\!R} H(\lambda, d(\omega-\lambda))\overline{H(\lambda', d(\omega'-\lambda))}E(Z(d\lambda)\overline{Z(d\lambda)}) \\ &= \int_{I\!R}\int_{I\!R} H(\lambda, d(\omega-\lambda))\overline{H(\lambda', d(\omega'-\lambda))}F(d\lambda, d\lambda') \end{aligned}$$

where $F(A,B) = E(Z(A)\overline{Z(B)})$ is of finite Vitali variation since $X(t)$ is strongly harmonizable.
Thus,

$$\begin{aligned}\int_{\mathbb{R}}\int_{\mathbb{R}} | \tilde{F}(d\omega, d\omega') | &= \int_{\mathbb{R}}\int_{\mathbb{R}} | \int_{\mathbb{R}}\int_{\mathbb{R}} H(\lambda, d(\omega-\lambda))\overline{H(\lambda', d(\omega'-\lambda'))}F(d\lambda, d\lambda') | \\ &\leq \int_{\mathbb{R}}\int_{\mathbb{R}}\int_{\mathbb{R}}\int_{\mathbb{R}} | H(\lambda, d(\omega-\lambda))\overline{H(\lambda', d(\omega'-\lambda'))}F(d\lambda, d\lambda') | \\ &< \infty\end{aligned}$$

since $| H | (\lambda, \mathbb{R})$ is bounded, proving the claim.
Now

$$\begin{aligned}\tilde{X}(t) &= \int_{\mathbb{R}} e^{i\omega t}\tilde{Z}(d\omega) = \int_{\mathbb{R}}\int_{\mathbb{R}} e^{i\omega t} H(\lambda, d(\omega-\lambda))Z(d\lambda) \\ &= \int_{\mathbb{R}} e^{i\lambda t}\int_{\mathbb{R}} e^{i(\omega-\lambda)t} H(\lambda, d(\omega-\lambda))Z(d\lambda) \\ &= \int_{\mathbb{R}} e^{i\lambda t} A(t,\lambda)Z(d\lambda) \\ &= X(t).\end{aligned}$$

So $X(t)$ is strongly harmonizable, which completes the proof of the theorem. □
An additional class of processes related to the oscillatory processes is given by:
DEFINITION 3.3 An oscillatory weakly harmonizable stochastic process $X : \mathbb{R} \to L_0^2(P)$ is *ε-slowly changing weakly harmonizable* if

$$B(\lambda) = \int_{\mathbb{R}} | x | | H | (\lambda, dx) \leq \epsilon, \ \forall \ \lambda \in \mathbb{R}.$$

Slowly changing stationary processes where first considered by Priestley (1981) and are of interest not only in engineering but also in economics. Priestley showed that it is possible to define a spectral measure for these processes. The class of slowly changing harmonizable processes introduced above extend the class of slowly changing stationary processes. The following corollary shows that it is possible to consider a similar concept for the slowly changing harmonizable class.
COROLLARY 3.1 *Slowly changing strongly harmonizable processes form a subclass of strongly harmonizable processes.*
Proof: The assumption is

$$\int_{\mathbb{R}} | x | | H | (\lambda, dx) \leq \varepsilon \ \forall \lambda \in \mathbb{R}.$$

Claim:

$$\int_{\mathbb{R}} | H | (\lambda, dx) < \infty.$$

In fact,

$$
\begin{aligned}
| H | (\lambda, I\!R) &= \int_{I\!R} | H | (\lambda, dx) \\
&= \int_{|x|<1} | H | (\lambda, dx) + \int_{|x|\geq 1} | H | (\lambda, dx) \\
&\leq K + \int_{|x|\geq 1} | H | (\lambda, dx) \leq K + \int_{|x|\geq 1} | x || H | (\lambda, dx) \\
&\leq K + \int_{I\!R} | x || H | (\lambda, dx) \leq K + \epsilon < \infty
\end{aligned}
$$

which is the claim. Now since K is finite, by the theorem $X(t)$ is strongly harmonizable, proving the corollary. □

4 AN OPERATOR CHARACTERIZATION

Using oscillatory harmonizable processes, it is possible to obtain a representation of a broader class of processes on $I\!R$.

DEFINITION 4.1 Let $\boldsymbol{S}$ be a locally compact space with $\mathcal{B}_0$ as the σ-ring generated by the bounded Borel sets of $\boldsymbol{S}$. If T is any index set, $\{X(t), t \in T\} \subset L_0^2(P)$ a second order process, r its covariance and $\tilde{\beta} : \mathcal{B}_0 \times \mathcal{B}_0 \to \mathbb{C}$, a bimeasure having locally bounded Fréchet variation then $X(\cdot)$ is said to be *(locally) weakly of class (C)* when $\tilde{\beta}$ is positive definite and

$$r(s,t) = \int_{\boldsymbol{S}} \int_{\boldsymbol{S}} g_s(\lambda)\overline{g_t(\lambda')}\beta(d\lambda, \lambda') \quad (s,t) \in T \times T \text{ (strict MT-integral)}$$

where $g_t : \boldsymbol{S} \to \mathbb{C}, t \in T$ a family of Borel functions for which the integral exists. If $\tilde{\beta}$ has locally finite Vitali variation, then the process is termed of *class (C)* relative to $\{g_s, S \in T\}$ and $\tilde{\beta}$.

Weak class (C) processes are considered extensively in Chang and Rao (1986). Oscillatory harmonizable processes affords this broad class of processes to have a simple representation on $I\!R$ as seen in the following

PROPOSITION 4.1 *The class of oscillatory weakly harmonizable processes* $\{X(t), t \in I\!R\} \subset L_0^2(P)$ *coincides with the class of weak class* (C) *processes indexed on* $I\!R$.

Proof: This follows by setting $g_s(\lambda) = e^{is\lambda}A_s(\lambda)$ in the definition of weak class (c) processes, since $F(\cdot,\cdot)$ always has finite Fréchet variation. □

Using this simple identification, an operator representation of weak class (C) processes indexed on $I\!R$ is possible. This result is an extension of that given in Chang and Rao (1988) for the oscillatory stationary class.

THEOREM 4.1 $X(\cdot)$ *is an oscillatory weakly harmonizable process iff it is representable as*

$$X(t) = a(t)T(t)Y(0), \quad t \in I\!R,$$

where

$$Y_0 = Y(0)$$

is some point in

$$H(X) = \overline{sp}\{X(t), t \in I\!R\}$$

with $a(t)$ a densely defined closed operator in $H(X)$ for each $t \in I\!R$ and

$$\{T(s), s \in I\!R\}$$

a weakly continuous family of positive definite contractive operators in $H(X)$ which commutes with each $a(t)$, $t \in I\!R$.

Proof: Suppose $X(t)$ is oscillatory weakly harmonizable, then

$$X(t) = \int_{I\!R} A(t,\lambda)e^{i\lambda t}dZ(\lambda)$$

where $Z(\cdot)$ is a stochastic measure satisfying

$$E(Z(B_1)\overline{Z(B_2)}) = F(B_1, B_2)$$

with $F(\cdot,\cdot)$ of bounded Fréchet variation.
Let

$$Y(t) = \int_{I\!R} e^{i\lambda t}dZ(\lambda)$$

then $Y(\cdot)$ is weakly harmonizable.

Now by a theorem of Rao (1982) there is a weakly continuous family of positive definite contractive operators

$$\{T(t), t \in I\!R\}$$

on

$$H(X) = \overline{sp}\{X(t), t \in I\!R\}$$

so that

$$Y(t) = T(t)Y_0.$$

Using the spectral theorem for this family of operators, (cf. Rao, 1982)

$$T(t) = \int_{I\!R} e^{i\lambda t}\tilde{E}(d\lambda), \quad t \in I\!R$$

where $\{\tilde{E}(\cdot), \boldsymbol{B}\}$ is the resolution of the identity of $\{T(t), t \in I\!R\}$ with $\boldsymbol{B}$ as the Borel σ-algebra of $I\!R$. So

$$Z(A) = \tilde{E}(A)Y_0, \quad A \in \boldsymbol{B}.$$

Now define

$$a(t) = \int_{I\!R} A(t,\lambda)\tilde{E}(d\lambda) \quad t \in I\!R.$$

It follows that $a(t)$ is closed and densely defined on $H(X)$ with its domain containing

$$\{Y(s), \ s \in I\!R\}.$$

Now since $T(t)$ and $\tilde{E}(D)$ commute for all t and D, then $a(t)$ and $\{\tilde{E}(D), \ D \in \boldsymbol{B}\}$ commute, so that $a(t)$ and $\{T(s), \ s \in I\!R\}$ commute for each t.

Thus

$$\begin{aligned} a(t)T(t)Y_0 &= a(t)\left(\int_{\mathbb{R}} e^{i\omega t}\tilde{E}(d\omega)Y_0\right) \\ &= \int_{\mathbb{R}}\int_{\mathbb{R}} A(t,\lambda)\tilde{E}(d\lambda)\left(\int_{\mathbb{R}} e^{i\omega t}E(d\omega)Y_0\right) \\ &= \int_{\mathbb{R}} A(t,\lambda)e^{i\lambda t}\tilde{E}(d\lambda)Y_0 \\ &= \int_{\mathbb{R}} A(t,\lambda)e^{i\lambda t}dZ(\lambda) \\ &= X(t) \end{aligned}$$

where (3) follows since

$$\tilde{E}(D)\int_{\mathbb{R}} e^{i\omega t}Y_0 = \int_D e^{i\omega t}\tilde{E}(d\omega)Y_0, \quad D \in \boldsymbol{B}.$$

Thus if $X(t)$ is oscillatory weakly harmonizable, then

$$X(t) = a(t)T(t)Y(0)$$

where

$$Y_0 = Y(0)$$

is some point in

$$H(X) = \overline{sp}\{X(t), t \in \mathbb{R}\}$$

$a(t)$ is a densely defined closed operator in $H(X)$ for each $t \in \mathbb{R}$ and

$$\{T(s), s \in \mathbb{R}\}$$

is a weakly continuous family of positive definite contractive operators in $H(X)$ which commutes with each $a(t)$, $t \in \mathbb{R}$.

Now suppose $X(t)$ can be represented as

$$X(t) = a(t)T(t)Y(0)$$

with $a(t)$, $T(t)$, and $Y(0)$ as stated in the theorem.

Then, using a classical result of von Neumann and F. Riesz (1990), $a(t)$ is a function $g(t)$ of $T(t)$ and further

$$a(t) = g(t)T(t) = \int_{\mathbb{R}} g(t,\lambda)\tilde{E}(d\lambda).$$

Thus

$$\begin{aligned} X(t) &= a(t)T(t)Y(0) \\ &= \int_{\mathbb{R}} g(t,\lambda)\tilde{E}(d\lambda)\int_{\mathbb{R}} e^{i\omega t}\tilde{E}(d\omega)Y_0 \end{aligned}$$

$$= \int_{I\!R} e^{i\lambda t} g(t,\lambda)\tilde{E}(d\lambda)Y_0$$

$$= \int_{I\!R} e^{i\lambda t} g(t,\lambda)Z(d\lambda)$$

but this is the representation of a oscillatory weakly harmonizable process. □

ACKNOWLEDGEMENTS

The author expresses his thanks to Professor M.M. Rao for his advice and encouragement during the work of this project. The author also expresses his gratitude to the Mathematics department at Western Kentucky University for release time during the Spring 1995 semester, during which this work was completed.

REFERENCES

1. D. K. Chang and M. M. Rao. (1986). Bimeasures and Nonstationary Processes. Real and Stochastic Analysis John Wiley and Sons, New York, p. 7.
2. D. K. Chang and M. M. Rao. (1988). Special Representations of Weakly Harmonizable Processes. Stoc. Anal. and Appl., 6(2):169.
3. I. I. Gihman and A. V. Skorohod. (1974). The Theory of Stochastic Processes I. Springer-Verlag, New York.
4. R. Joyeux. (1987). Slowly Changing Processes and Harmonizability. J. Time Series Anal.. 8, No.4.
5. Y. Kakihara. Multidimensional Second Order Stochastic Processes. World Scientific, In preparation.
6. M. H. Mehlman. (1992). Prediction and Fundamental Moving Averages for Discrete Multidimensional Harmonizable Processes. J. Multiv. Anal., 43, No.1.
7. M. B. Priestley. (1981). Spectral Analysis and Time Series. Vol. 1 and 2, Academic Press, London.
8. M. M. Rao. (1978). Covariance Analysis of Non Stationary Time Series, Developments in Statistics. 1, p. 171.
9. M. M. Rao. (1982). Harmonizable Processes: Structure Theory. L'Enseign Math, 28, p. 295.
10. M. M. Rao. (1989). Harmonizable Signal Extraction, Filtering and Sampling. Topics in Non-Gaussian Signal Processing. (E.J. Wegman, S.C. Schwartz, J.B. Thomas, eds.), Springer-Verlag, New York.
11. M. M. Rao. (1991). Sampling and Prediction for Harmonizable Isotropic Random Fields. J. Comb., Info. and Sys. Sci.. 16 No. 2- 3 p. 207.
12. M. M. Rao. (1994). Harmonizable processes and inference: unbiased prediction for stochastic flows. J. Stat. Plan. and Infer.. 39 p. 187.
13. F. Riesz & B. Sz-Nagy. (1990). Functional Analysis. Dover, New York.
14. R. Swift. (1994). The Structure of Harmonizable Isotropic Random Fields. Stoch. Anal. and Appl., 12, No. 5, p. 583.

15. R. Swift. (1995). Representation and Prediction for Locally Harmonizable Isotropic Random Fields. J. Appl. Math. and Stoch. Anal.. VIII, p. 101.
16. R. Swift. (1996a). A Class of Harmonizable Isotropic Random Fields. J. Comb., Info. and Sys. Sci., (to appear).
17. R. Swift. (1996b). Almost Periodic Harmonizable Processes. Georgian Math. J., (to appear).
18. R. Swift. (1996c). Stochastic Processes with Harmonizable Increments. J. Comb., Info. and Sys. Sci., (to appear).
19. A.M. Yaglom. (1987). Correlation Theory of Stationary and Related Random Functions. Vol. 1 and 2, Springer-Verlag, New York.

Operator Algebraic Aspects for Sufficiency

MAKATO TSUKADA, Department of Information Sciences, Toho University, Funabashi City, Chiba 274, Japan

0. Introduction. Sufficiency is one of the most important concepts in mathematical statistics. In the measure theoretic context ([Halmos and Savage, 1949]), it is specified with a measurable space $(\Omega, \mathcal{F})$, a set of probability measures $\mathcal{P}$ and a σ-subfield $\mathcal{G}$ of $\mathcal{F}$. However for a technical reason it is often assumed that $\mathcal{P}$ is dominated by some σ-finite measure. If not so, several pathological difficulties occur (see for example [Burkholder, 1960]). More general property than domination was, for example, introduced by [Pitcher, 1965]. On the other hand, [LeCam, 1964, 1986] discussed sufficiency in an abstract framework, namely, the theory of Banach lattices. Including these, several attempts have been made to remedy such difficulties. Some of these are also related to an axiom of set theory, that is, existence of measurable cardinality ([Ramamoothi and Yamada, 1981],[Luschgy and Mussmann, 1985], etc.). Also see, [Fujii and Morimoto, 1986],[Luschgy, 1988], [Luschgy, Mussmann and Yamada, 1988].

In this note, we give another definition of sufficiency in the view of operator algebras and apply it to the theory of Gibbs states on countable sets.

1. Basic spaces. Let $(\Omega, \mathcal{F})$ be a measurable space. We denote by $ca(\Omega, \mathcal{F})$ (resp. $pr(\Omega, \mathcal{F})$) the set of all countably additive bounded complex-valued measures (resp. probability measures) on $(\Omega, \mathcal{F})$.

Now let $\{(\Omega_\mu, \mathcal{F}_\mu) : \mu \in pr(\Omega, \mathcal{F})\}$ be a family of disjoint copies of $(\Omega, \mathcal{F})$. A bimeasurable bijection from $(\Omega, \mathcal{F})$ to $(\Omega_\mu, \mathcal{F}_\mu)$ is denoted by ι_μ for each $\mu \in$

$pr(\Omega, \mathcal{F})$. Put

$$\oplus\Omega = \bigcup_{\mu \in pr(\Omega,\mathcal{F})} \Omega_\mu,$$

$$\oplus\mathcal{F} = \{A \subseteq \oplus\Omega : A \cap \Omega_\mu \in \mathcal{F}_\mu \quad (\forall \mu \in pr(\Omega, \mathcal{F})\},$$

$$m(A) = \sum_{\mu \in pr(\Omega,\mathcal{F})} \mu({\iota_\mu}^{-1}(A \cap \Omega_\mu)) \quad (A \in \oplus\mathcal{F}).$$

Since $(\oplus\Omega, \oplus\mathcal{F}, m)$ is a direct sum of $\{(\Omega, \mathcal{F}, \mu) : \mu \in pr(\Omega, \mathcal{F})\}$, it is a localizable measure space, and the Banach space $L^p(\oplus\Omega, \oplus\mathcal{F}, m)$ of the set of all m-equivalence classes of p-th power integrable complex-valued functions on $(\oplus\Omega, \oplus\mathcal{F}, m)$ can be identified with the Banach space

$$\bigoplus_{\mu \in pr(\Omega,\mathcal{F})} L^p(\Omega, \mathcal{F}, \mu) = \{\{f_\mu\}_{\mu \in pr(\Omega,\mathcal{F})} : f_\mu \in L^p(\Omega, \mathcal{F}, \mu) \quad (\forall \mu \in pr(\Omega, \mathcal{F}))$$

$$\text{and} \quad \sum_{\mu \in pr(\Omega,\mathcal{F})} \int |f_\mu|^p d\mu < \infty\}$$

for all $1 \leq p < \infty$ and $L^\infty(\oplus\Omega, \oplus\mathcal{F}, m)$ the set of all m-equivalence classes of essentially bounded complex-valued functions on $(\oplus\Omega, \oplus\mathcal{F}, m)$ with

$$\bigoplus_{\mu \in pr(\Omega,\mathcal{F})} L^\infty(\Omega, \mathcal{F}, \mu) = \{\{f_\mu\}_{\mu \in pr(\Omega,\mathcal{F})} : f_\mu \in L^\infty(\Omega, \mathcal{F}, \mu) \quad (\forall \mu \in pr(\Omega, \mathcal{F}))$$

$$\text{and} \quad \sup_{\mu \in pr(\Omega,\mathcal{F})} \mu\text{-ess.}\sup |f_\mu| < \infty\}.$$

$L^\infty(\oplus\Omega, \oplus\mathcal{F}, m)$ is the dual Banach space of $L^1(\oplus\Omega, \oplus\mathcal{F}, m)$. On the other hand $L^\infty(\oplus\Omega, \oplus\mathcal{F}, m)$ can also be identified with a commutative von Neumann algebra as the multiplicative operator algebra on $L^2(\oplus\Omega, \oplus\mathcal{F}, m)$. The weak* topology and the weak operator topology on $L^\infty(\oplus\Omega, \oplus\mathcal{F}, m)$ coincide because

$$L^1(\oplus\Omega, \oplus\mathcal{F}, m) = \{g\bar{h} : g, h \in L^2(\oplus\Omega, \oplus\mathcal{F}, m)\}.$$

Let $B(\Omega, \mathcal{F})$ be the set of all bounded measurable complex-valued functions on $(\Omega, \mathcal{F})$. We define

$$\pi(f) = \{[f]_\mu\}_{\mu \in pr(\Omega,\mathcal{F})} \quad (f \in B(\Omega, \mathcal{F}))$$

where $[f]_\mu$ denotes the μ-equivalence class of f in $L^\infty(\Omega, \mathcal{F}, \mu)$. Let $\mathcal{M}(\Omega, \mathcal{F})$ be the weak* closure of $\operatorname{Im}\pi$ in $L^\infty(\oplus\Omega, \oplus\mathcal{F}, m)$.

Proposition 1. *$\mathcal{M}(\Omega, \mathcal{F})$ is a von Neumann algebra and its predual is isometrically isomorphic to $ca(\Omega, \mathcal{F})$, which is equipped with the total variation norm.*

Proof. It is trivial that $\mathcal{M}(\Omega, \mathcal{F})$ is a von Neumann algebra. Let $\mathcal{M}_0$ be the polar of $\mathcal{M}(\Omega, \mathcal{F})$. That is,

$$\mathcal{M}_0 = \{f \in L^1(\oplus\Omega, \oplus\mathcal{F}, m) : \int fg dm = 0 \quad (\forall g \in \mathcal{M}(\Omega, \mathcal{F}))\}.$$

$L^1(\oplus\Omega, \oplus\mathcal{F}, m)/\mathcal{M}_0$ can be identified with the predual of $\mathcal{M}(\Omega, \mathcal{F})$. Suppose $F \in L^1(\oplus\Omega, \oplus\mathcal{F}, m)/\mathcal{M}_0$ and $f \in F$. Let

$$\nu_F(A) = \int_A f dm \qquad (\forall A \in \mathcal{F}).$$

Then ν_F does not depend on the choice of f and is a countably additive bounded complex-valued measure on $(\Omega, \mathcal{F})$. Conversely if $\nu \in ca(\Omega, \mathcal{F})$ then there exist $\alpha_1, \alpha_2, \alpha_3, \alpha_4 \geq 0$ and $\mu_1, \mu_2, \mu_3, \mu_4 \in pr(\Omega, \mathcal{F})$ such that

$$\nu = \alpha_1\mu_1 - \alpha_2\mu_2 + i(\alpha_3\mu_3 - \alpha_4\mu_4).$$

Put

$$f_\mu = \begin{cases} \alpha_j, & \text{if } \mu = \mu_j \text{ for some } j = 1,2,3,4\ ; \\ 0, & \text{otherwise.} \end{cases}$$

Then $F \in L^1(\oplus\Omega, \oplus\mathcal{F}, m)/\mathcal{M}_0$ such that $\{f_\mu\}_{\mu \in pr(\Omega,\mathcal{F})} \in F$ satisfies $\nu_F = \nu$. It is also straightforward that the mapping $F \mapsto \nu_F$ is an isometric isomorphism. $\square$

By the above proposition, for each $\nu \in ca(\Omega, \mathcal{F})$ there exists a unique weak* linear functional φ_ν on $\mathcal{M}(\Omega, \mathcal{F})$ such that

$$\varphi_\nu(\pi(f)) = \int f d\nu \quad (\forall f \in B(\Omega, \mathcal{F})).$$

If $\nu \in pr(\Omega, \mathcal{F})$ and $\{f_\mu\}_{\mu \in pr(\Omega,\mathcal{F})} \subset \mathcal{M}(\Omega, \mathcal{F})$ then

$$\varphi_\nu(\{f_\mu\}_{\mu \in pr(\Omega,\mathcal{F})}) = \int f_\nu d\nu.$$

2. Experiments and observables. Let $\mathcal{P}$ be a non-empty subset of $pr(\Omega, \mathcal{F})$ and we call it a set of *experiments*. Define

$$\mathcal{P}^0 = \{F \in \mathcal{M}(\Omega, \mathcal{F}) : \varphi_\nu(|F|) = 0 \quad (\forall \nu \in \mathcal{P})\}.$$

Then

$$\mathcal{P}^0 = \{\{f_\mu\}_{\mu \in pr(\Omega,\mathcal{F})} \in \mathcal{M}(\Omega, \mathcal{F}) : f_\nu = 0 \quad (\forall \nu \in \mathcal{P})\}$$

because

$$\begin{aligned} f_\nu = 0 \quad &\Leftrightarrow \quad \int |f_\nu| d\nu = 0 \\ &\Leftrightarrow \quad \varphi_\nu(\{|f_\mu|\}_{\mu \in pr(\Omega,\mathcal{F})}) = 0. \end{aligned}$$

We can easily see that $\mathcal{P}^0$ is a weak* closed ideal of $\mathcal{M}(\Omega, \mathcal{F})$. Therefore there exists $E \in \mathcal{M}(\Omega, \mathcal{F})$ such that

$$E = \overline{E} = E^2$$

(i.e., E is $\{0,1\}$-valued and it is an orthogonal projection as a multiplicative operator) and

$$\mathcal{P}^0 = \{EF : F \in \mathcal{M}(\Omega, \mathcal{F})\}.$$

Let E be $\{e_\mu\}_{\mu \in pr(\Omega,\mathcal{F})}$. Put

$$\overline{\mathcal{P}} = \{\mu \in pr(\Omega, \mathcal{F}) : e_\mu = 0\}.$$

Lemma 2.

$$\overline{\mathcal{P}} = \{\mu \in pr(\Omega, \mathcal{F}) : \varphi_\mu(F) = 0 \quad (\forall F \in \mathcal{P}^0)\}.$$

Proof. ($\subseteq$)

$$\begin{aligned} \nu \in \overline{\mathcal{P}} \quad &\Rightarrow \quad e_\nu = 0 \\ &\Rightarrow \quad [f_\mu = 0 \quad (\forall \mu \in \mathcal{P}) \quad \Rightarrow \quad f_\nu = 0] \\ &\Rightarrow \quad [|f_\mu| = 0 \quad (\forall \mu \in \mathcal{P}) \quad \Rightarrow \quad f_\nu = 0] \\ &\Rightarrow \quad [\varphi_\mu(|F|) = 0 \quad (\forall \mu \in \mathcal{P}) \quad \Rightarrow \quad \varphi_\nu(F) = 0] \\ &\Rightarrow \quad \varphi_\nu(F) = 0 \quad (\forall F \in \mathcal{P}^0). \end{aligned}$$

($\supseteq$) Because $F \in \mathcal{P}^0$ implies $|F| \in \mathcal{P}^0$,

$$\begin{aligned} &\varphi_\nu(F) = 0 \quad (\forall F \in \mathcal{P}^0) \\ \Rightarrow \quad &\varphi_\nu(|F|) = 0 \quad (\forall F \in \mathcal{P}^0) \\ \Rightarrow \quad &[\varphi_\mu(|F|) = 0 \quad (\forall \mu \in \mathcal{P}) \quad \Rightarrow \quad \varphi_\nu(|F|) = 0] \\ \Rightarrow \quad &[f_\mu = 0 \quad (\forall \mu \in \mathcal{P}) \quad \Rightarrow \quad f_\nu = 0] \\ \Rightarrow \quad &e_\nu = 0 \\ \Rightarrow \quad &\nu \in \overline{\mathcal{P}}. \quad \square \end{aligned}$$

Let $\mathcal{L}(\mathcal{P})$ be the linear span of $\overline{\mathcal{P}}$. Then we have the following.

Lemma 3.

$$\begin{aligned} \mathcal{L}(\mathcal{P}) &= \{\nu \in ca(\Omega, \mathcal{F}) : \varphi_\nu(F) = 0 \quad (\forall F \in \mathcal{P}^0)\} \\ &= \{\nu \in ca(\Omega, \mathcal{F}) : \exists \mu_1, \ldots, \mu_n \in \mathcal{P}, \\ &\qquad \exists c_1, \ldots, c_n > 0, |\nu| \leq c_1\mu_1 + \cdots + c_n\mu_n\}^-, \end{aligned}$$

where $|\nu|$ *means the total variation measure of* ν *and* $^-$ *is the closure in norm topology.*

Proof. The first equality follows from the previous lemma and the Hahn-Jordan decomposition.

¿From this, $\mathcal{L}(\mathcal{P})$ is a closed subspace of $ca(\Omega, \mathcal{F})$. Put

$$\mathcal{X} = \{\nu \in ca(\Omega, \mathcal{F}) : \exists \mu_1, \ldots, \mu_n \in \mathcal{P}, \exists c_1, \ldots, c_n > 0, |\nu| \leq c_1\mu_1 + \cdots + c_n\mu_n\}.$$

$\mathcal{L}(\mathcal{P}) \supseteq \mathcal{X}$ is trivial. Now suppose that there exists $\nu \in \mathcal{L}(\mathcal{P}) \setminus \overline{\mathcal{X}}$. By the Hahn-Banach theorem and Proposition 1 there exists $F \in \mathcal{M}(\Omega, \mathcal{F})$ such that

$$\varphi_\nu(F) \neq 0 \qquad \text{and} \qquad \varphi_\mu(F) = 0 \quad (\forall \mu \in \mathcal{L}(\mathcal{P})).$$

This contradicts the first equality. $\square$

Proposition 4. *$\mathcal{M}(\Omega, \mathcal{F})/\mathcal{P}^0$ is a von Neumann algebra and its predual is isometrically isomorphic to $\mathcal{L}(\mathcal{P})$.*

Proof. This is a direct consequence of

$$\mathcal{P}^0 = \{F \in \mathcal{M}(\Omega, \mathcal{F}) : \varphi_\nu(F) = 0 \quad (\forall \nu \in \mathcal{L}(\mathcal{P}))\}. \quad \square$$

We denote $\mathcal{M}(\Omega, \mathcal{F})/\mathcal{P}^0$ by $\mathcal{M}(\Omega, \mathcal{F}, \mathcal{P})$ and call it a set of *observables*. This space can also be constructed as follows. The direct sum

$$\bigoplus_{\mu \in \mathcal{P}} L^\infty(\Omega, \mathcal{F}, \mu) = \{\{f_\mu\}_{\mu \in \mathcal{P}} : f_\mu \in L^\infty(\Omega, \mathcal{F}, \mu) \quad (\forall \mu \in \mathcal{P}) \text{ and } \sup_{\mu \in \mathcal{P}} \mu\text{-ess.}\sup |f_\mu| < \infty\}$$

is a von Neumann algebra and its predual is

$$\bigoplus_{\mu \in \mathcal{P}} L^1(\Omega, \mathcal{F}, \mu) = \{\{f_\mu\}_{\mu \in \mathcal{P}} : f_\mu \in L^1(\Omega, \mathcal{F}, \mu) \quad (\forall \mu \in \mathcal{P}) \text{ and } \sum_{\mu \in \mathcal{P}} \int |f_\mu| d\mu < \infty\}.$$

We define a mapping $\pi_\mathcal{P}$ from $B(\Omega, \mathcal{F})$ into $\bigoplus_{\mu \in \mathcal{P}} L^\infty(\Omega, \mathcal{F}, \mu)$ by

$$\pi_\mathcal{P}(f) = \{[f]_\mu\}_{\mu \in \mathcal{P}} \qquad (\forall f \in B(\Omega, \mathcal{F})).$$

Then $\pi_\mathcal{P}$ is *-homomorphism and

$$\ker \pi_\mathcal{P} = \{f \in B(\Omega, \mathcal{F}) : f = 0 \quad \mu\text{-a.e.} \quad (\forall \mu \in \mathcal{P})\}.$$

Since $\ker \pi_\mathcal{P}$ is a closed ideal of $B(\Omega, \mathcal{F})$, $B(\Omega, \mathcal{F})/\ker \pi_\mathcal{P}$ is a C*-algebra with norm

$$\|F\| = \inf_{f \in F} \|f\| \qquad (F \in B(\Omega, \mathcal{F})/\ker \pi_\mathcal{P}).$$

Moreover this space is *-isomorphic to $\operatorname{Im} \pi_\mathcal{P}$. Hence

$$\|[f]_\mathcal{P}\| = \sup_{\mu \in \mathcal{P}} \|f\|_{\infty,\mu} \qquad (\forall f \in B(\Omega, \mathcal{F}))$$

where

$$[f]_\mathcal{P} = \{g \in B(\Omega, \mathcal{F}) : f = g \quad \mu\text{-a.e.} \quad (\forall \mu \in \mathcal{P})\}$$

and $\|\cdot\|_{\infty,\mu}$ is the norm of $L^\infty(\Omega, \mathcal{F}, \mu)$.

Theorem 5. *$\mathcal{M}(\Omega, \mathcal{F}, \mathcal{P})$ is the closure of $\operatorname{Im} \pi_{\mathcal{P}}$ in $\bigoplus_{\mu \in \mathcal{P}} L^\infty(\Omega, \mathcal{F}, \mu)$ in the weak* topology. In particular, $\operatorname{Im} \pi_{\mathcal{P}}$ is weakly* closed if and only if there exists a localizable measure λ on $(\Omega, \mathcal{F})$ such that each $\mu \in \mathcal{P}$ has density $d\mu/d\lambda$.*

Proof. The former assertion is easy and we only prove the latter. Suppose $\operatorname{Im} \pi_{\mathcal{P}}$ is weak* closed, namely, it is a commutative von Neumann algebra. Since commutative von Neumann algebras are semi-finite, there exists a faithful normal semi-finite trace τ on it. Put

$$\lambda(A) = \tau(\pi_{\mathcal{P}}(1_A)) \qquad (A \in \mathcal{F}).$$

This is a localizable measure which we want.

Conversely if there exists a localizable measure λ on $(\Omega, \mathcal{F})$ such that each $\mu \in \mathcal{P}$ has density $d\mu/d\lambda$. We define

$$\mathcal{I} = \{f \in L^\infty(\Omega, \mathcal{F}, \lambda) : f = 0 \quad \mu\text{-a.e.} \quad (\forall \mu \in \mathcal{P})\}.$$

Then it can be easily seen that $\mathcal{I}$ is a weakly* closed ideal of $L^\infty(\Omega, \mathcal{F}, \lambda)$ and that $L^\infty(\Omega, \mathcal{F}, \lambda)/\mathcal{I}$ is *-isomorphic to $B(\Omega, \mathcal{F})/\ker \pi_{\mathcal{P}}$ and then to $\operatorname{Im} \pi_{\mathcal{P}}$. □

Remark. The topology on $B(\Omega, \mathcal{F})/\ker \pi_{\mathcal{P}}$ induced from the weak* topology coincides with the topology defined by [Pitcher,1965]. An analogous theorem is also proved by [Luschgy and Mussman, 1985].

Example 1. Let Ω be $[0, 1]$ and $\mathcal{F}$ the Borel field. Suppose $\mathcal{P}$ is the set of all Dirac measures on $(\Omega, \mathcal{F})$. Then $\bigoplus_{\mu \in \mathcal{P}} L^\infty(\Omega, \mathcal{F}, \mu)$ is identified with $l^\infty[0, 1]$ (the set of all bounded complex-valued functions on $[0, 1]$) and so is $\mathcal{M}(\Omega, \mathcal{F}, \mathcal{P})$. It is really bigger than $\operatorname{Im} \pi_{\mathcal{P}}$. However we can modify $\mathcal{F}$ such as $\mathcal{M}(\Omega, \mathcal{F}, \mathcal{P})$ is identified with $\operatorname{Im} \pi_{\mathcal{P}}$. Namely, let $\mathcal{F}$ be the power set of Ω.

Example 2. Let $(\Omega, \mathcal{F})$ be the same as the above. Suppose $\mathcal{P}$ is the set of all Dirac measures and the Lebesgue measure. Then

$$\begin{aligned} \bigoplus_{\mu \in \mathcal{P}} L^\infty(\Omega, \mathcal{F}, \mu) &= \mathcal{M}(\Omega, \mathcal{F}, \mathcal{P}) \\ &= l^\infty[0, 1] \oplus L^\infty[0, 1]. \end{aligned}$$

In this example, we can not make any kind of modification like the above.

3. σ-subfields and sufficiency. Let $\mathcal{G}$ be a σ-subfield of $\mathcal{F}$ and

$$\mathcal{P}|\mathcal{G} = \{\mu|\mathcal{G} : \mu \in \mathcal{P}\}.$$

However if there is no ambiguity, $\mathcal{P}|\mathcal{G}$ is merely denoted by $\mathcal{P}$ like $\mathcal{M}(\Omega, \mathcal{G}, \mathcal{P})$ rather than $\mathcal{M}(\Omega, \mathcal{G}, \mathcal{P}|\mathcal{G})$. It is also true for $\mu \in pr(\Omega, \mathcal{F})$ like $L^p(\Omega, \mathcal{G}, \mu)$ rather than $L^p(\Omega, \mathcal{G}, \mu|\mathcal{G})$. $\mathcal{M}(\Omega, \mathcal{G})$ and $\mathcal{M}(\Omega, \mathcal{G}, \mathcal{P})$ are considered as von Neumann subalgebras of $\mathcal{M}(\Omega, \mathcal{F})$ and $\mathcal{M}(\Omega, \mathcal{F}, \mathcal{P})$ respectively.

The conditional expectation of $f \in L^p(\Omega, \mathcal{F}, \mu)$ with respect to $\mathcal{G}$ is denoted by $E_\mu(f|\mathcal{G})$ for each $\mu \in pr(\Omega, \mathcal{F})$. The mapping $E_\mu(\cdot|\mathcal{G})$ is a projection of norm-one from $L^p(\Omega, \mathcal{F}, \mu)$ onto the subspace $L^p(\Omega, \mathcal{G}, \mu)$ for every $1 \le p \le \infty$. Now we define

$$E(\{f_\mu\}_{\mu \in pr(\Omega, \mathcal{F})}|\mathcal{G}) = \{E_\mu(f_\mu|\mathcal{G})\}_{\mu \in pr(\Omega, \mathcal{F})} \; (\{f_\mu\}_{\mu \in pr(\Omega, \mathcal{F})} \in L^p(\oplus\Omega, \oplus\mathcal{F}, m)).$$

Then $E(\cdot|\mathcal{G})$ is a projection of norm-one from $L^p(\oplus\Omega, \oplus\mathcal{F}, m)$ onto $L^p(\oplus\Omega, \oplus\mathcal{G}, m)$ for every $1 \leq p \leq \infty$. $E(\cdot|\mathcal{G})$ naturally induces a projection of norm-one from $\bigoplus_{\mu\in\mathcal{P}} L^p(\Omega, \mathcal{F}, \mu)$ onto $\bigoplus_{\mu\in\mathcal{P}} L^p(\Omega, \mathcal{G}, \mu)$. Is the range of $\mathcal{M}(\Omega, \mathcal{F}, \mathcal{P})$ contained in $\mathcal{M}(\Omega, \mathcal{G}, \mathcal{P})$? This containment is not always true. If it is true, then we say that $\mathcal{G}$ is *sufficient* for $\mathcal{P}$. Namely, $\mathcal{G}$ is sufficient for $\mathcal{P}$ if and only if $\{E_\mu(f|\mathcal{G})\}_{\mu\in\mathcal{P}}$ belongs to $\mathcal{M}(\Omega, \mathcal{G}, \mathcal{P})$ for all $f \in B(\Omega, \mathcal{F})$. In general, this condition is really weaker than that for any $f \in B(\Omega, \mathcal{F})$ there exists $g \in B(\Omega, \mathcal{G})$ such that $E_\mu(f|\mathcal{G}) = g$ μ-a.e. for all $\mu \in \mathcal{P}$. Let us consider Example 1 in the previous section and let $\mathcal{G}$ be the σ-subfield generated by all singletons contained in Ω. Then $\mathcal{G}$ is sufficient for $\mathcal{P}$ because $\mathcal{M}(\Omega, \mathcal{G}) = \mathcal{M}(\Omega, \mathcal{F})$ and $E(\cdot|\mathcal{G})$ is identity on $\mathcal{M}(\Omega, \mathcal{F})$. For any $f \in B(\Omega, \mathcal{F})$, $E_{\delta_x}(f|\mathcal{G}) = f(x)$, and no $g \in B(\Omega, \mathcal{G})$ satisfies $f = g$. Note that $\mathcal{P}$ is dominated by the semi-finite counting measure λ and $d\delta_x/d\lambda$ is $\mathcal{G}$-measurable. In Example 2, $\mathcal{G}$ is also sufficient for $\mathcal{P}$, because $\mathcal{M}(\Omega, \mathcal{G}, \mathcal{P})$ is equal to $l^\infty[0,1] \oplus \mathbb{C}$ and $E(\cdot|\mathcal{G})$ maps $f \in B(\Omega, \mathcal{F})$ to $f \oplus \int_0^1 f(x)dx$.

4. Gibbs states on a countable set. Let S be a countable set and Ω the power set of S (the set of all subsets of S). For each $s \in S$, we define a $\{0,1\}$-function σ_s on Ω by

$$\sigma_s(X) = 1_X(s) \qquad (X \in \Omega)$$

where 1_X is the indicator function of X on S . It is well known that the weakest topology on Ω induced by $\{\sigma_s\}_{s\in S}$ is totally disconnected, compact, and metrizable. The space $C(\Omega)$ of all complex valued continuous functions defined on Ω is a C*-algebra. The Borel field on Ω is denoted by $\mathcal{F}$, which is the smallest σ-field on Ω generated by $\{\sigma_s\}_{s\in S}$. It coincides with the Baire field on Ω, which is generated by $C(\Omega)$. Every probability measure is identified with a state on $C(\Omega)$ (i.e., positive linear functional μ with $\mu(1) = 1$). $\bigoplus_{\mu\in ca(\Omega,\mathcal{F})} L^\infty(\Omega, \mathcal{F}, \mu)$ is known as the enveloping von Neumann algebra which is the second dual Banach space $C(\Omega)^{**}$ of $C(\Omega)$.

For any $\Lambda \subseteq S$, we denote by $\mathcal{F}_\Lambda$ the σ-subfield of $\mathcal{F}$ which is generated by $\{\sigma_s\}_{s\in\Lambda}$. Clearly $\mathcal{F}_\emptyset = \{\emptyset, \Omega\}$ and $\mathcal{F}_S = \mathcal{F}$. We put

$$[A, \Lambda] = \{Y \in \Omega : Y \cap \Lambda = A\} \qquad (A \subseteq X).$$

Then $\mathcal{F}_\Lambda$ is the smallest σ-subfield containing $\{[A, \Lambda] : A \subseteq \Lambda\}$.

Let $\mathcal{C}$ be the set of all finite subsets of S . A subset $\{f^\Lambda\}_{\Lambda\in\mathcal{C}}$ of $C(\Omega)$ indexed by $\mathcal{C}$ is called a *local specification* if:

$$f^\Lambda(X) \geq 0 \qquad (X \in \Omega),$$
$$\sum_{A\subseteq\Lambda} f^\Lambda(A \cup B) = 1 \qquad (B \subseteq \Lambda^c)$$

for all $\Lambda \in \mathcal{C}$, and

$$f^{\Lambda_2}(A \cup B) = f^{\Lambda_1}(A \cup B) \sum_{A'\subseteq\Lambda_1} f^{\Lambda_2}(A' \cup B)$$

for all $A \subseteq \Lambda_1 \subseteq \Lambda_2 \in \mathcal{C}$ and $B \subseteq {\Lambda_1}^c$. We say that $\mu \in pr(\Omega, \mathcal{F})$ is a *Gibbs state* with specification $\{f^\Lambda\}_{\Lambda\in\mathcal{C}}$ if

$$E_\mu(1_{[A,\Lambda]}|\mathcal{F}_{\Lambda^c})(X) = f^\Lambda(A \cup (X \cap \Lambda^c))$$

for all $A \subseteq \Lambda \in \mathcal{C}$ and $X \in \Omega$.

Let $\mathcal{P}$ be the set of Gibbs states with specification $\{f^\Lambda\}_{\Lambda \in \mathcal{C}}$. It is known that $\mathcal{P}$ is a non-empty compact convex subset of $pr(\Omega, \mathcal{F})$ in the vague topology (see, for example, [Preston,1974]). For any $\mu \in \mathcal{P}$, if $\Lambda \subseteq \Lambda'$,

$$\begin{aligned} E_\mu(1_{[A,\Lambda']}|\mathcal{F}_{\Lambda^c})(X) &= E_\mu(1_{[A\cap\Lambda,\Lambda]\cap[A\cap(\Lambda'\setminus\Lambda),\Lambda'\setminus\Lambda]}|\mathcal{F}_{\Lambda^c})(X) \\ &= E_\mu(1_{[A\cap\Lambda,\Lambda]}1_{[A\cap(\Lambda'\setminus\Lambda),\Lambda'\setminus\Lambda]}|\mathcal{F}_{\Lambda^c})(X) \\ &= 1_{[A\cap(\Lambda'\setminus\Lambda),\Lambda'\setminus\Lambda]}(X)E_\mu(1_{[A\cap\Lambda,\Lambda]}|\mathcal{F}_{\Lambda^c})(X) \\ &= 1_{[A\cap(\Lambda'\setminus\Lambda),\Lambda'\setminus\Lambda]}(X)f^\Lambda((A\cap\Lambda)\cup(X\cap\Lambda^c)) \end{aligned}$$

for all $A \subseteq \Lambda'$ and $X \in \Omega$. This says that $\mathcal{F}_{\Lambda^c}$ is sufficient for $\mathcal{P}$. We put $\mathcal{F}_\infty = \bigcap_{\Lambda\in\mathcal{C}} \mathcal{F}_{\Lambda^c}$. Since $\mathcal{M}(\Omega, \mathcal{F}_\infty, \mathcal{P}) = \bigcap_{\Lambda\in\mathcal{C}} \mathcal{M}(\Omega, \mathcal{F}_{\Lambda^c}, \mathcal{P})$, using the martingale convergence theorem on von Neumann algebras (see,[Tsukada,1985]), we conclude the following theorem.

Theorem. *$\mathcal{F}_\infty$ is sufficient for $\mathcal{P}$.*

REFERENCES

1. P.R. Halmos & L.J. Savage, *Application of the Radon-Nikodym theorem to the theory of sufficient statistics*, Ann. Math. Statist. **20** (1949), 225–241.
2. F. Hiai, M. Ohya & M. Tsukada, *Sufficiency, KMS condition and relative entropy in von Neumann algebras*, Pacific J. Math. **96** (1981), 99-109.
3. F. Hiai, M. Ohya & M. Tsukada, *Sufficiency and relative entropy in *-algebras with applications in quantum systems*, Pacific J. Math. **107** (1983), 117–140.
4. L. LeCam, *Asymptotic Methods in Statistical Decision Theory*, Springer, 1986.
5. H. Luschgy & D. Mussmann, *Equivalent properties and completion of statistical experimaents*, Sankya: Indian J. Stat. **47** (1985), 174–195.
6. T.S. Pitcher, *A more general property than domination for sets of probability measures*, Pacific J. Math. **15** (1965), 597–611.
7. D. Petz, *Sufficient subalgebras and the relative entropy of states on a von Neumann algebra*, Commun. Math. Phys. **105** (1986), 123–131.
8. C. Preston, *Gibbs States on Countable Sets*, Cambridge Univ. Press, 1974.
9. M. Tsdukada, *Convergence of closed convex sets and σ-fields*, Z. Wahrsch. verw. Geb. **62** (1983), 137–146.
10. ———, *The strong limit of von Neumann subalgebras with conditional expectations*, Proc. Amer. Math. Soc. **94** (1985), 259–264.
11. H. Umegaki, *Conditional expectation in an operator algebra III*, Kodai Math. Sem. Rep. **11** (1959), 51–64.

Nonlinear Parabolic Equations, Favard Classes, and Regularity

GISÈLE RUIZ GOLDSTEIN † Department of Mathematics, Louisiana State University, Baton Rouge, LA 70803, and CERI and Department of Mathematical Sciences, University of Memphis, Memphis, TN 38152

1. INTRODUCTION

Let A be an m-dissipative operator (not necessarily linear) on a Banach space X. By the Crandall-Liggett theorem A determines a contraction semigroup T on $(\overline{\mathcal{D}(A)})$. The *Favard class* (or the *generalized domain*) $\widehat{\mathcal{D}}(A)$ is defined to be

$$\widehat{\mathcal{D}}(A) = \{f \in (\overline{\mathcal{D}(A)}) : \lim_{\lambda \downarrow 0} \|A_\lambda f\| < \infty\}.$$

Here A_λ is the Yosida approximation of A, namely $A_\lambda = \lambda^{-1}(I - (I - \lambda A)^{-1})$ for $\lambda > 0$. It is not difficult to show that the Favard class can be equivalently defined to be

$$\begin{aligned}\widehat{\mathcal{D}}(A) &= \{f \in (\overline{\mathcal{D}(A)}) : \|T(t)f - f\| \leq M_f t \text{ for some } M_f > 0 \text{ and } 0 < t < 1\} \\ &= \{f \in (\overline{\mathcal{D}(A)}) : \text{ for some sequence } \{g_n\} \in \mathcal{D}(A) \\ &\qquad \text{with } g_n \to f, Ag_n \text{ is bounded as } n \to \infty\}.\end{aligned}$$

Clearly,

$$\mathcal{D}(A) \subset \widehat{\mathcal{D}}(A) \subset (\overline{\mathcal{D}(A)}),$$

† Partially supported by an NSF grant.

and one can show $\widehat{\mathcal{D}}(A) = \mathcal{D}(A)$ if X is reflexive. From our perspective, the most important aspect of the Favard class is the property

$$T(t)(\widehat{\mathcal{D}}(A)) \subset \widehat{\mathcal{D}}(A) \text{ for each } t > 0,$$

that is, the Favard class is an invariant set for the semigroup. Hence, the Favard class contains information on spatial regularity of a problem. For example if we can show that $W_0^{k,p}(\Omega) \subset \widehat{\mathcal{D}}(A) \subset W^{k,p}(\Omega)$, (1.1) says the solution $u(t)$ will have spatial derivatives up to and including order k, each of which is in $L^p(\Omega)$. The problem with this method is that $\widehat{\mathcal{D}}(A)$ is very difficult to compute explicitly.

Our purpose in this paper is to calculate the Favard class explicitly in the case of a nonlinear parabolic problem with degeneracy and to draw some conclusions about regularity.

The problem of calculating the Favard class for this problem in the case $\psi \equiv 0$ with either Dirichlet or nonlinear boundary conditions was studied in [4], [5]. In this paper we consider a more general operator with several different types of boundary conditions, so that even in the case where no lower order terms are present this paper gives new results.

The main result is Theorem 2. It is stated in Section 2 and proved in Section 3. Section 4 contains some extensions, while Section 5 contains concluding remarks and directions for future research.

2. A SINGULAR NONLINEAR PARABOLIC PROBLEM

We consider the problem

$$u_t(t,x) = \varphi(x, u_x(t,x))u_{xx}(t,x) + \psi(x, u(t,x), u_x(t,x)) \tag{2.1}$$

for $x \in [0,1]$ and $t \in [0,\infty]$. Let $X = C[0,1]$; we assume the initial condition

$$u(0,x) = u_0(x).$$

We allow several types of boundary conditions at $j = 0,1$.

$$u(t,j) = 0 \tag{$(BC_j)^D$}$$

$$(-1)^j u_x(t,j) \in \beta_j(u(t,j)) \tag{$(BC_j)^N$}$$

Here β_j is a strictly increasing maximal monotone graph in $\mathbb{R}^2$ containing the origin. Thus $0 \in \beta_j(0)$, and if $y_i \in \beta(x_i)$ for $i = 1,2$ and $x_1 < x_2$, then $y_1 < y_2$. Note that $(BC_j)^N$ includes the linear boundary conditions

$$u(t,j) = (-1)^j \alpha_j u'(t,j) \tag{(BC_j)}$$

for $\alpha_j > 0$. We also allow for periodic-like boundary conditions

$$\begin{aligned} u(t,0) &= u(t,1) \\ u_x(t,0) &= u_x(t,1). \end{aligned} \tag{$(BC)^P$}$$

Regarding φ and ψ we assume

$$\begin{aligned} &\varphi(x,q) > 0 \text{ for } 0 < x < 1, \varphi(x,q) \geq \varphi_0(x) \text{ where} \\ &\varphi_0 \in C[0,1], \varphi_0(x) > 0 \text{ for } x \in [0,1] \setminus \mathcal{S} \text{ and} \\ &\varphi_0^{-1} \in L^1[0,1]. \end{aligned} \tag{2.2}$$

Here

$$\mathcal{S} = \{x \in [0,1] : \varphi_0(x) = 0\}.$$

(Hence meas $\mathcal{S} = 0$.)

There exist positive constants L, M, and N such that

$$|\psi(x,p,q) - \psi(x,\tilde{p},q)| \leq L|\, p - \tilde{p} \,| \tag{2.3}$$

$$|\psi(x,p,q)| \leq \mathcal{M}(|p|)(1 + \varphi(x,q))M(1 + |q|) \tag{2.4}$$

and

$$|\psi(x,p,0)| \leq N(1 + |p|). \tag{2.5}$$

where $\mathcal{M} : [0,\infty) \to [0,\infty)$ is a continuous nondecreasing function.

In fact the constant L can be replaced by a continuous nondecreasing function $\mathcal{L}(|q|)$, so that (2.3) holds only locally, and our theorems still remain valid. For such extensions see [2].

Let X be the Banach space $C[0,1]$ with the sup norm. We define the operator A on X by

$$Au = \varphi(\cdot, u')u'' + \psi(x, u, u').$$

Choose ***one*** boundary condition at $j = 0$ and ***one*** at $j = 1$ from the conditions $(BC_j), (BC_j)^N$, or choose $(BC)^P$. Then we define the set

$$Y_{BC} = \begin{cases} C[0,1] & \text{if } (BC_0^D), (BC_1)^D \text{ hold} \\ C[0,1] \cap C^1[0,1) & \text{if } (BC)_0^N \text{ and } (BC_1)^D \text{ hold} \\ C[0,1] \cap C^1(0,1] & \text{if } (BC_0)^N \text{ and } (BC_1)^D \text{ hold} \\ C^1[0,1] & \text{if } (BC_j)^N \text{ holds for } j = 1,2, \text{ or if } (BC)^P \text{ hold}. \end{cases}$$

We define the domain $\mathcal{D}(A)$ of the operator A by

$$\mathcal{D}(A) = \{u \in Y_{BC} \cap C^2(0,1) : Au \in C[0,1] \text{ and } u \text{ satisfies the chosen boundary conditions at } x = 0,1\}.$$

Theorem 1: *A is m-dissipative on X.*

This result is due to J.A. Goldstein and C.Y. Lin [8] in the special case $\psi \equiv 0, \varphi_0 > 0$ and $\varphi_0^{-1} \in L^2(0,1)$. J.R. Dorroh and G. R. Rieder [1] proved the result for the boundary conditions $(BC_j)^D$ and the linear version of $(BC_j)^N$, G. R. Goldstein [3], [12] obtained this result for the boundary conditions $(BC_j)^N, (BC_j)^P$.

We shall require one additional hypothesis for our result.

There exist $\varepsilon_0, C_0 > 0$ such that

(2.6)
$$\varphi(x,\xi + a) \leq C_0\varphi(x,\xi)$$
$$\psi(x,y,\xi + a) \leq C_0\psi(x,y,\xi)$$

for all $x \in [0,1]$, all $\xi \in \mathbb{R}$ and all $a \in \mathbb{R}$ with $|a| < \varepsilon_0$.

The following theorem is the main result.

Theorem 2: *Define A and $\mathcal{D}(A)$ as in the preceding discussion, and assume (2.2)-(2.6). Then the Favard class of A is*

$$\widehat{\mathcal{D}}(A) = \{u \in C^2(0,1) \cap Y_{BC} : u' \in AC[0,1]$$

$$\varphi(x,u')u'' + \psi(x,u,u') \in L^\infty(0,1)$$

and u satisfies the boundary conditions at $j = 0,1\}$.

3. PROOF OF THEOREM 2

Proof: Let

$$D = \{u \in C^2(0,1) \cap Y_{BC} : u' \in AC[0,1], Au \in L^\infty(0,1)$$

and u satisfies the boundary conditions at $j = 0,1\}$.

Our goal is to show D is precisely the Favard class of A.

First, we shall show that $\widehat{\mathcal{D}}(A) \subseteq D$. Let $u \in \widehat{\mathcal{D}}(A)$, and choose $u_n \in \mathcal{D}(A)$ such that $\|u_n - u\|_\infty \to 0$ and $\|Au_n\|_\infty \leq M_1$ for some $M_1 > 0$ and for all $n = 1, 2 \cdots$. Let $M_0 = \|\varphi_0^{-1}\|_1$. Then for all $n \geq 1$,

$$\|u_n''\|_1 \leq \int_0^1 \frac{|\varphi(x,u_m')u_n'' + \psi(x,u_n,u_n')\,|}{|\varphi_0(x)|}dx \qquad (3.1)$$
$$\leq M_0 M_1.$$

Define two quantities

$$\omega_L(f;\delta) = \sup\{\int_E \mid f(x)|dx : E \text{ is a subinterval of } [0,1] \text{ with } |E| < \delta\},$$

and

$$\omega_C(f;\delta) = \sup\{|f(x) - f(y)| : x, y \in [0,1] \text{ with } |x-y| \le \delta\}.$$

It is not difficult to show (cf. [1]) that for $\delta \in (0, \frac{1}{2}]$, and if $f \in Y_{BC} \cap C^2(0,1)$ with $f'' \in L^1[0,1]$, we have

$$\|f'\|_\infty \le 4\|f\|_\infty + \|f''\|_1 \tag{3.2}$$

$$\|f'\|_\infty \le \frac{2}{\delta}\|f'\|_\infty + \omega_L(f'';\delta). \tag{3.3}$$

Also, notice that if $\{f_n\} \subseteq L^1[0,1]$, then the statement that for every $\varepsilon > 0$ there is a $\delta > 0$ such that $\omega_L(f_n;\delta) < \varepsilon$ for all n is equivalent to saying that the sequence $\{f_n\}$ is uniformly integrable on $[0,1]$. Similarly, the statement that for each $\varepsilon > 0$ there is a $\delta > 0$ such that $\omega_C(f_n;\delta) < \varepsilon$ for all n is equivalent to saying that the sequence $\{f_n\}$ is equicontinuous on $[0,1]$.

In our problem if we write

$$u_n''(x) \le \frac{\varphi(x, u_n'(x))u_n''(x) + \psi(x, u_n(x), u_n'(x))}{\varphi_0(x)}$$

it follows that, given $\varepsilon > 0$,

$$\omega_L(u_n'';\delta_\varepsilon) \le M_1\omega_L(\varphi_0^{-1};\delta_\varepsilon) < \varepsilon \tag{3.4}$$

and

$$\omega_C(u_n';\delta_\varepsilon) \le \omega_L(u_n'';\delta_\varepsilon) \tag{3.5}$$

for some $\delta_\varepsilon > 0$ and all n since φ_0^{-1} is integrable. Together the estimates (3.1) and (3.2) show that $\{u_n'\}$ is a pointwise bounded sequence in $C[0,1]$, while (3.4) and (3.5) show that $\{u_n\}$ is an equicontinuous sequence in $C^1[0,1]$. Hence by the Arzela-Ascoli theorem there is a subsequence, which we again denote by $\{u_n\}$ which converges uniformly to a function $u \in C^1[0,1]$. Using the boundedness of $\{Au_n\}$, we have, at least for some subsequence,

$$u_n \to u \text{ in } X,$$

$$u_n' \to u' \text{ in } X,$$

$$u_n'' \to u'' \text{ a.e.}$$

$$\omega_L(u'',\delta) \le M_1\omega_L(\varphi_0^{-1},\delta) \le M_1M_0$$

and

$$\omega_L(u'',\delta) \le \varepsilon$$

for some δ sufficiently small. It also follows that

$$\|\widetilde{A}u\|_\infty \leq M,$$

where $\widetilde{A}u = \varphi(x, u')u'' + \psi(x, u, u')$. Thus it remains to show that the boundary conditions hold. In the case of Dirichlet $(BC_j)^D$ or periodic-like boundary conditions $(BC)^P$, the result follows by the uniform convergence of u_n to u. In the case of the nonlinear boundary conditions $(BC_j)^N$, the result follows from the closedness of the graph β_j.

This completes half of our proof. Next, we show that $D \subseteq \widehat{\mathcal{D}}(A)$. Let $u \in D$. Clearly, u can be written uniquely in the form

$$u(x) = a + bx + \int_0^x \int_0^y u''(s)dsdy. \tag{3.6}$$

Since the continuous functions are dense in $L^1[0,1]$, we can choose a sequence $\{f_n\} \subseteq C[0,1]$ with

(a) $f_n \to u''$ a.e. and in $L^1(0,1)$

(b) $|f_n(x)| \leq 2|u''(x)| + 1$ a.e. for $n \geq 1$

(c) $\sup \|\varphi(x, u_n')u_n'' + \psi(x, u_n, u_n')\|_\infty \leq N < \infty$

where $u_n \in C^2[0,1]$ is defined by

$$u_n(x) = a_n + b_n x + \int_0^x \int_0^y f_n(s)dsdy. \tag{3.7}$$

Clearly,

$$u_n'(x) = b_n + \int_0^x f_n(s)ds \quad \text{for all } x \in (0,1),$$

$$u_n''(x) = f_n(x) \quad \text{a.e.}$$

The constants a and b in (3.6) are uniquely determined. Specifically, $a = u(0)$ and $b = u'(0)$. In (3.7), the definition of u_n, choose a_n and b_n so that as $n \to \infty$

$$a_n \to a, b_n \to b.$$

The fact that $\sup_n \|\varphi(x, u_n')u_n'' + \psi(x, u_n, u_n')\| < \infty$ follows from (2.3), and (2.4) and the assumption that $u \in D$.

In order to complete the proof that $u \in \widehat{\mathcal{D}}(A)$, it remains to show that the boundary conditions hold for each u_n, so that $\{u_n\} \subseteq \mathcal{D}(A)$. This amounts to choosing a_n and b_n appropriately.

Recall that $a = u(0)$ and $b = u'(0)$, and ***define*** the constants c_n, c, d_n and d by

$$c_n = \int_0^1 f_n(s)ds \tag{3.8}$$

$$c = \int_0^1 u''(s)ds, \tag{3.9}$$

$$d_n = \int_0^1 \int_0^y f_n(s)dsdy \tag{3.10}$$

$$d = \int_0^1 \int_0^y u''(s)dsdy. \tag{3.11}$$

It follows from the choice of $\{f_n\}$ that

$$c_n \to c \text{ and } d_n \to d \text{ as } n \to \infty.$$

We consider the different cases based on the boundary conditions chosen.

Case 1: $(BC_0)^D, (BC_1)^D$

Under these boundary conditions $a = 0$ and $b = -d$. Hence, the boundary conditions will be satisfied by u_n if we choose $a_n = 0$ and $b_n = -d_n$. (The latter holds since $d_n \to d$.)

Case 2: $(BC)^P$.

With periodic like boundary conditions we see $a = a+b+d$ and $b = b+c$: whence $c = 0$ and $b = -d$. Choosing $b_n = -d_n$ and $a_n = a$, we see that the sequence $(a_n, b_n) \to (a, b)$ as $n \to \infty$ and that $u_n(0) = u_n(1), u_n'(0) = u_n'(1)$. Thus, u_n satisfies $(BC)^P$.

Case 3: $(BC_0)^N, (BC_1)^N$.

This is the most difficult case. With these nonlinear boundary conditions, we must have $b \in \beta_0(a)$ and $-(b + c) \in \beta_1(a + b + d)$. For $u_n \in \mathcal{D}(A)$, we need $b_n \in \beta_0(a_n), -(b_n + c_n) \in \beta_1(a_n + b_n + d_n)$ to hold for all n.

Define the maximal monotone graphs, γ_n, γ on $\mathbb{R}$ by

$$\gamma_n(s) := \beta_1((I + \beta_0)(s) + d_n) + \beta_0(s)$$
$$\gamma(s) := \beta_1((I + \beta_0)(s) + d) + \beta_0(s).$$

By the strict monotonicity of β_0 and β_1, Range (γ) = Range $(\gamma_n) = J$ where J is the open interval

$$J = \left(\inf_{\mathbb{R}} \beta_1 + \inf_{\mathbb{R}} \beta_0, \sup_{\mathbb{R}} \beta_1 + \sup_{\mathbb{R}} \beta_0 \right).$$

Note that J is independent of n.

For both boundary conditions to hold we need $-c_n \in \gamma_n(a_n)$. Since u satisfies the boundary conditions, we have $-c \in \gamma(a)$; in particular, $c \in J$. Since $c_n \to c$, it follows that $c_n \in J$ for n sufficiently large. Hence, by the strict monotonicity there exists a unique a_n

with $-c_n \in \gamma(a_n)$. Also, for such n, there is a (uniquely determined if β_1 is single valued) b_n with $b_n \in \beta_0(a_n)$ such that

$$-c_n \in \beta_1(a_n + b_n + d_n) + b_n.$$

Even if b_n is not uniquely determined, from the facts that $b \in \beta_0(a)$ and $a_n \to a$, we see that we can choose $b_n \in \beta_0(a_n)$ for sufficiently large n in such a way that

$$-(b_n + c_n) \in \beta_1(a_n + b_n + d_n).$$

and $b_n \to b$ as $n \to \infty$.

Case 4: $(BC_0)^0, (BC_1)^N$.

Case 5: $(BC_0)^N, (BC_1)$.

The proofs in Cases 4 and 5 are similar. We omit the details.

4. FURTHER RESULTS

Let $Y = L^\infty(0,1)$. We define the natural extension $\widehat{A}$ of A from $C[0,1]$ to $L^\infty(0,1)$ by $(\widehat{A}u)(x) = \varphi(x, u'(x))u''(x) + \psi(x, u(x), u'(x))$ for $0 < x < 1$. The domain of $\widehat{A}$ is defined to be

$$\mathcal{D}(\widehat{A}) := \widehat{\mathcal{D}}(A) = \{u \in C^1(0,1) \cap Y_{BC} : u' \in AC[0,1], \widehat{A}u \in Y\},$$

that is, the natural maximal domain of $\widehat{A}$ is the Favard class of A.

In [5] G. Goldstein, J. Goldstein and S. Oharu prove the following theorem in the case $\psi \equiv 0$.

Theorem 3: *The operator $\widehat{A}$ is dissipative, and for all $\lambda > 0, \overline{\mathcal{D}(\widehat{A})} \subset \mathcal{R}(I - \lambda\widehat{A})$, that is, $\widehat{A}$ satisfies the hypotheses of the Crandall-Liggett theorem.*

Thus, $\widehat{A}$ generates a contraction semigroup $\widehat{T} = \{\widehat{T}(t) : t \geq 0\}$ on $\overline{\mathcal{D}(\widehat{A})} = \overline{\mathcal{D}(A)} \subset C[0,1]$. It can be shown that

$$\overline{\mathcal{D}(A)} = \begin{cases} C[0,1] & \text{if } (BC_0)^N, (BC_1)^N \text{ hold} \\ \{u \in C[0,1] : u(0) = 0\} & \text{if } (BC_0)^D, (BC_1)^N \text{ hold} \\ \{u \in C[0,1] : u(1) = 0\} & \text{if } (BC_0)^N, (BC_1)^D \text{ hold} \\ \{u \in C[0,1] : u(0) = u(1) = 0\} & \text{if } (BC_0)^D, (BC_1)^D \text{ hold} \\ \{u \in C[0,1] : u(0) = u(1)\} & \text{if } (BC)^P \text{ hold}. \end{cases}$$

For each $u_0 \in \overline{\mathcal{D}(A)}$, the semigroup $\widehat{T}$ gives a unique mild solution $u(t,x) = (\widehat{T}(t)u_0)(x)$ of (2.1) satisfying $u(0,x) = u_0(x)$.

For each $\lambda > 0$

$$\begin{aligned} \overline{\mathcal{D}(\widehat{A})} = \overline{\widehat{\mathcal{D}}(A)} = \overline{\mathcal{D}(A)} \subseteq C[0,1] &= \mathcal{R}(I - \lambda A) \\ &= \mathcal{R}(I - \lambda\widehat{A}) \end{aligned}$$

so the range condition in Theorem 3 follows easily from Theorem 1 and the fact that $\widehat{A}$ is an extension of A. The difficult part of Theorem 3 is the dissipative estimate. One must find an analogue of the second derivative test on $L^\infty(0,1)$. Heuristically, evaluation at a point can be viewed a linear functional on $L^\infty(0,1)$, but it is not a "good" linear functional. Application of the Hahn-Banach theorem, which requires a careful study of the duality map of $L^\infty(0,1)$, leads quite naturally to using finitely (but not countably) additive set functions on the Borel sets in $[0,1]$ which take values in $[0,1]$. The important facts about the duality map in ℓ^∞ and $L^\infty(0,1)$ are contained in [10] and [11], respectively.

In [5] we prove Theorem 3 for $\psi \equiv 0$; we also use Theorem 3 in a critical way to prove the next theorem in that case. This theorem which can be extended to the present situation, is important since it gives us information on regularity in time of solutions of (2.1).

Theorem 4: *Let $\widehat{A}$ be the extension of A on $L^\infty(0,1)$ and $\widehat{\mathcal{D}}(A)$ be as above. Then for all $u_0 \in \widehat{\mathcal{D}}(A)$, there is a unique mild solution $u(t) = \widehat{T}(t)u_0$ of*

$$\begin{cases} u'(t) = \widehat{A}u \\ u(0) = u_0 \end{cases}$$

satisfying

$$wk^*\frac{d}{dt}u(t) = \widehat{A}u(t) \tag{4.1}$$

for $t \geq 0$.

The statement (4.1) means that for every $h \in L^1(0,1)$,

$$\langle u(t), h\rangle = \int_0^1 u(t,x)h(x)dx \in AC_{\text{loc}}\,[0,\infty)$$

as a function of t, and

$$\frac{d}{dt}\langle u(t), h\rangle = \langle \widehat{A}u(t), h\rangle \text{ a.e.}$$

Notice that we cannot hope for a similar result on the space $C[0,1]$; $L^\infty(0,1)$ is a dual space whereas $C[0,1]$ is not.

5. FUTURE DIRECTIONS

We plan to investigate Favard classes for operators of the type we have been considering but with Wentzel boundary conditions rather than the ones used here. Let A be the operator $u \to \varphi(x,u')u'' + \psi(x,u,u')$ acting on a subset of $C[0,1]$. The general Wentzel boundary condition associate with the operator A at the endpoint $j (j = 0,1)$ is $a_jAu(j) + b_ju(j) + c_ju'(j) = 0$ where $V_j = (a_j, b_j, c_j)$ is a nonzero vector in $\mathbb{R}^3$. The case of $V_0 = V_1 = (1,0,0)$ is treated in [9] as far as existence is concerned. That is, in [9] it was shown that certain realizations of A are m-dissipative, but no Favard classes were computed. Here we give one sample calculation. (Questions of this nature are being pursued in collaboration with Jerry Goldstein and Silvia Romanelli.) Consider the boundary conditions defined by $V_j = (1, b_j, 0)$ for $j = 0, 1$ where $b_0 \geq 0 \geq b_1$ with $b_0 - b_1 > 0$. Let $u_i - \lambda A u_i = h_i$ for $i = 1, 2$ when $\lambda > 0$. To prove dissipativity we must show $\|u_1 - u_2\|_\infty \leq \|h_1 - h_2\|_\infty$. Choose $x_0 \in [0, 1]$ such that $\|u_1 - u_2\|_\infty = (u_1 - u_2)(x_0)$. When $0 < x_0 < 1$, the proof proceeds as usual by the first and second derivative tests. Now consider the case $x_0 = 0$. (The case $x_0 = 1$ is similar.) Evaluate $u_j - \lambda A u_i = h_i$ at 0 and use the boundary condition $Au_j(0) + b_j u_i(0) = 0$ where $b_0 \geq 0$. Then $u_i(0)(1 + \lambda b_0) = h_i(0)$; whence,

$$\begin{aligned} \|u_1 - u_2\|_\infty &= (u_1 - u_2)(0) = (1 + \lambda b_0)^{-1}(h_1(0) - h_2(0)) \\ &\leq \|h_1 - h_2\|_\infty. \end{aligned}$$

This implies the dissipativity of the operator.

Favard classes associated with this type of boundary condition seem quite difficult to classify. We plan to study these objects in the future.

References

1. J.R. Dorroh, and G.R. Rieder, A singular quasilinear parabolic problem in one space dimension, ***J. Diff. Eqns.*** 91 (1991), 1-23.

2. J.R. Dorroh, and G.R. Goldstein, Existence and regularity for singular parabolic problems, in preparation.

3. G. R. Goldstein, Nonlinear singular diffusion with nonlinear boundary conditions, ***Math. Meth. Appl. Sci.*** 20 (1993), 1-20.

4. G.R. Goldstein, J.A. Goldstein, and S. Oharu, The Favard class for a nonlinear parabolic problem, in ***Evolution Equations*** (ed. by A. C. McBride and G.F. Roach), Longman, Pitman Notes, Harlow (1995), 134-147.

5. G.R. Goldstein, J.A. Goldstein and S. Oharu, in preparation.

6. G.R. Goldstein, J.A. Goldstein and S. Romanelli, in preparation.

7. J.A. Goldstein, ***Semigroups of Nonlinear Operators***, in preparation.

8. J. A. Goldstein and C. Y. Lin, Singular nonlinear parabolic boundary value problems in one space dimension, ***J. Diff. Eqns.*** 68 (1987), 429-43.

9. J. A. Goldstein and C. Y. Lin, Highly degenerate parabolic boundary value problems, ***Diff. Int. Eqns.*** 2 (1989), 216-227.

10. I. Hada, K. Hashimoto and S. Oharu, On the duality map of ℓ^∞, ***Tokyo J. Math*** 2 (1979), 71-97.

11. K. Hashimoto and S. Oharu, On the duality mapping of $L^\infty(0, 1)$, to appear.

12. G. R. Rieder, Spatially degenerate diffusion with periodic-like boundary conditions, in ***Differential Equations with Applications in Biology, Physics, and Engineering*** (J. A. Goldstein, F. Kappel, W. Schappacher, eds.), Lecture Notes in Pure and Applied Math., Marcel Dekker, New York (1991), 301-312.

Index